Mechanical Properties and Performance of Engineering Ceramics and Composites

T0318965

Mechanical Properties and Performance of Engineering Ceramics and Composites

A collection of papers presented at the 29th International Conference on Advanced Ceramics and Composites, January 23-28, 2005, Cocoa Beach, Florida

Editor
Edgar Lara-Curzio

General Editors
Dongming Zhu
Waltraud M. Kriven

The American Ceramic Society
www.ceramics.org

Published by

The American Ceramic Society
735 Ceramic Place
Suite 100
Westerville, Ohio 43081
www.ceramics.org

Mechanical Properties and Performance of Engineering Ceramics and Composites

ISSN 0196-6219

ISBN 1-57498-232-X

Contents

Fracture and Damage of Ceramics and Composites

Materials Characterization, NDE and Novel Techniques

Processing and Properties of Ceramics

vii

Hardness and Wear Resistance of Ceramics

Properties of Fiber-Reinforced Composites

Laminated Ceramics and Particulate-Reinforced Ceramics

Joining

Preface

The "Cocoa Beach Meeting" has become the premier international forum for the dissemination of information related to the processing, properties and behavior of structural and multifunctional ceramics, ceramic composites, emerging ceramic technologies and applications of engineering ceramics. While symposia addressing bioceramics, solid-oxide fuel cells, ceramic coatings, and armor materials have found a home in the Cocoa Beach meeting, the Symposium on Mechanical Behavior and Design of Engineering Ceramics and Composites, continues to be the backbone of this meeting. As part of the 29th International Conference & Exposition on Advanced Ceramics & Composites, which was held in Cocoa Beach, Florida, January 24-28, 2005, this symposium was organized to address the following topics:

- Fibers, Interphases and Interfaces
- Fracture and Damage of Ceramics and Composites
- Materials Characterization, NDE and Novel Techniques
- Processing and Properties of Ceramics
- Hardness and Wear Resistance of Ceramics
- Properties of Fiber-reinforced Composites
- Laminated Ceramics and Particulate-Reinforced Ceramics
- Joining

This volume contains papers presented in these areas by scientists and engineers in government, industrial and academic organizations from around the world. The tradition of the Cocoa Beach meeting to address core fundamentals as well as timely topics on ceramic science and technology is demonstrated very well by the contents in this volume.

We would like to thank and express appreciation for the many volunteers and the staff of The American Ceramic Society, who make possible the Cocoa Beach meeting. In particular we are indebted to those who attended and participated in the meeting, to the session chairs and organizers, and to those who helped us in the review of the manuscripts contained in this volume.

Edgar Lara-Curzio

General

CREATING A VISION FOR THE FUTURE OF ADVANCED CERAMICS

Frank Kuchinski
Triton Systems, Inc.
200 Turnpike Road
Chelmsford, MA, 01824

Karen Hall
United States Advanced Ceramics Association
1800 M Street NW, Suite 300
Washington, DC, 20036-5802

ABSTRACT

The United States Advanced Ceramics Association (USACA) has supported programs to insert advanced ceramics in government and targeted end-user applications since 1989. These have included a variety of programs with DOE, DOD, NIST, and NASA, as well as investigating the application of advanced ceramics in energy-intensive industries.

In 1999, USACA and the U.S. Department of Energy co-sponsored a workshop to bring together a broad range of ceramic manufacturers and end-user companies. The 1999 workshop produced an Advanced Ceramics Technology Roadmap that set forth the research, development and demonstrations needed for improving advanced structural ceramics. In April 2004, USACA conducted a workshop to revisit the Advanced Ceramics Roadmap to assess progress and ongoing needs, as well as to focus on the manufacturing and fabrication requirements to insert advanced ceramics into the marketplace.

These activities led to the creation of three new USACA-led initiatives: The Ceramic Materials for Energy Independence Initiative; The Technology Transition Initiative; and The Ultra High Temperature Ceramics Initiative.

BACKGROUND

The United States Advanced Ceramics Association

USACA is the premier association that champions the common business interests of the advanced ceramic producer and end-user industries. USACA was formed in 1985 to facilitate the commercialization of the United States' advanced ceramics industry and quickly became the leading voice of the advanced ceramics industry before the U.S. Congress and federal agencies.

USACA recognizes that the key to the industry's survival is the identification of new commercial market opportunities for advanced ceramics. USACA has a long-standing commitment to promoting the use of advanced ceramics as the foundation for a new generation of high-efficiency and high-performance products for surface transportation, aerospace, defense, energy, and industrial applications.

The objectives of USACA are to develop and maintain close working relationships between its membership and the industries they serve; provide liaison between its members and Congress, government agencies, and allied interest industrial organizations and associations; advocate continued and increased funding of research and development on advanced ceramics with congressional leaders and their staffs; provide a mechanism for dissemination of advanced ceramics information; and to promote the benefits of advanced ceramics to Congress, government agencies, and targeted industrial end-users. USACA accomplishes these objectives through a committee structure that guides and supports the staff.

USACA continues to support the needs of the advanced ceramics community through workshops, support of the Annual Conference on Composites, Materials, and Structures (Cocoa Beach, FL), and a quarterly electronic newsletter. In addition, USACA has formed close relationships with similarly aligned organizations, such as the American Ceramic Society, with whom USACA has a Memorandum of Understanding to facilitate information exchange.

Advanced Ceramics Technology Roadmap

In 1999, USACA and the U.S. DOE co-sponsored a workshop to bring together a broad range of ceramic manufacturers and end-user companies. This roadmap, which summarizes the insights of those 40 workshop participants, set forth the research, development and demonstrations needed for improving advanced structural ceramics. Achievement of the RD&D will significantly improve energy efficiency and productivity in many industries and help them reach their performance targets for 2020.

USACA conducted a workshop in April 2004 sponsored by DOE to revisit the Advanced Ceramics Technology Roadmap. It was important to assess progress and ongoing needs, as well as to focus on the manufacturing and fabrication requirements to insert advanced ceramics into the marketplace.

Advanced ceramic materials being incorporated into hot-sections of distributed energy technologies are allowing them to meet strict emission standards while increasing energy efficiency by operating at higher temperatures. Future efficiency and emissions improvements will significantly benefit from the continued research and development of advanced ceramics.

Key challenges for advanced ceramics include improvements in silicon nitride durability (impact resistance), engine integration technology (attachments), design and analysis tools for life prediction, coatings to reduce environmental degradation and leakage of gas flows, and cost reduction. Additional challenges include utilizing new materials, operational maintainability (repair), and scale-up (manufacturing).

The supply of advanced ceramics for application to gas turbines and engines has become of increasing concern because of industry restructuring. It has resulted in fewer suppliers and many of the remaining suppliers being captive to gas turbine manufacturers, which may lead to higher costs or limited product availability.

A decline in government funding for advanced ceramic materials R&D has occurred due to shifts in priorities and rising interest in alternative energy sources. Hydrogen energy systems and fuel cells are receiving funding because they are less mature and viewed by many as being in greater need of federal R&D investment. The workshop provided a set of ideas that could be used to strengthen research planning and raise awareness about the need for and benefits from the development of advanced ceramics for energy conversion equipment such as engines and turbines.

Strengthening existing public-private partnerships in advanced ceramics is one of the keys, including stronger industry, university, and national laboratory teams to accelerate technology development and enhance the chances for market acceptance.

Forming Industry-led Initiatives

USACA's three most recent workshops looked in various ways at the advanced ceramics needs in aerospace and federal agencies. The result of these workshops was the formation of three initiatives in December 2004 to focus on high-payoff applications and stimulate champions. The first workshop, held in the Fall of 2003, reviewed the status of aerospace hot structures and gas turbine materials progress and needs. In the spring of 2004, USACA explored the status of progress being made in meeting mission needs of the federal agencies. In the fall of 2004 USACA held a workshop to investigate initiatives in energy, technology transition, and ultra-high temperature materials.

Key thoughts that came out of these workshops were as follows:
- The advanced ceramics community should think & strategize in capability terms.
- Performance parameters need to be translated into capability parameters.
- Capability based on new (or novel applications of) materials & structures needs to be demonstrated.
- The advanced ceramics community should rethink and perhaps redefine the boundary between material & structure.
- Future discussions and planning should include a viable industrial base and a solid technology base.

It was clear that there was a real opportunity for USACA to lead initiatives to facilitate insertion of advanced ceramics in key areas. Advanced ceramics are wear-resistant, corrosion-resistant, lightweight, and more stable than other materials in high-temperature environments. Because of this combination of properties, advanced ceramics have an especially high potential to resolve a wide number of today's material challenges in process industries, power generation, aerospace, transportation, and military applications. Such applications are vital to maintaining global competitiveness, decreasing energy consumption, and minimizing pollution. These initiatives could provide the framework for a stronger partnership between an alliance of the advanced ceramics industry and the U.S. federal government with the goal of establishing a viable and sustainable advanced ceramics industrial base.

The DOE and DOD could encourage these initiatives by strengthening collaborative efforts in advanced ceramics research. By combining their expertise in materials, components, and manufacturing processes these agencies can help to accelerate the timetable for developing light, strong, corrosion-resistant ceramics capable of performing effectively under an array of environmental conditions.

The advanced ceramics industry needs to expand its scope of thinking and planning horizons to develop leap-frog ideas for distributed energy applications including opportunities for waste heat recovery, heat transfer, and thermal storage. These applications could potentially highlight advanced ceramics in a nationally recognized research and design agenda.

THREE NEW INITIATIVES

Ceramic Materials for Energy Independence

This initiative is focused on specific needs of the US DOE. It is chaired by Jay Morrison, Siemens Westinghouse. USACA created this initiative to accelerate development of advanced ceramic materials that are wear-resistant, corrosion-resistant, lightweight, and more stable than other materials in high-temperature environment in critical energy-related applications. The formation of this initiative is based upon the following:

- There is a strong desire on the part of industry to develop advanced materials for power generation end users of conventional technologies, advanced central and advanced distributed resources.
- An important element of a secure energy future is the use of distributed power production resources to provide more options for serving critical loads in times of power disturbances and terrorism. The DOE has identified the need for advanced materials to better perform these critical functions and goals for fuel flexibility, and efficient and durable central and distributed energy generation.
- Industry needs to see a potential market for it to invest in continued development of advanced ceramics.

USACA's goal for this initiative is to achieve advanced ceramic material insertion into the energy marketplace. To accomplish this goal, the USACA Board of Directors set the following objectives: Create an energy security vision for advanced material requirements that can provide durable, efficient affordable, fuel-flexible generation in central and small distributed resource packages to complement the existing base of power generation assets; Develop programmatic metrics that can lead to the vision; and Develop strategies to address barriers to market insertion and propose plans/initiatives of action.

The Committee will include USACA members, DOE stakeholders, national laboratories and other interested parties that can contribute to the Initiative. An invitation to participate will be extended to relevant industry associations, such as the American Ceramics Society, the Gas Turbine Association, US Fuel Cell Council, and Engine Manufacturers Association.

Beyond the above objectives, desired outcomes include increased communications between the U.S. DOE and the advanced ceramics industry; and a sustainable advanced materials infrastructure that satisfies energy security requirements.

Advanced Ceramics Technology Transition Initiative

This initiative will mainly support critical needs of the U.S. DOD. It is chaired by Bob Licht, Saint-Gobain High Performance Materials.

USACA created this initiative to develop pathways to affordable advanced ceramic materials that are wear-resistant, corrosion-resistant, lightweight, and more stable than other materials in high-temperature environment in critical military applications. The emphasis in this initiative is in the application of available ceramics through efforts to reduce manufacturing costs. The formation of this initiative is based upon the following:

- There is a strong desire on the part of industry to develop advanced materials for military end users.
- The Department of Defense requires advanced materials to better perform critical missions. Applications include:
  o Mirrors
  o Space transportation vehicles
  o Ceramic body, aircraft and vehicle armor
  o Improved transparent armor
  o Small military engines

6

- o Gas turbine
- o Propulsion systems
- o Erosion/corrosion resistant components
- There is national interest in assuring a domestic supply base for military-critical technologies.
- Industry needs to see a potential market to warrant continued development of advanced ceramics.

USACA's goal for this initiative is to achieve advanced ceramic material insertion into the military marketplace. To accomplish this goal, the USACA Board of Directors set the following objectives: Increase the technology readiness level and affordability of advanced ceramics through advances in manufacturing technology; Demonstrate enhanced war fighter capability by the application of advanced ceramics; and Establish a sustainable advanced ceramics industrial base through the use of the Defense Production Act.

The Committee will include USACA members, DOD stakeholders, laboratories and other interested parties that can contribute to the Initiative.

Beyond the objectives above, desired outcomes include: Development of industry partnerships where necessary, with government, to accelerate the application of high performing ceramics; Better communications between government agencies and the advanced ceramics industry; and Increased availability of advanced materials for the DOD.

Ultra High Temperature Ceramics Initiative

This initiative will mainly support key needs of the U.S. DOD and National Aeronautics & Space Administration. It is chaired by Frank Kuchinski, Triton Systems.

USACA created this initiative to accelerate development of advanced ceramic materials that are wear-resistant, corrosion-resistant, lightweight, and more stable than other materials in high-temperature environment in critical defense and space applications. The formation of this initiative is based upon the following:

- There is a strong desire on the part of industry to develop advanced materials for military end and aerospace users.
- The DOD requires advanced materials to better perform critical missions and to serve evolving applications including:
  - o High velocity flight
  - o Digital solid propulsion for kill vehicles
  - o Mirrors
  - o Aerospace nozzles
  - o Scram jet missiles
  - o Hypersonic propulsion
  - o Space-based vehicle hot structures
  - o Electromagnetic transparency materials
- There is national interest in assuring a domestic manufacturing base for military and space-critical technologies.
- Industry needs to see a potential market to warrant continued development of advanced ceramics.

USACA's goal for this initiative is to achieve advanced ceramic material insertion into the military and aerospace marketplace. To accomplish this goal, the USACA Board of Directors set the following objectives: Perform an analysis of the impact on cost and mission capability by not having access to high temperature materials necessary to perform future critical functions; Create awareness of the needs among users of advanced ceramic materials; and Create ceramic technologies and a manufacturing infrastructure to support important requirements for ultra high temperature materials.

The Committee will include USACA members, Department of Defense and NASA stakeholders, laboratories and other interested parties that can contribute to the Initiative. The concept of developing a users consortium will be explored to help prepare the prime contractors and other users for specifying these new materials.

Beyond the above objectives, desired outcomes include: Better communications among government agencies, systems developers, and the advanced ceramics industry; Increased availability of advanced materials for the Department of Defense and NASA; and Generate the funding to implement resulting initiatives.

CONCLUSIONS

- Advanced Ceramics play a key role in multiple US Government applications
- Needs in advanced materials are similar across DOE, DOD, and NASA
- Joint and/or collaborative R&D programs will promote faster product transition and leverage available limited resources.
- USACA hopes to create a vision and strategy to ensure product insertion in critical US applications through its 3 new initiatives.

REFERENCES

[1]USACA, US Department of Energy, Energetics, Richerson & Associates, "Advanced Ceramics Technology Roadmap: Charting Our Course," *Joint USACA/DOE Publication,* (2000).

Fibers, Interphases and Interfaces

MEASUREMENT OF FIBER COATING THICKNESS VARIATION

Randall S. Hay, Geoff Fair
U.S. Air Force Research Laboratory
Materials and Manufacturing Directorate
WPAFB, OH 45433-6533

Pavel Mogilevsky, Emmanuel Boakye
UES, Inc.
4401 Dayton-Xenia Road,
Beavercreek, OH 45432

ABSTRACT

The thickness of a fiber coating may vary significantly along a fiber, and the average thickness may vary between fibers. Coating bridges and fins between close packed fibers and crusting of coating around the perimeter of a fiber tow or cloth are some features that contribute to thickness nonuniformity. Such features have generally been described subjectively. However, quantitative and reproducible thickness nonuniformity measurements are desirable so that the effects of such nonuniformity on properties of ceramic matrix composites can be evaluated. Parameters that may sufficiently characterize coating thickness nonuniformity and methods for measuring those parameters are presented and discussed.

INTRODUCTION

The importance of fiber coatings to the mechanical properties of ceramic fiber-matrix composites (CMCs) is well known.[1] The optimal thickness of such coatings is less well known. Some studies have been done for carbon and BN fiber coatings in SiC-SiC and SiC-BMAS CMCs, respectively.[2,3] Development of oxidation resistant coatings for oxide-oxide CMCs is less mature, and coating thickness optimization for these CMCs is lacking.

When fiber coatings are uniform, coating thickness measurement is trivial. This can be the case for fiber coatings made by CVD; however, such coatings are often thicker on the perimeter of a fiber tow or cloth than they are in the interior. When filaments are close together the coating often bridges the filaments, creating a "crust" around the tow perimeter. Oxide fiber coatings deposited by sol-gel and solution methods also have this nonuniformity. Other oxide coating flaws include "fins" that form from during drying of precursor liquids that bridge filaments, and coating segments that crack off because of excessive thickness or rough handling.[4] These types of coatings are usually evaluated subjectively because there are no accepted methods for quantifying such nonuniformity, and the relevant thickness metrics are not obvious.

Coating non-uniformity may affect CMC mechanical properties. A significant fraction of uncoated fiber surface is certainly a major concern. The fraction and distribution of coatings that are too thin to dissipate roughness induced stresses during fiber pullout should also be of concern.[1,5] Coating that is excessively thick replaces matrix and should affect matrix dominated properties. Coating bridges that span many fibers may function as flaws that degrade interlaminar properties and promote early matrix cracking in CMCs with 2 and 3-D fiber architectures.[6-8] Coating bridges and fins may also impede matrix impregnation during processing and consequently affect CMC properties. These coating nonuniformities can be minimized by multiple coating with low concentration, low viscosity precursors.[4,9] However, this increases the time and expense of fiber coating. Evaluation of the effects of coating thickness nonuniformity on CMC mechanical properties requires methods to quantify the nonuniformity, which allows the effort to improve fiber coating uniformity to be optimized. A preliminary attempt to develop methods that quantify coating thickness nonuniformity is presented.

METHOD

Sample Preparation

Nextel™ 610 fibers were coated with monazite ($LaPO_4$) by methods discussed elsewhere.[9,17] Two types of monazite coatings were characterized; a fiber tow with 3 and 6 coatings (A-3Cts and A-6Cts, respectively), and a cloth coating in close-packed, intermediate-packed, and loose-packed areas (B-CP, B-Int, and B-LP, respectively). These coated fibers were chosen to illustrate the image analysis and measurement process for this paper, and are not necessarily representative of most coated fibers that have been analyzed. Polished cross-sections of the coated fibers were prepared and imaged by SEM (Leica FEG). SEM images of sufficient quality for image analysis required several steps for preparation of the polished cross-sections: 1) The coatings had to be dense. Otherwise there was ambiguity in the grayscale contrast between the coating and fiber. 2) Final polish was done with diamond films (1 μm, 3M) to minimize surface relief. Otherwise charging created a "halo" around the fiber perimeter, which made it difficult to distinguish fiber from coating by the pixel grayscale. 3) The fiber cross-sections had to be perpendicular to the fiber axis, so that they were circular rather than elliptical.

Image Analysis

Digital SEM images were typically saved at ~500 × 500 and ~1000 × 1000 pixels at 8-bit grayscale, with resolutions of ~4 and ~8 pixels/μm, respectively. The five coated fibers described here used a resolution of 0.14 pixels/μm. The images were edited in *Adobe Photoshop*™ to remove fibers intersected by the edge of the image, and the grayscale was adjusted so that the monazite coating was white (grayscale 255-200), the fibers were gray (grayscale 100-155), and the epoxy mounting material was black (grayscale 0-50). A thin black line was inserted where adjacent fibers touched each other so that the touching fibers were not identified as a single fiber by the image analysis program. An example of this image editing is in figure 1.

The edited images were imported into an image analysis program written in *Mathematica*™ 5.0. Grayscale thresholds were used to identify fiber, coating, and mounting material. Typically grayscales of 128 ± 35 were identified as fiber and changed to 128. Anything above this threshold was identified as coating and changed to 255, and anything below it was identified as mounting material and changed to 0. The threshold of ± 35 was varied until the best correspondence between the grayscale threshold and image phases was found by visual inspection of the original and processed images. The result is an image with only three grayscales – 0 for coating, 128 for fiber, and 255 for mounting material.

The centers of each fiber were found. The code used for this was general and time consuming for large images. It was much faster to scale down the image by a factor of 8, then find fiber centers, magnify the center coordinates by 8×, and superimpose them on the original. In this way 10,000 × 10,000 pixel images could potentially be analyzed. Voronoi polyhedra (also called Dirichlet cells) for each fiber were calculated from the fiber centers (Fig. 1).[13,14] The coating for a particular fiber was defined as the coating inside the Voronoi polyhedra for that fiber. Coating thickness was measured radially from the fiber centers in 1° increments by counting and scaling pixels with 255 grayscale up to the Voronoi polyhedra boundary. Coating that radially contiguous with the fiber and coating that was separated from the fiber by 0 grayscale pixels (usually from cracks or coating overhang) were both measured. Coating that bridged fibers was defined as coating that intersected the Voronoi polyhedra boundary. Uncoated fiber surface was defined as that surface with a step from 128 grayscale (fiber) to 0 grayscale (epoxy).

Coating Thickness Variables

Typically several images were analyzed for each particular coating, which meant that >100 fibers were measured and >36,000 coating thickness measurements were made. The data was saved for each fiber in columns in a spreadsheet, from which plots of coating thickness along the circumference of a particular fiber and histograms of that thickness could be

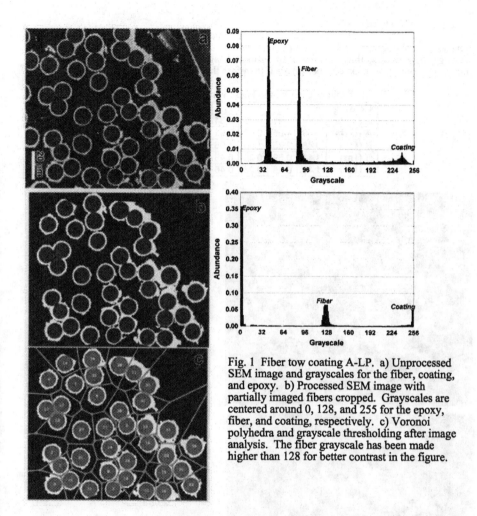

Fig. 1 Fiber tow coating A-LP. a) Unprocessed SEM image and grayscales for the fiber, coating, and epoxy. b) Processed SEM image with partially imaged fibers cropped. Grayscales are centered around 0, 128, and 255 for the epoxy, fiber, and coating, respectively. c) Voronoi polyhedra and grayscale thresholding after image analysis. The fiber grayscale has been made higher than 128 for better contrast in the figure.

conveniently made. We chose to calculate the following parameters for each fiber: 1) Average coating thickness (δ). 2) Standard deviation of δ for all fibers considered together, for the average δ of all fibers, the average of the standard deviation of δ for each individual fiber, and the standard deviation of the standard deviations of δ for each fiber (λ, λ_{av}, λ_e, λ_λ, respectively). These standard deviations were all normalized by δ. 3) Fraction of fiber that is uncoated (ζ). 4) Fraction of fiber that is bridged by coating (η). In some cases the fiber spacing distribution was calculated from the Voronoi polyhedra area. Many other quantities such as the fractal dimension, roughness frequency, etc., could potentially be calculated. Problems with the parameter definitions are discussed in the next section.

RESULTS AND DISCUSSION

Parameter measurements for the five coated fibers are shown in Table 1, and processed images with Voronoi polyhedra are shown in figures 1-5. All fibers are ~12 μm in diameter. Histograms of coating thickness for all fibers in an image are shown in figure 6. These histograms omit the uncoated fiber (ζ) area; therefore the histogram areas differ, since some

Table I. Coating thickness variation parameters

Coating	δ	λ	λ_{av}	λ_e	λ_λ	ζ	η
A (3 cts)	0.24	1.51	0.53	1.34	0.47	0.53	0.011
A (6 cts)	0.59	1.05	0.40	0.91	0.36	0.24	0.013
B (CP)	0.48	2.25	1.56	1.06	1.28	0.63	0.095
B (Int)	1.11	0.59	0.29	0.48	0.20	0.045	0.26
B (LP)	1.48	0.55	0.23	0.46	0.20	0.049	0.20

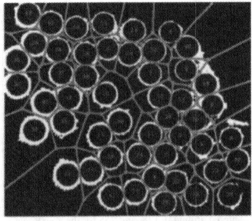

Fig. 2 Processed image with Voronoi polyhedra for B-Int.

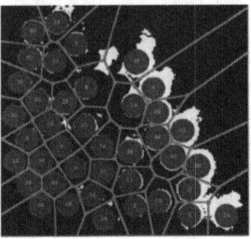

Fig. 3 Processed image with Voronoi polyhedra for B-CP.

14

coated fibers (A-3 Cts and B-CP) have very large ζ (Table I). Note that fiber defined as "uncoated" includes all coating thinner than the image resolution of 0.14 μm, and ζ must therefore be larger than the actual fraction of fiber that is uncoated. The measurements use

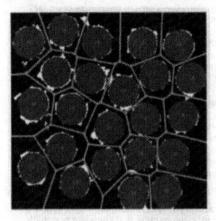

Fig. 4 Processed image with Voronoi polyhedra for A-3cts.

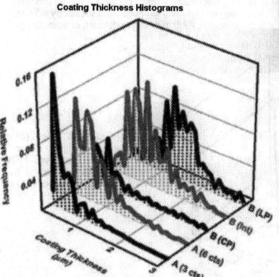

Coating Thickness Histograms

Fig. 6. Histograms of coating thickness for the five fibers in Table I.

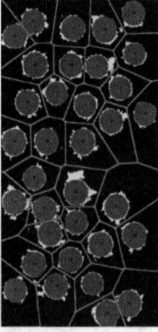

Fig. 5 Processed image with Voronoi polyhedra for A-6cts.

all coating inside a Voronoi polyhedra, regardless of whether it was radially contiguous. Coating overhangs, cracked coating, and isolated blocks of coating were not abundant for the five coatings, so there was not a significant difference between measurements for radially contiguous and non-contiguous coating.

Bridged coating was sometimes found erroneously. If a fiber with a thick coating was next to a fiber with a thin and the fibers were close together, the Voronoi polyhedra boundary sometimes ran over the thick coating and it was counted as bridged, even though the coating did not connect the fibers. Isolated blocks of coating unconnected with any fiber could sometimes cause the same problem. However, since fibers with thick coatings tended to be clumped together, these types of errors were not judged to be large. The clumping of similar types of coating (thick, bridged, uncoated, etc.), and the correlation between coating characteristics to the location of a fiber in tow (e.g. perimeter or interior of tow) might be another important parameter to measure.

Preliminary attempts to classify coating quality are shown in figures 7 and 8. Figure 7 shows the relationship between coating thickness (δ) and the standard deviation of the average thicknesses for all fibers and the standard deviation in thickness for each fiber (λ_{av} and λ_e, respectively). Higher quality coatings are assumed to be those with smaller λ_{av} and λ_e. The optimal thickness is unknown but clearly the thicker coatings tended to have smaller normalized standard deviations in their thickness. Figure 8 is a bubble plot. The bubble size scales with the coating thickness (δ). Bubbles are plotted for the standard deviation in thickness for all coatings considered together (λ), the uncoated fiber fraction (ζ), and the bridged fraction (η). Coating quality is assumed to be higher for lower values of λ, ζ, and η. As expected, coatings with high average thicknesses had more bridging and smaller fractions of uncoated fiber. δ, λ, ζ, and η are the simplest parameters to calculate but may not be the parameters that correlate best with CMC mechanical properties. Correlation of these proposed coating quality metrics with CMC mechanical properties is necessary to verify the subjective assessments of "better" and "worse" that are used in the plots. Even after such correlation, it will be unclear whether metrics are causal. For example, is high variation in coating thickness "worse" because it has an inherent effect on mechanical functionality of the coating, or is it "worse" because it indirectly affects CMC mechanical properties through affects on matrix precursor impregnation, or something else?

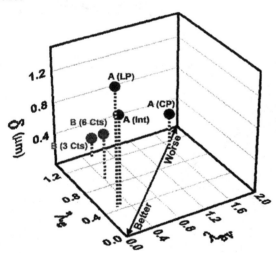

Fig. 7 Coating thickness (δ) vs. the standard deviation of the average thicknesses for all fibers and the standard deviation in thickness for each fiber (λ_{av} and λ_e, respectively).

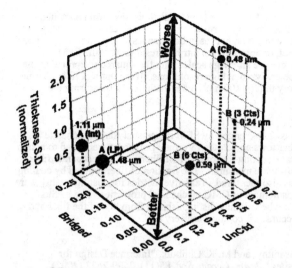

Fig. 8 A bubble plot of the coating thickness standard deviation for all fibers (λ) vs. fraction of coating that is bridged (η) and the fraction of fiber that is uncoated (ζ) (Table I). The bubble size scales with the average coating thickness (δ).

Other aspects of image analysis may affect data quality. There is a tendency to slightly over sample closely packed fibers because they take up less image space. Sampling many images can partially correct for this.[15] Otherwise, digitization at high resolution had no apparent affect on data quality. Cracked coating and coating "overhangs" were not abundant in the particular images used for analysis. There was not a significant difference between the measurements for radially contiguous and non-contiguous coating, so we chose to count all coating that was radially non-contiguous with the fiber. Circumferentially cracked coating may be mechanically functional if it sinters back on to the fiber during matrix processing. However, it is not clear how coating "overhangs" would function mechanically at the fiber-matrix interface. They may have a tendency to break off during matrix processing, and therefore should not be counted. A more general method for handling such anomalies is desirable.

Coating variation was measured in only one dimension – along the fiber circumference. Variation along the fiber axis was not measured. From SEM observation of coated fiber surfaces it is usually obvious that coating irregularities such as "fins" persist at length scales that are significantly larger than the fiber diameter. Variation along the axial direction would also have to be measured if variation along this larger length scale is comparable or more important to CMC mechanical properties than circumferential variation. Another possibility is that parameters defining the axial and circumferential coating thickness variations are highly correlated with each other, so only one would have to be measured.

Other information about relationships between CMC fiber architecture and coating thickness can be extracted by image analysis. For example, the area of Voronoi polyhedra and the average spacing between fibers can be correlated with coating thickness. This makes sense for cloth coatings, but not for coated fiber tows. For coated tows, the imaged fiber spacing has no obvious relationship to the spacing present when the fiber was coated, or what the spacing will be when incorporated in a CMC. The Voronoi polyhedra were not calculated correctly for fibers with elliptical cross-sections or non-uniform diameter. This was not a problem for coated Nextel[TM] 610 tows and simple cloth weaves, but the image analysis would have to be modified for complex fiber architectures or irregular fibers.

The dependence of the image analysis method on the ability to use grayscale to unambiguously identify fiber, coating, and matrix phases is critical. At some scale this breaks down. Slight surface relief can cause charging "halos" around the fiber perimeter that introduce grayscale ambiguity. The intrinsic image resolution depends on the pixel scale of the analyzed images (~140 nm). Coatings thinner than 140 nm could not be distinguished from uncoated fiber. This is a severe limitation for fibers with uniform coatings that are very thin. Analysis of coating with δ_{av} of 20 nm would require much higher resolution images, which in turn might be affected by charging from surface relief, or approach the intrinsic resolution of the SEM instrument used. The time it takes to analyze an image scales with the squared reciprocal of the pixel resolution, so fewer fibers can be analyzed at high resolutions. If the optimal fiber coating thickness for CMCs turns out to be very thin (< 100 nm), other methods for measuring the thickness variation may be necessary.

SUMMARY AND CONCLUSIONS

Coating thickness variation was characterized by SEM image analysis for monazite ($LaPO_4$) coatings on Nextel™ 610 fiber tows and cloths. Voronoi polyhedra were used to assign coating to a particular fiber. The average coating thickness of all fibers, for each separate fiber, and their respective standard deviations were measured. The fraction of fiber bridged by coating and the fraction that was uncoated was also measured. Tentative metrics for coating quality were defined that require validation by measurement of CMC mechanical properties. The method requires that the grayscale of the coating be uniquely distinguishable from that of the fiber and matrix, and may have unacceptable accuracy if coatings are very thin.

REFERENCES
[1]R. J. Kerans, R. S. Hay, T. A. Parthasarathy, and M. K. Cinibulk, "Interface Design for Oxidation Resistant Ceramic Composites," *J. Am. Ceram. Soc.* **85**(11), 2599-2632 (2002).
[2]J. P. Singh, D. Singh, and R. A. Lowden, "Effect of Fiber Coating on Mechanical Properties of Nicalon Fibers and Nicalon-Fiber/SiC Matrix Composites," *Ceram. Eng. Sci. Proc.* **15**(4), 456-464 (1994).
[3]E. Y. Sun, S. R. Nutt, and J. J. Brennan, "Fiber Coatings for SiC-Fiber-Reinforced BMAS Glass-Ceramic Composites," *J. Am. Ceram. Soc.* **80**(1), 264-266 (1997).
[4]E. Boakye, R. S. Hay, and M. D. Petry, "Continuous Coating of Oxide Fiber Tows Using Liquid Precursors: Monazite Coatings on Nextel 720," *J. Am. Ceram. Soc.* **82**(9), 2321-2331 (1999).
[5]R. J. Kerans, "Viability of Oxide Fiber Coatings in Ceramic Composites for Accommodation of Misfit Stresses," *J. Am. Ceram. Soc.* **79**(6), 1664-1668 (1996).
[6]G. N. Morscher, F. I. Hurwitz, and A. M. Calomino, "C-Coupon Studies of SiC/SiC Composites Part I: Acoustic Emission Monitoring," *Ceram. Eng. Sci. Proc.* **23**(3), 379-386 (2002).
[7]F. I. Hurwitz, A. M. Calomino, T. R. McCue, and G. N. Morscher, "C-Coupon Studies of SiC/SiC Composites Part II: Microstructural Characterization," *Ceram. Eng. Sci. Proc.* **23**(3), 387-393 (2002).
[8]T. Hinoki, E. Lara-Curzio, and L. L. Snead, "Evaluation of Transthickness Tensile Strength of SiC/SiC Composites," *Ceram. Eng. Sci. Proc.* **24**(4), 401-406 (2003).
[9]E. E. Boakye, R. S. Hay, P. Mogilevsky, and L. M. Douglas, "Monazite Coatings on Fibers: II, Coating without Strength Degradation," *J. Am. Ceram. Soc.* **84**(12), 2793-2801 (2001).
[10]E. E. Boakye and P. Mogilevsky, "Fiber Strength Retention of La and Ce-PO$_4$ Coated Nextel™ 720," *J. Am. Ceram. Soc.* **87**(2), 314-316 (2004).
[11]E. E. Boakye, P. Mogilevsky, and R. S. Hay, "Synthesis of Spherical Rhabdophane Particles," *J. Am. Ceram. Soc.* (submitted).
[12]G. Fair, "Precipitation Coating of Monazite on Ceramic Fiber Cloths and Tows", Unpublished Results.
[13]D.-S. Kim, Y.-C. Chung, J. J. Kim, D. Kim, and K. Yu, "Voronoi Diagram as an Analysis Tool for Spatial Properties for Ceramics," *Ceram. Proc. Res.* **3**(3), 150-152 (2002).
[14]H. V. Atkinson and G. Shi, "Characterization of Inclusions in Clean Steels; A Review Including the Statistics of Extremes Methods," *Prog. Mater. Sci.* **48**, 457-520 (2003).
[15]D. A. Coker and S. Torquato, "Extraction of Morphological Quantities from a Digitized Medium," *J. Appl. Phys.* **77**(12), 6087-6099 (1995).

MODEL OF DEVIATION OF CRACKS AT INTERFACES OR WITHIN INTERPHASES

J. Lamon and S. Pompidou
Laboratoire des Composites ThermoStructuraux
UMR 5801 : CNRS-Snecma-CEA-UB1
Domaine Universitaire
3, allée de la Boétie
33600 Pessac
France

ABSTRACT

An approach to crack deflection at interfaces or within interphases is proposed on the basis of the Cook and Gordon's mechanism[1]: a crack is nucleated along an interface, ahead of a propagating crack; deflection of this crack then results from coalescence with the interfacial crack.

The stress state induced by a crack was computed in a cell of bimaterial using the finite element method. The cell represents a matrix and a fiber, or an interphase and a fiber or two layers in a multilayered matrix. A master curve was established. It is based on a debonding depending on strengths and elastic moduli of constituents. The master curve allows deflection of cracks to be predicted with respect to constituent properties. It was used to evaluate the strength of various fiber/matrix interfaces and interphases in ceramic matrix composites. It is discussed with respect to experimental data and crack observations.

INTRODUCTION

The fiber/matrix interfacial domain is a critical part of ceramic matrix composites because load transfers from the matrix to the fiber and vice versa occur through the interface. Fiber/matrix interfaces thus exert a profound influence on the mechanical behavior and the lifetime of composites. Experimental efforts have been directed towards optimization of interface properties. Fiber/matrix interfaces in CVI SiC/SiC composites consist of a thin layer (less than 1 µm thick) of one or several materials deposited on the fiber (interphase). CVI SiC/SiC composites with rather strong interfaces have been obtained using fibers that had been treated in order to increase the fiber/coating bond. The concept of strong interfaces has been established on CVI SiC/SiC composites with PyC and multilayered (PyC/SiC)$_n$ fiber coatings[2].

In the presence of weak fiber/coating bonds, matrix cracks generate a single long debond at the surface of fibers[2]. In the presence of stronger fiber/coating bonds, the matrix cracks are deflected into the coating, involving short and branched multiple cracks[2]. The following properties are significantly higher when comparing to those conventional composites with classical weak interfaces: strength, fracture toughness, resistance to fatigue and creep at high temperatures, etc. Therefore a sound model of crack deflection taking into account characteristics of interfaces located ahead of matrix crack, is to be sought to predict deviation of cracks.

Approaches to crack deflection can be grouped into two families. First, those that consider a stationary crack lying at the interface[3-5]. These approaches are interesting essentially from a qualitative point of view. However they may not reflect the true mechanism of crack deviation. They cannot be applied to a layered interphase such as PyC when the fiber/coating bond is strengthened.

Second, the approach based on the mechanism proposed by Cook and Gordon[1]: a crack nucleates first ahead of the propagating one, and it initiates interface fracture (debonding). The deflection then results from coalescence of both cracks. This mechanism of debonding has been observed on several material combinations[6-7].

Cook and Gordon have computed the stress field induced by a semi-elliptical crack (tip radius ρ) placed in a homogeneous material subjected to uniaxial tension (Fig. I). They have shown that the stress state at crack tip is multiaxial, although a uniform uniaxial tensile stress is applied. In particular, they evidenced that the stress component σ_{rr}, parallel to the crack plane, reaches a maximum at a distance from the crack tip in the order of ρ. When an interface is placed perpendicular to crack extension direction, a crack may nucleate if $\sigma_{rr} \geq \sigma_i^c$, where σ_i^c is the interfacial strength. Cook and Gordon considered a single material with an interface. In the present paper, this approach is extended to combinations of different materials so that it can be applied to composites and multilayered structures.

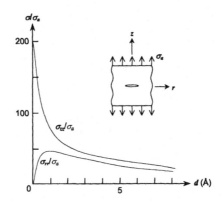

Fig. I. Stress components along crack axis, parallel and perpendicular to the load axis, versus distance to crack tip[1]

MODEL

Let's consider the cell shown on figure II. It consists of two different materials perfectly bonded. These materials are considered to be homogeneous, isotropic, elastic and brittle. A transverse crack is lying in the first phase, denoted 1 (Young's modulus E_1, Poisson's ratio ν_1). The crack is supposed to extend perpendicular to the interface, towards the second constituent denoted 2 (E_2, ν_2, strength σ_2^c). The ligament (l) is defined as the distance from crack tip to the interface. The cell is subjected to a uniform uniaxial tensile stress (σ_a) perpendicular to crack plane.

20

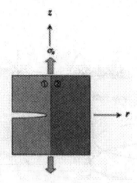

Fig. II. Bimaterial cell subjected to a uniform tensile stress (σ_a)

Debonding occurs when the stress in the interface exceeds the bond strength and the stress in material 2 remains smaller than the resistance to failure of material 2:

$$\sigma_{rr}(r = l) \geq \sigma_i^c \tag{1}$$

$$\sigma_{zz}(r > l, z) < \sigma_2^c \tag{2}$$

where σ_{rr} and σ_{zz} are stress components (coordinates are shown in figure I). r is the distance to crack tip. σ_2^c is the resistance to failure of material 2. σ_i^c is the bond strength: resistance of interface to opening (mode I). A debonding criterion can be summarized by the following condition derived from (1) and (2):

$$\frac{\sigma_i^c}{\sigma_2^c} \leq \frac{\sigma_{rr}(r = l)}{\sigma_{zz}(r > l, z)} \tag{3}$$

The perpendicular value of l that can be extracted from (3) characterizes the position of crack tip when debonding occurs. Alternately, in the presence of various interfaces and an interphase, the particular values of l for which condition (3) is satisfied indicates the location of debonding (figure III). Strengths of interfaces and interphase can be taken into account. For instance, debonding within the interphase at a distance l_l from crack tip corresponds to the following conditions:

• in the matrix/interphase interface (interface 1):

$$\frac{\sigma_{i1}^c}{\sigma_I^c} > \frac{\sigma_{rr}(r = l_1)}{\sigma_{zz}(r > l_1, z)} \tag{4}$$

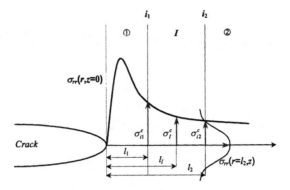

Fig. III. Schematic diagram showing the stress component σ_{rr}
at crack tip in a cell with an interphase

- in the interphase (material I):

$$\frac{\sigma_{iI}^c}{\sigma_I^c} \leq \frac{\sigma_{rr}(r=l_I)}{\sigma_{zz}(r>l_I,z)} \tag{5}$$

- in the interphase/fiber interface (interface 2):

$$\frac{\sigma_{i2}^c}{\sigma_2^c} > \frac{\sigma_{rr}(r=l_2)}{\sigma_{zz}(r>l_2,z)} \tag{6}$$

Indexes 1, 2 and I designate respectively interface 1 (equation (4)), interface 2 and material 2 (equation (6)) and the interphase (equation (5)).

However, the strength of interface expressed in terms of an opening stress is generally not available, essentially because experimental tests have not been developed up to now. Therefore, l cannot be derived from equation (3). Considering the peak stress σ_{rr}^{max} allows this shortcoming to be overcome:

$$\sigma_{rr}^{max} = \sigma_{rr}(r=\rho) \geq \sigma_i^c \tag{7}$$

Thus, when condition (1) is fulfilled, it is obvious that condition (7) is also satisfied and vice versa. The debonding condition is now:

$$\frac{\sigma_i^c}{\sigma_2^c} \leq \frac{\sigma_{rr}^{max}}{\sigma_{zz}^{max}(r>l)} \tag{8}$$

22

where $\sigma_{zz}^{max}(r > l)$ is the maximum of σ_{zz} in material 2. Expression (8) is convenient because σ_{rr}^{max} and σ_{zz}^{max} can be determined. However, note that it provides a necessary but not sufficient condition.

Determination of stress state

The stresses in the cell shown on figure II were computed using a finite element method (CASTEM 2000 computer code). A bidimensional finite element mesh was constructed. An elliptical crack (tip radius ρ) is placed at a distance $l = \rho$ from the interface: this is the critical configuration[1,8], $\sigma_{rr}(l = \rho) = \sigma_{rr}^{max}$. It has been shown that ρ does not influence the results[8]. Thus, a minimum ligament length beyond which the main crack will not be able to propagate without reaching the interface can be introduced. This ensures at last the convergence of numerical results.

Computations were run for various constituent properties: Young's modulus E_2/E_1 ratio ranging from 0.01 to 100. Poison's ratios were taken to be identical ($\nu_1 = \nu_2 = 0.2$).

RESULTS AND DISCUSSION

Ratios of maximum stresses σ_{rr}^{max} and $\sigma_{zz}^{max}(r > l)$ that were derived from finite element analysis were plotted versus ratios of Young's moduli (E_2/E_1) (Fig. IV). The resulting master curve $\sigma_{rr}^{max}/\sigma_{zz}^{max}(r > l) = f(E_2/E_1)$ allows debonding to be predicted. Condition (8) is satisfied in the domain under the curve: debonding domain. This domain contains the pertinent values of $(\sigma_i^c/\sigma_2^c, E_2/E_1)$ for which debonding will occur. Condition (8) is not satisfied in the domain located above the curve; debonding cannot occur. However note that to get failure of material 2, condition (2) must be also satisfied, since, as mentioned above, condition (8) is necessary but not sufficient.

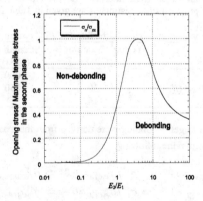

Fig. IV. Criterion of debonding ahead of a matricial crack with respect to constituent properties

The master curve reaches the maximum of 1 for $E_2/E_1=3.86$. Debonding is thus enhanced for this typical material combination.

For a single material: $\sigma_i^c/\sigma_2^c = 1$, $E_2/E_1 = 1$. It can be noticed that the (1,1) point is located in the non-debonding domain (Fig. IV). This result is consistent with logical expectation since it is well known that a homogeneous and isotropic material under tension cannot fail parallel to load direction.

When the modulus ratio decreases to zero, debonding becomes quite impossible. This may be attributed to compliance of material 2 which can deform easily. As a consequence, σ_{rr} is very small, as shown by computations.

The master curve is not symmetrical. The higher E_2/E_1, the higher resistance of material 2 to deformations induced by crack in material 1. As a consequence, σ_{rr} can be much higher than in the former case and debonding is possible.

Applications of criterion

When mechanical properties $\left(E_1, E_2, \sigma_i^c, \sigma_2^c\right)$ are all known, modulus and strength ratios have to be introduced in the diagram shown on figure IV. Depending on position of point, debonding or non-debonding is predicted. Furthermore, location of debonding can be determined using equation (3) : within interphase, or at interfaces. If non-debonding is predicted, condition (2) needs to be checked for fracture in material 2.

As mentioned above, bond strength σ_i^c is generally unknown. For those combinations of materials for which debonding is observed, an upper bound for σ_i^c can be estimated: σ_i^{ct}. σ_i^{ct} is derived from the particular value of stress ratio on the master curve, that corresponds to the actual value of E_2/E_1.

Example of SiC/SiC composite with a pyrocarbon interphase

Condition (8) was examined at each interface: fiber/interphase, interphase/matrix and within interphase. Mechanical properties of constituents are given in Table I.

Table I. Mechanical properties of the SiC/PyC/SiC constituents

Constituent	Young's Modulus E (GPa)	Strength σ^c (MPa)
SiC fiber	280	3000
PyC interphase (//)*	30	150
SiC matrix	415	

Values of σ_i^c required to get debonding were derived from the master curve using ratios of elastic moduli, as summarized in the following:

- PyC$_i$/SiC$_m$: $E_{PyC,i}/E_{SiC,m} = 0.07 \Rightarrow \sigma_i^{ct}/\sigma_{PyC_{//},i}^c = 0.01 \Rightarrow \sigma_{PyC,i/SiC,m}^{ct} = 0.3$ Mpa
- PyC$_i$/PyC$_i$: $E_{PyC_{//},i}/E_{PyC_{//},i} = 1 \Rightarrow \sigma_i^{ct}/\sigma_{PyC_{//},i}^c = 0.46 \Rightarrow \sigma_{PyC,i/PyC,i}^{ct} = 69.5$ MPa
- SiC$_f$/PyC$_i$: $E_{SiC,f}/E_{PyC_{//},i} = 9.33 \Rightarrow \sigma_i^{ct}/\sigma_{SiC,f}^c = 0.79 \Rightarrow \sigma_{SiC,f/PyC,i}^{ct} = 2.2$ GPa

24

Results show that σ_i^c must be smaller than 0.3 MPa so that debonding can occur at the matrix/interphase interface. Debonding at matrix/interphase interface thus requires a very weak bond. Higher σ_i^c values are logically expected. Furthermore debonding is not observed at the matrix/interphase interface in SiC/SiC composites.

Debonding at the interphase/fiber interface seems to be easier, since the maximum bond strength is estimated to be 2.2 GPa. Debonding is generally observed at this interface, in conventional SiC/PyC/SiC composite, i.e. when this interface is not strengthened via fiber treatment.

Debonding seems also to be possible within the PyC interphase, as indicated by the stress ratio and the master curve. This can be obtained when conditions (4) to (6) are satisfied, which requires that σ_{12}^c be increased according to equation (6). Experimental observations on SiC/PyC/SiC composites reinforced with treated fibers (strong interface concept) show that crack deviation occurs within the interphase.

CONCLUSIONS

A criterion for crack deviation at interfaces was established on the basis of the mechanism proposed by Cook and Gordon. It reduces to a master curve relating interface strength to strength and Young's modulus of constituents. The master curve was produced using finite element computations of the stress state at crack tip. This curve allows prediction of debonding when properties of interface and constituents are known. Alternately, it provides an upper bound of bond strength required to obtain debonding. Predictions were found to be in agreement with available experimental data. The model can explain and predict debonding within a PyC interphase in the presence of a strong fiber/interphase bond. It allows determination of location of debonding when bond strengths data are available.

ACKNOWLEDGEMENTS

This work has been supported by the French Ministry of Education and Research.

FOOTNOTES
* //: along the graphitical plane direction

REFERENCES
[1] J. Cook, J.E. Gordon, "A mechanism for the control of crack propagation in all-brittle systems", *Proc. Roy. Soc.*, **28A**, 508-520 (1964)

[2] C. Droillard, J. Lamon, "Fracture toughness of 2D woven SiC/SiC CVI composites with multilayered interphases", Vol. 79, N°4, 849-858 (1996)

[3] M.Y. He, J.W. Hutchinson, "Crack deflection at an interface between dissimilar elastic materials", *Int. J. Solids Structures*, **25** [9], 1053-1067 (1989)

[4] A.K. Kaw, N.J. Pagano, "Axisymmetric thermoelastic response of a composite cylinder containing an annular matrix crack", *Journal of Composite Materials*, Vol. 27, No. 6, 540-571 (1993)

[5] N. Carrère, E. Martin, J. Lamon, "The influence of the interphase and associated interfaces on the deflection of matrix cracks in ceramic matrix composites", *Composites: Part A*, 31, 1179-1190 (2000)

[6] W. Lee, S.J. Howard, W.J. Clegg, "Growth of interface defects and its effect on crack deflection and toughening criteria", *Acta mater.*, 44 [10], 3905-3922 (1996)

[7] A.H. Barber, E. Wiesel, H.D. Wagner, "Crack deflection at a transcrystalline junction", *Composites Science and Technology*, 62, 1957-1964 (2002)

[8] S. Pompidou, "Déviation des fissures par une interface ou une interphase dans les composites et les multicouches", Thèse de l'Université Bordeaux 1, N° 2694 (11 juillet 2003)

PREPARATION AND CHARACTERIZATION OF MAGNESIUM-SILICON BASED OXIDE COATING ON HIGH-CRYSTALLINE SiC FIBER AS AN INTERPHASE IN SiC/SiC COMPOSITE

Naoki Igawa, Tomitsugu Taguchi, Reiji Yamada, Yoshinobu Ishii and Shiro Jitsukawa
Japan Atomic Energy Research Institute
Tokai-mura, Ibaraki-ken, 319-1195, Japan

ABSTRACT

Mg-Si-O coating on the advanced SiC fibers as an interphase of SiC/SiC composites was prepared by the alkoxide method. Almost all fiber surface was coated completely by 5 times dipping into the coating solution of $[Mg]_T$=0.50 mol/dm^3 and $[Si]_T$=0.25 mol/dm^3. The phases of the coating were a mixture of MgO, MgSiO$_3$, Mg$_2$SiO$_4$ and SiO$_2$ up to 1400 °C in air. The tensile strengths of the coated fibers were retained those of as-received fiber below 1400 °C, while those were slightly degraded at 1400 °C.

INTRODUCTION

SiC fiber reinforced SiC composites (SiC/SiC) are being considered as candidate materials for structural applications of fusion reactors[1-3] because of their low activation under the fusion-neutron irradiation and excellent mechanical properties at high temperatures. In SiC/SiC, the interphase between SiC fibers and SiC matrix is very important for their excellent mechanical properties; the roles of the interphase are to allow debonding between fiber and matrix during the deformation to branch the cracks to increase fracture toughness, and to prevent chemical reactions between fiber and matrix for reduction of fiber-degradations during the processing. Several interphases were investigated in the previous studies[4-8]; BN is one of the best materials to allow for fiber pull-out.[7] For fusion reactor component, however, BN interphase is not favorable because of its chemical reactivity with oxygen and water, and its residual radioactivity after neutron irradiation.[9] Lee et al. reported SiO$_2$/ZrO$_2$/SiO$_2$ multilayered oxide interface for conventional Hi-Nicalon SiC fiber reinforced SiC mini-composites.[8] Those composites were exhibited good mechanical behavior with evidence of multiple matrix cracking by the tensile test at room temperature due to significant coefficient of thermal expansion mismatch between ZrO$_2$ and SiO$_2$, which results in "weakening" of the interface. However their maximum load capabilities were decreased after heating even at 960 °C in comparison to the as-prepared specimens. Recently, the advanced SiC fibers have been developed, including Hi-Nicalon Type S[10] and Tyranno SA[11] which possess superior mechanical, thermal properties and oxidation-resistance as well as neutron radiation-resistance. Therefore oxide interphases become a potential candidate to improve the thermal stability of SiC/SiC under oxidation and/or fusion-neutron environment.

In this study, we fabricated Mg-Si-O coating on the advanced SiC fiber using the alkoxide method. Induced activity of the Mg-Si-O coating is expected to be low and resistance for irradiation damage is also estimated to be high because of the rather high threshold energy for the displacement damage. This coating is probably a weak interphase due to the coefficient of thermal expansion mismatch between the coating and SiC. The interaction between SiC fiber and the interphase was examined and the tensile property of coated fiber was evaluated. Hi-Nicalon SiC fiber/SiC matrix composite with the Mg-Si-O interphase was fabricated by the chemical vapor infiltration method and carried out the bending test as a preliminary study.

EXPERIMENTAL PROCEDURE

The SiC fibers used in the present work were Hi-Nicalon Type S (Nippon Carbon Co. Ltd., Tokyo, Japan) and Tyranno SA (Ube Industries, Ube, Japan). The fiber diameter and tensile strength of Hi-Nicalon type S are 12 μm and 2.6 GPa and these of Tyranno SA are 7.5 μm and 2.5 GPa, respectively.

The flow chart of Mg-Si-O coating process in this study is shown in Fig. 1. The coating solution was prepared by the dissolving the magnesium ethoxide, $Mg(OC_2H_5)_2$ and tetraethyl orthosilicate, $Si(OC_2H_5)_4$ (Wako Pure Chemical Industries, Ltd., Osaka, Japan) into 2-propanol. The magnesium and silicon concentrations of the coating solution were 0.50 and 0.25 mol/dm^3, respectively. The SiC fibers were dipped into the coating solution and hydrolyzed the alkoxides in air. Dip-and-hydrolysis process was repeated 5 times. Finally the fibers were heated up to 1400 °C in air: the heating and cooling rate was 7 °C/min. The phase of the coating was analyzed by using X-ray diffractometer. CuKα radiation of 40 KV and 20 mA was used with a step-scanning technique in the 2θ range from 25 to 80°. The surface of the coating was examined by using a scanning electron microscopy (SEM). The tensile strength of the coated fibers was also measured at room temperature by a single filament method using the Instron-type tensile testing machine with the gauge length of 25 mm and the cross-head speed of 0.5 mm/min.

As a preliminary test for this coating, SiC/SiC with the Mg-Si-O interphase was fabricated by a forced flow chemical vapor infiltration method (FCVI). Two dimensional-plain weave of Hi-Nicalon fabrics was used. After 5 times dip-and-hydrolysis process, the coated fabric was heated at 700 °C in air for 1 h. Nine layers of the coated fabrics with a fabric layer orientation of [0°/90°] were restrained in a furnace. The preform was 55 mm in diameter and 2.5 mm in thickness. The precursor for the SiC matrix was methyltrichlorosilane (MTS, CH$_3$SiCl$_3$, Shin-Etsu Chemicals Co., Ltd., Tokyo, Japan). The SiC matrix was formed at 1050 °C with the MTS flow rate of ca. 0.5 g/min carried by 1 dm^3/min of H$_2$ under 0.05 MPa for 7 h. The perform was flipped twice through the FCVI process for matrix uniformity. Three-point bending test was carried out at room temperature with a span length of 20 mm and the cross-head speed of 0.1 mm/min. The size of the specimen for the bending test was about 25 x 5 x 2 mm^3.

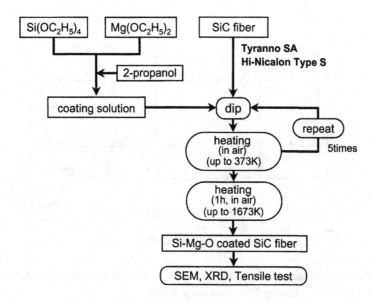

Fig. 1. Flow chart of fabrication process of Mg-Si-O coating on the SiC fibers.

RESULTS AND DISCUSSION

Figure 2 shows the SEM photographs of uncoated and Mg-Si-O coated SiC fibers after heating at 1000 °C. The uncoated fibers have a smooth surface. From photos, it was found that the surface of the fibers were fully coated and obtained relatively excellent smooth surface: it was found that the surface of coated Tyranno SA SiC fiber was smoother than that of Hi-Nicalon Type S SiC fiber.

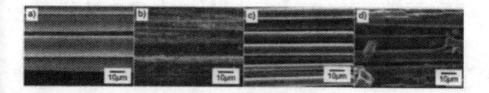

Fig. 2. SEM microphotographs of a) uncoated, b) coated Hi-Nicalon TypeS, c) uncoated and d) coated Tyranno SA SiC fiber after heating at 1000 °C in air.

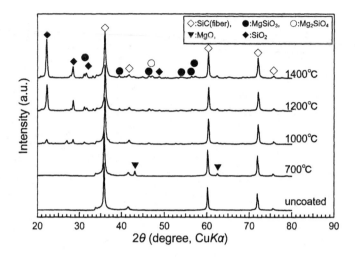

Fig. 3. X-ray diffraction patterns of Mg-Si-O coated Tyranno SA SiC fiber as a function of coating fabrication temperature.

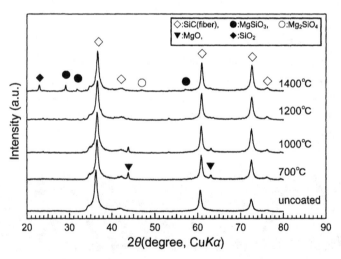

Fig. 4. X-ray diffraction patterns of Mg-Si-O coated Hi-Nicalon Type S SiC fiber as a function of coating fabrication temperature.

Figures 3 and 4 show the X-ray diffraction patterns of the coated Tyranno SA and Hi-Nicalon Type S SiC fibers as a function of coating fabrication temperature. For comparison, as-received fibers were also shown in those figures. The XRD patterns of the uncoated fibers showed five peaks around the 2θ of 35.7°, 41.4°, 60.4°, 71.8° and 76° (d = 0.251, 0.218, 0.154, 0.131 and 0.125 nm) which were assigned to the 111, 200, 220, 311 and 222 reflections of β-SiC. In addition to these β-SiC peaks, the phase of MgO was observed in the coated fiber at 700 °C. Although the peaks attributed to MgO phase were decreased with increasing temperature, these peaks not completely disappeared during the heating. The peaks attributed to $MgSiO_3$, Mg_2SiO_4 and SiO_2 were increased with the temperature. Unfortunately, it is not determined that these SiO_2 peaks were attributed to whether the oxidation of SiC fiber or the decomposition and crystallization of $Si(OC_2H_5)_4$. The intensity ratio of coating to the SiC fiber was different between Tyranno SA and Hi-Nicalon Type S. It is probably due to the difference of the fiber diameter: the diameter of Tyranno SA was smaller than that of Hi-Nicalon Type S, therefore the volume ratio of coating layer to fiber of Tyranno SA was larger than that of Hi-Nicalon Type S. The widths of diffraction peaks attributed to Hi-Nicalon Type S was much larger than those of Tyranno SA, since the crystallite size of Hi-Nicalon Type S is smaller than that of Tyranno SA. While the widths of diffraction peaks attributed to Hi-Nicalon SiC fiber was narrowed by the heating[12], no change of the peak shape and position of both Hi-Nicalon Type S and Tyranno SA SiC fiber was observed in the present work. This suggests that these advanced SiC fibers are more stable than the conventional fibers under those conditions.

Figure 5 shows the tensile strength of the coated Hi-Nicalon Type S and Tyranno SA SiC fibers as a function of coating fabrication temperature. For the comparison, those of uncoated fiber as a function of heating temperature are also shown in those figures. The tensile strengths of as-received Hi-Nicalon Type S and Tyranno SA ($S_{0,H}$ and $S_{0,T}$) were 2.6 and 2.5 GPa, respectively. While the tensile strengths, S of the coated fibers were degraded to 85 % of $S_{0,H}$ for Hi-Nicalon Type S and 80 % of $S_{0,T}$ for Tyranno SA after heating to 1400 °C, those were retained the $S_{0,H}$ and $S_{0,T}$ below the 1400 °C. Since the strength of the coated conventional Hi-Nicalon SiC fiber was remarkably degraded to 55 % of those of as received fiber even at 1000 °C[12], it is indicated that those advanced SiC fibers have much higher oxidation resistance. Among the various fabrication processes of SiC/SiC, FCVI is one of the best ones to produce a highly pure, stoichiometric, crystalline β-SiC matrix with a low thermal stress. Moreover process temperature (1100-1200 °C) is relatively lower compared with other fabrication process, which minimizes the fiber damage.[13] The present work shows the degradation of the tensile strength of both Mg-Si-O coated Hi-Nicalon Type S and Tyranno SA SiC fibers below 1400 °C was relatively small enough to apply a new oxygen resistance interphase.

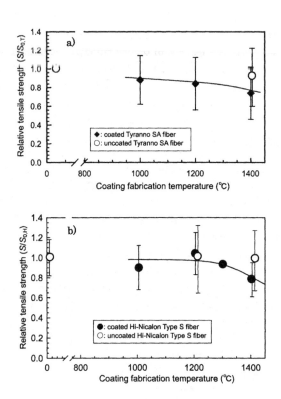

Fig. 5. Relative tensile strength of a) Tyranno SA and b) Hi-Nicalon Type S SiC fiber with Mg-Si-O coating as a function of coating fabrication temperature.

Figure 6 shows a typical load-displacement curve of the bending test for the coated Hi-Nicalon/SiC composite. The specimen showed some ductile failure behavior after the maximum load. The load did not become zero after the large deformation since the specimen was not turn off due to the tolerance of fibers. The average bending strength was about 70 MPa. This low bending strength was due to conventional SiC fiber and low matrix density (ca.60 %TD): the oxygen-resistance of Hi-Nicalon fiber is lower than that of advanced SiC fibers described above. Figure 7 shows a typical fracture surface of the composite after bending test. The cracking between fiber and interphase was observed, therefore the fiber pull-out was occurred in the figure. Those results are indicated that the Mg-Si-O layer has a potential to good properties as an interphase of SiC/SiC.

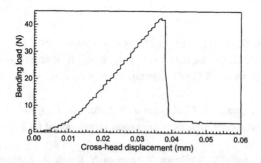

Fig. 6. Typical bending behavior of Hi-Nicalon/SiC composite with Mg-Si-O interphase.

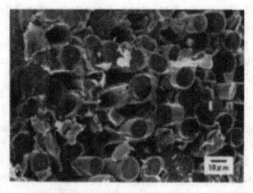

Fig. 7. Fracture surface of Hi-Nicalon/SiC composite with Mg-Si-O interphase.

CONCLUSION

Mg-Si-O coating on the advanced SiC fibers, Hi-Nicalon Type S and Tyranno SA, as an interphase of SiC/SiC composites was prepared by the alkoxide method. The surface on the fiber was coated completely by 5 times dipping in the coating solution of $[Mg]_T$=0.50 mol/dm^3 and $[Si]_T$=0.25 mol/dm^3. The coating was consisted of the mixture of MgO, SiO$_2$ and Mg$_x$Si$_z$O at 1400 °C in air. Compare with uncoated fiber, the tensile strength of coating fiber was retained below 1400 °C but slightly degraded at 1400 °C. Hi-Nicalon SiC fiber/SiC composite with the Mg-Si-O interphase was fabricated by FCVI process for a preliminary test. The fiber pull-out effect was observed in the fracture surface of the composites. Those results indicate that the Mg-Si-O layer has a potential to good properties as an interphase of SiC/SiC.

REFERENCES

[1]P. Fenici, A.J.F. Rebelo, R.H. Jones, A. Kohyama and L.L. Snead, "Current Status of SiC/SiC Composites R&D," *J. Nucl. Mater.*, **258-263**, 215-225(1998).

[2]A. Hasegawa, A. Kohyama, R.H. Jones, L.L. Snead, B. Riccardi, and P. Fenici, "Critical Issues and Current Satus of SiC/SiC Composites for Fusion," *J. Nucl. Mater.*, **283-287**, 128-137(2000).

[3]B. Riccardi, L. Giancarli, A. Hasegawa, Y. Katoh, A. Kohyama, R.H. Jones, and L.L. Snead, "Issues and Advances in SiC$_f$/SiC Composites Development for Fusion," *J. Nucl. Mater.*, **329-333**, 56-65(2004).

[4]T. Taguchi, N. Igawa, S. Jitsukawa, T. Nozawa, Y. Katoh, A. Kohyama, L.L. Snead, and J.C. McLaughlin, "Optimizing the Fabrication Process for Excellent Mechanical Properties in Stoichiometric SiC fiber/ FCVI SiC Matrix composites," *Ceram. Trans.*, 144, 69-79(2002).

[5]T. Taguchi, N. Igawa, R. Yamada, M. Futakawa, and S. Jitsukawa, "Fracture Behavior of High Densified SiC/SiC Composites Fabricated by Reaction Bonding: Effect of SiC Coating as Interface Layer," *Ceram. Eng. Sci. Proc.*, **22A**, 533-538(2000).

[6]T. Taguchi, T. Nozawa, N. Igawa, Y. Katoh, S. Jitsukawa, A. Kohyama, T. Hinoki, and L.L. Snead, "Fabrication of Advanced SiC Fiber/F-CVI SiC Matrix Composites with SiC/C Multi-layer Interphase," *J. Nucl. Mater.*, **329-333**, 572-576(2004).

[7]S. Suyama, Y. Itoh, T. Kameda and A. Sayano, "Development of SiC Fiber Reinforced Reaction-Sintered SiC Matrix Composite"; pp.89-96, in *Proc. of the 2nd IEA/JUPITER Joint International Workshop on SiC/SiC Ceramic Composites for Fusion Applications*, Edited by A. Kohyama, R.H. Jones and P. Fenici. The Japanese Society of Materials for Advanced Energy Systems, 1997.

[8]W.Y. Lee, E. Lara-Curzio, and K.L. More, "Multilayered Oxide Interphase Concept for Ceramic-matrix Composites," *J. Am. Ceram. Soc.*, **81**, 717-720(1998).

[9]Y. Seki, I. Aoki, N. Yamano and T. Tabara, "Preliminary Evaluation of Radwaste in Fusion Power Reactors," *Fusion Technology*, **30**,1624-30(1996).

[10]Takeda, A. Urano, J. Sakamoto and Y. Imai, "Microstructure and Oxidative Degradation Behavior of Silicon Carbide Fiber Hi-Nicalon Type S," *J. Nucl. Mater.*, **258-263**, 1594-1599(1998).

[11]T. Ichikawa, Y. Kohtoku, K. Kumagawa, T. Yamanura, and T. Nagawasa, "High-strength Alkali-resistant Sintered SiC Fibre Stable to 2200 °C," *Nature*, **391**, 773-775(1998).

[12]N. Igawa, T. Taguchi, R. Yamada, and S. Jitsukawa, "Mg-Si-Al-O Coating on Hi-Nicalon SiC Fiber by Alkoxide Method," *Ceram. Eng. Sci. Proc.*, **21B**, 237-242(2000).

[13]N. Igawa, T.Taguchi, L.L. Snead, Y. Katoh, S. Jitsukawa, and A. Kohyama, "Optimizing the Fabrication Process for Superior Mechanical Properties in the FCVI SiC Matrix/stoichiometric SiC Fiber Composite System," *J. Nucl. Mater.*, **307-311**, 1205-1209(2002).

POLYBOROSILAZANE-DERIVED CERAMIC FIBERS IN THE Si-B-C-N QUATERNARY SYSTEM FOR HIGH-TEMPERATURE APPLICATIONS

Samuel Bernard, David Cornu, Philippe Miele
Laboratoire des Multimatériaux et Interfaces (UMR CNRS 5615)
University Claude Bernard - Lyon 1
43 Bd du 11 novembre 1918
Villeurbanne, France, 69622

Markus Weinmann, Fritz Aldinger
Max-Planck-Institut für Metallforschung and Institut für Nichtmetallische Anorganische Materialien,
University Stuttgart, Pulvermetallurgisches Laboratorium
Heisenbergstrasse 5,
Stuttgart, Germany, 70569

ABSTRACT

Amorphous Si-B-C-N ceramic fibers prepared at 1000°C from a melt-processable boron-modified polysilazane $[B(C_2H_4SiCH_3NCH_3)_3]_n$ were annealed in the temperature range 1000-1800°C in a nitrogen atmosphere to identify the changes in the thermal and structural stability as well as in the related fiber strength. Fibers were shown to be extremely stable up to 1600°C without decomposition and measurable changes in their amorphous structure. At higher temperatures, X-ray diffraction and thermogravimetry analyses indicated that the structural and thermal properties of fibers were probably controlled by the carbothermal decomposition of a minor part of the silicon nitride phase providing Si_3N_4/SiC nanograins in the material at high temperature. The excellent strength retention after heat-treatment at 1600°C (1.3 GPa) is clearly related to the high structural and thermal stability of fibers. Between 1600°C and 1700°C, the fiber strength decreased to 0.9 GPa then dropped to about one-quarter the original value at 1800°C while structural changes were evident. With an excellent stability in air at 1300°C, these Si-B-C-N fibers are potential candidates for Continuous Fiber-reinforced Ceramic-matrix Composites (CFCCs).

INTRODUCTION

There is long interest and demand of aerospace industry for ceramic matrix composites because of their superior properties compared to monolithic ceramic materials. In a general way, such lightweight structural materials are made from a variety of continuous fibers and matrix combinations in which fibers, embedded in the ceramic matrix, act as the reinforcing phase in Continuous-reinforced Ceramic matrix Composites (CFCCs).[1] Performances of CFCCs for high-temperature aerospace and energy-related areas depend upon judicious selection of fibers with the proper chemical and physical properties of the ceramic which is employed.

Non-oxide ceramic fibers are particularly attractive for their light weight, thermal shock resistance, creep resistance and relatively high tensile strength and modulus values making them useful for aerospace applications.[2] Among them, those fibers prepared from organometallic/inorganic polymers are of particular interest because of their small diameter and flexibility to allow weaving and braiding. Furthermore, they exhibit desired properties via the control of the composition, structure and processability of the starting polymer. The silicon carbide Nicalon® fibers, produced via the curing and subsequent pyrolysis of polycarbosilane-based fibers, are the most popular polymer-derived ceramic fibers since they were developed by Yajima and co-workers about 28 years ago.[3] Nicalon® fibers were efficiently used in CFCCs because of their high tensile strength and potentially high oxidation resistance at moderated temperature.[4-5] Unfortunately, these first generation of commercially available ceramic fibers are known to be structurally, mechanically and chemically unstable above 1200°C in both oxidative and non-oxidative atmospheres making them inappropriate nowadays for aerospace applications which require thermal stability above 1500°C.[2]

For high temperature operation, the most desired critical fiber properties are high strength and stiffness and the reliable retention of these properties throughout the service life of the application. As a result, much research has been done to develop a new generation of fibers with improved mechanical, thermal and structural properties.[6-12] In particular, we have synthesized a boron-modified polysilazane which displays desired polymer properties to develop Si-B-C-N ceramic fibers.[12] For validating their potential in CFCCs, an understanding of the high-temperature stability and corrosion behavior is of critical interest. The present paper reports the changes in the microstructure and the tensile strength of developmental Si-B-C-N ceramic fibers which have been monitored as functions of heat treatments in a nitrogen atmosphere at various temperatures from 1000 to 1800°C. Their thermal stability in air is briefly discussed.

EXPERIMENTAL SECTION
General comments

All reactions leading to the preparation of the polymer $[B(C_2H_4SiCH_3NCH_3)_3]_n$ were carried out in a purified argon atmosphere using standard Schlenk techniques as previously reported.[12] Argon, nitrogen and ammonia with 99.999% purity were used during the specimen preparation.

Tensile tests and diameter values were obtained at room temperature from single filaments with a gauge length of 10 mm. 50 single filaments were analyzed for each test. The diameter ϕ of each filament was measured by laser interferometry and an average diameter was calculated from the 50 as-obtained values. Single filament tensile properties were determined using a standard tensile tester (Adamel Lhomargy DY 22). The strength distribution was described by Weibull statistics[13] and the average room-temperature value of fibers was estimated for a failure probability $P = 0.632$.

High-temperature thermogravimetric analysis (HT-TGA, Netzsch STA 501 equipment) of fibers was performed in a nitrogen atmosphere (1000-1800 °C; heating rates: 5 °C/min (T < 1400 °C) and 2 °C/min (T > 1400 °C)) using graphite crucibles.

XRD of amorphous Si-B-C-N fibers was achieved using a Philipps PW 3040/60 X'Pert PRO X-ray diffraction system. Fibers were crushed prior characterization.

Fiber morphology was observed by scanning electron microscopy (SEM) with field emission equipment, Hitachi S800.

Fiber preparation

The synthesis and characterization of the melt-processable boron-modified polysilazane $[B(C_2H_4SiCH_3NCH_3)_3]_n$ were previously described.[12]

The polymer was continuously spun around 80°C and drawn via a take-up drum into an endless fiber (monofilament) using a lab-scale melt-spinning apparatus set up in a dry glove-box filled with nitrogen. The as-spun fiber, wound on the drum to prevent crimping of fibers during conversion, was transferred to a curing and pyrolysis furnace equipped with a silica glass tube directly connected to the glove-box in order to minimize contamination by oxygen (air) and moisture. It was first cured in a mixture of ammonia and nitrogen (70:30) atmosphere (25°C-200°C, atmospheric pressure) at a heating rate of 0.5°C/min and the following heat-treatment was carried out in a pure nitrogen atmosphere (P_{N2} = 1atm) to 1000°C: 1°C/min for hold times of 1h. After such treatments, the fine-diameter fiber was no longer sensitive to oxygen and moisture and could be handled in the open air. As-prepared fibers were black colored and of flexible form. Elemental compositions showed that contamination with oxygen (O < 2wt%) was negligible.[12] Typical chemical formula of as-pyrolyzed fibers was $Si_{3.0}B_{1.0}C_{5.0}N_{2.4}$.

Additional heat-treatments (5°C/min) were performed in a high-temperature graphite furnace in a nitrogen atmosphere (P_{N2} = 1atm) in the temperature range 1400-1800°C for hold times of 2h at each intermediate temperature.

Oxidation tests (5°C/min) in air were conducted in a silica furnace by exposing the as-prepared Si-B-C-N fibers to temperatures up to 900°C for hold times of 10h then in an alumina furnace for exposures at 1300°C and 1500°C for hold times 15min.

RESULTS AND DISCUSSION
Thermal stability and microstructural changes from 1000 to 1800°C in a nitrogen atmosphere

The thermal stability of fibers has been measured through their tensile strength retention after annealing in a nitrogen atmosphere for 2h at temperatures in the range 1400°C-1800°C (Fig. 1).

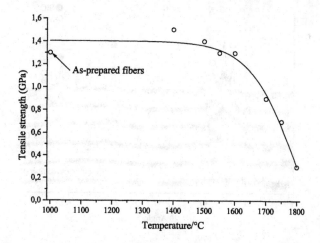

Fig. 1. Tensile strength of the Si-B-C-N fibers after 2h exposure at elevated temperature in N_2.

Without optimizing conditions, the as-prepared Si-B-C-N fibers showed an average room-temperature tensile strength of 1.3 GPa (Table I) due to their homogeneous amorphous structure and small average diameters (Table I) due to a tailored spinnability of the starting polymer.

The average room-temperature tensile strength was maintained around 1.3 GPa after thermal treatments at 1600°C. These results show that the Si-B-C-N fibers exhibit excellent mechanical stability at high temperature (Fig. 1). Although different atmosphere were applied (nitrogen for SiBCN fibers and argon for SiC fibers), these fibers have greater retention of their original strength at high temperature compared to the strength retention reported for Hi-Nicalon® and Hi-Nicalon type S® SiC fibers. The latter demonstrated a loss of ~50% after exposure in argon at 1600°C.[14] Studies of thermal stability in an argon environment are in progress.

Table I. Tensile strength and diameter values of the Si-B-C-N fibers annealed at different temperatures from 1400 to 1800°C in a nitrogen atmosphere for hold times of 2h.

As-prepared fibers	Temperature/°C	1400	1500	1550	1600	1700	1750	1800
1.3	Average strength/GPa	1.5	1.4	1.3	1.3	0.9	0.7	0.3
12.0	Diameter/μm	12.4	11.7	12.0	10.9	12.6	11.5	11.0

Above 1600°C, a continuous decrease in strength occurred and the average tensile strength of fibers dropped to 0.3 GPa as fibers were exposed at 1800°C.

In order to understand the tensile strength behavior, XRD analyses were investigated. XRD patterns of the as-prepared Si-B-C-N fibers and those annealed at different temperatures from 1400 to 1800°C for 2h in N_2 are reported in Figure 2.

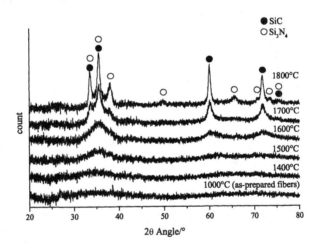

Fig. 2. XRD patterns of the Si-B-C-N fibers after annealing to 1800°C in N_2 for hold times of 2h.

38

As-prepared Si-B-C-N fibers are predominantly amorphous since XRD analysis showed no diffraction peaks. Fibers retained their amorphous structure through heat-treatments at 1600°C despite of the appearance of broad peaks due to the nucleation of nanosized SiC crystals at 1600°C. It is clear that, in presence of additional elements such as boron, the onset of crystallization in Si-B-C-N materials is greatly retarded compared to that of polymer-derived SiC and Si_3N_4 materials which starts around 1000°C.[15] In this way, Si-B-C-N fibers formed predominantly SiC crystals at 1700°C. Raising the annealed temperature to 1800°C resulted clearly in the presence, besides SiC, of Si_3N_4 as distinct part of the crystalline fraction. We can therefore postulated that the drop of the fiber strength starts while the amorphous-to-crystalline transition occurs.

The HT-TGA measurements between 1000°C and 1800°C supported these interpretations since they showed that the thermal stability extended up to 1600°C (Fig. 3). The high thermal and structural stability of the Si-B-C-N fibers in a nitrogen atmosphere is documented by the low continuous mass loss starting from 1600°C up to 1800°C.

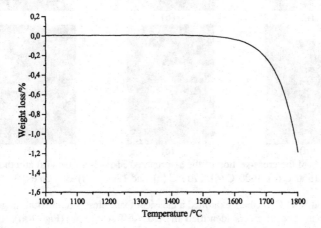

Fig. 3. Relative change in mass of the as-prepared Si-B-C-N fibers annealed up to 1800°C in N_2.

No weight loss was detected during the heat-treatment of fibers up to 1600°C, whereas it reached ~30 wt% for Si-C-O fibers and ~6.6 wt% for low oxygen SiC fibers at 1600°C.[16] Above 1600°C, the degradation of fibers slowly started and the measured weight loss was as low as -1.2 wt% at 1800°C.

Based on XRD and thermodynamic data, the observed degradation of Si-B-C-N materials at high temperature often results from the carbothermal decomposition of the Si_3N_4 phase which is formed by in situ crystallization of the amorphous Si-B-C-N network providing SiC and elemental nitrogen gas according to the following equation (1)[17]:

$$Si_3N_4 + 3C \longrightarrow 3SiC + 2N_2 \quad (1)$$

In the present paper, we cannot conclude that the above reaction occurs in the SiBCN fibers simply based on XRD data (nucleation of nanosized Si_3N_4/SiC grains) and TGA experiments. Raman/FTIR spectroscopies and TEM and chemical analyses are in progress to determine the exact composition and the phase evolution of the complex nanostructure of these fibers at high temperature. Previous observations are nicely reflected in SEM images (Fig. 4).

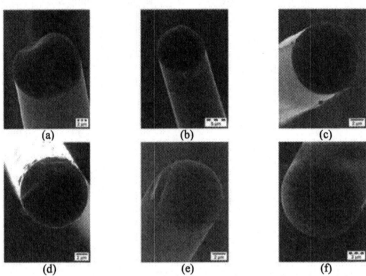

Fig. 4. SEM images of the cross-section of the as-prepared Si-B-C-N fibers (a) and those annealed in N_2 at 1400°C (b), 1500°C (c), 1600°C (d), 1700°C (e), and 1800°C (f) after 2h exposure.

As-prepared Si-B-C-N fibers exhibited a classical featureless microtexture in good agreement with the amorphous state of fibers identified in XRD diffractograms (Fig. 4(a)). The glassy-like microtexture of the as-prepared Si-B-C-N fibers was preserved through heat-treatments at 1400°C (Fig. 4(b)), 1500°C (Fig. 4(c)) and 1600°C (Fig. 4(d)). Due to the amorphous-to-crystalline conversion, fiber cross-section changed from glassy to granular between 1600°C (Fig. 4(d)) and 1700°C (Fig. 4(e)). When fibers were exposed at 1800°C (Fig. 4(f)), a coarse-grained microtexture was clearly observed. It reflected the polycristalline nature of the fibers which, in the end, corroborated the conclusions drawn from TGA and XRD data.

Thermal stability in air up to 1500°C

It is known that oxidation in air (P_{air} = 1atm) of Si-based ceramics proceeds by a passive oxidation mechanism in which diffusion of oxygen through a growing silica layer is often the rate-determining step according to Eqs (2) and (3)[18]:

$$2SiC + 3O_2 \longrightarrow SiO_2 + 2CO \quad (2)$$
$$Si_3N_4 + 3O_2 \longrightarrow 3SiO_2 + 2N_2 \quad (3)$$

Typical cross-section of fibers exposed at 500°C and 900°C for hold times of 10h and 1300°C as well as 1500°C for hold times of 15 min are shown in Figs. 5. As-prepared fibers oxidized in air up to 900°C for aging of 10h did not change in appearance (Figs. 5(a), and (b)). In particular, their surface remained smooth without defects. At 1300°C for 15min (Fig. 5(c)), the fiber morphology seemed also similar to that of as-prepared fibers whereas the formation of deleterious pores and bubbles in a glassy SiO_2 layer was pronounced as the exposure temperature reached 1500°C (Fig. 5(d)). The formation of the SiO_2 layer can be correlated to the weight gain (+ 2.9 $_{wt}$%) measured by thermogravimetry after a thermal treatment at 1500°C in a previous study.[12] Further works are in progress to measure the influence of the oxidation on the tensile strength of fibers.

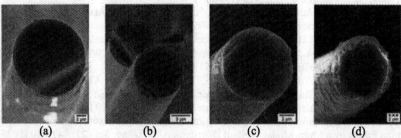

(a)   (b)   (c)   (d)

Fig. 5. SEM images of the as-prepared Si-B-C-N ceramic fibers oxidized at 500°C (a) and 900°C (b) for 10h, 1300°C (c), and 1500°C (d) for 15min.

CONCLUSION

In the present paper, we have studied the changes in properties of developmental Si-B-C-N fibers on exposure at elevated temperatures in a nitrogen atmosphere. We found that the fibers were extremely stable up to 1600°C without decomposition and crystallization and these fibers were even able to retain their initial tensile strength at such temperatures. Above 1600°C, the tensile strength became to decrease slowly but its value was maintained above 1GPa before it dropped at 1700°C as the nucleation of SiC and Si_3N_4 grains occurred and elemental nitrogen gas was probably lost. When exposed in air, Si-B-C-N fibers showed a good stability up to 1300°C. A glassy SiO_2 layer with pores and bubbles was deposed when fibers were exposed at 1500°C.

REFERENCES

[1]J. R. Strife, J. J. Brennan, and K. M. Prewo, "Status of Continuous Fiber-Reinforced Ceramic Matrix Composite Processing Technology," *Ceram. Eng. Sci. Proc.*, **11**, 871-919 (1990).

[2]K. Okamura, "Ceramic Fibres from Polymer Precursors," *Composites*, **18**, 107-20 (1987).

[3]S. Yajima, J. Hayashi, M. Omari, and K. Okamura, "Development of a SiC Fibre with High Tensile Strength," *Nature*, **261**, 683-85 (1976).

[4]C.-H. Andersson, and R. Warren, "Silicon Carbide Fibres and Their Potential for Use in Composite Material. Part I," *Composites*, **15**, 16-24 (1984).

[5]R. Warren, and C.-H Andersson, "Silicon Carbide Fibres and Their Potential for Use in Composite Material. Part II," *Composites*, **15**, 101-11 (1984).

[6]M. Takeda, Y. Imai, H. Ichikawa, and T. Ichikawa, "Properties of the Low Oxygen Content SiC Fiber on High Temperature Heat Treatment," *Ceram. Eng. Sci. Proc.*, **12**, 1007-18 (1991).

[7]J. Lipowitz, J. A. Rabe, and G. A. Zank, "Polycristalline SiC Fibers from Organosilicon Polymers," *Ceram. Eng. Sci. Proc.*, **12**, 1819-31 (1991).

[8]M. Takeda, J.-I. Sakameto, Y. Imai, H. Ichikawa, and T. Ichikawa, "Properties of Stoichiometric silicon carbide Fiber Derived from Polycarbosilane," *Ceram. Eng. Sci. Proc.*, **15**, 133-41 (1994).

[9]M. Sacks, W. Toreki, C. D. Batich, and G. J. Choi, "Preparation of Boron-Doped Silicon Carbide Fibers," *U.S. Patent* No 5,792,416 (1998).

[10]H. P. Baldus, M. Jansen, and D. Sporn, "Ceramic Fibers for Matrix Composites in High-Temperature Engine Applications," *Science*, **285**, 699-703 (1999).

[11]D. Schawaller, and B. Clauss, "Preparation of Non-Oxide Ceramic Fibers in the System Si-C-N and Si-B-C-N," in *High Temperature Ceramic Matrix Composites*, W. Krenkel, R. Naslain, and H. Schneider eds., Wiley-VCH, Weinheim, p. 56-61 (2001).

[12]S. Bernard, M. Weinmann, P. Gerstel, P. Miele, and F. Aldinger, "Boron-Modified Polysilazane as a Novel Single-Source Precursor for SiBCN Ceramic Fibers: Synthesis, Melt-spinning, Curing and Ceramic Conversion, " *J. Mater. Chem.*, **15**, 289-299 (2005).

[13]K. Goda, and H. Fukunaga, "The Evaluation of the Strength Distribution of Silicon Carbide and Alumina Fibers by a Multi-Modal Weibull Distribution," *J. Mater. Sci.*, **21**, 4475-80 (1986).

[14]M. Takeda, A. Urano, J.-I. Sakameto, and Y. Imai, "Microstructure and Oxidation Behavior of Silicon Carbide Fibers Derived from Polycarbosilane," *J. Am. Ceram. Soc.*, **83**, 1171-76 (2000).

[15] R. Riedel, "Advanced Ceramics from Inorganic Polymers," in *Materials Science and Technology: A Comprehensive Treatment*, R. W. Cahn, P. Haasen, and E. J. Kramer eds., Wiley-VCH, Weinheim, p. 1-50 (1996).

[16]Z. H. Chu, Y. C. Song, C. X. Feng, Y. S. Xu, and Y. B. Fu, "A Model SiC-Based Fiber with a Low Oxygen Content Prepared from a Vinyl-Containing Polycarbosilane Precursor," *J. Mater. Sci. Let.*, **20**, 585-87 (2001).

[17]H. J. Seifert, J. Peng, J. Golczewski, and F. Aldinger, "Phase Equilibria of Precursor-Derived Si-(B-)C-N Ceramics," *Appl. Organometal. Chem.*, **15**, 794-808 (2001).

[18]W. L. Vaughn, and H. G. Moahs, "Active-to-Passive Transition in the Oxidation of Silicon Carbide and Silicon Nitride in Air," *J. Am. Ceram. Soc.*, **73**, 1540-43 (1990).

MICROTEXTURAL AND MICROSTRUCTURAL EVOLUTION IN POLY[(ALKYLAMINO)BORAZINE]-DERIVED FIBERS DURING THEIR CONVERSION INTO BORON NITRIDE FIBERS

Samuel Bernard, Fernand Chassagneux, David Cornu, and Philippe Miele
Laboratoire des Multimatériaux et Interfaces (UMR CNRS 5615)
University Claude Bernard - Lyon 1
43 Bd du 11 novembre 1918
Villeurbanne, France, 69622

ABSTRACT

Boron nitride (BN) fibers were efficiently prepared from a B-aminoborazine-based polymer according to the Polymer-Derived Ceramic (PDC) route *via* melt-spinning and heat-treatments up to 1800°C in a controlled atmosphere. The microtextural and microstructural changes in the material during the polymer-to-ceramic conversion were investigated by means of electron microscopy and XRD observations. The microtexture of the fibers was featureless as glassy-like materials when fibers were exposed at temperatures below 1400°C. Mechanical properties of such amorphous fibers were poor at these temperatures. Upon further heating to 1500°C, the microstructure changed from disordered nanocrystals embedded into an amorphous phase to a turbostratic phase. This amorphous-to-crystalline transition was accompanied with a large increase in the mechanical properties. At 1800°C, the microtexture of the fibers was coarse-grained and was correlated to the identification of a "meso-hexagonal" BN phase with basal layers almost aligned along the *fiber*-axis in the material. Polycrystalline BN fibers exhibited high mechanical properties (σ = 1.4 GPa, E = 360 GPa) after curing of the polymer fibers at 400°C and subsequent pyrolysis of cured fibers at 1800°C.

INTRODUCTION

Inorganic/organometallic preceramic polymers play a major role in the preparation of shape-controlled non-oxide ceramics due to their adjustable processability.[1-2] One significant advantage of the Polymer-Derived Ceramic (PDC) route is that flexible and small-diameter ceramic fibers with desired properties can be prepared via spinning of tractable polymers, curing of the as-spun fibers and pyrolysis of the resulting cured fibers according to the processing scheme presented in Fig. 1. These small-diameter and flexible fibers are generally ideally suited for Continuous Fibers-reinforced Ceramic matrix Composites (CFCCs) since they are easily weavable to produce net-shape fiber preforms that are subsequently infiltrated by a matrix.[3]

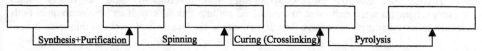

Fig.1. Processing scheme for the preparation of ceramic fibers from preceramic polymers

The preparation of amorphous silicon carbide fibers from polycarbosilane clearly illustrated this method.[4] This process is also related to that used for the preparation of carbon fibers from polyacrilonitrile.[5] These examples highlight the utility of the polymer precursor approach to fiber research. Unfortunately, such fibers are either mechanically and chemically unstable above 1000°C in both oxidative and non-oxidative environment (SiC fibers) or poorly resistant to oxidation in air above 400°C (carbon fibers). For CFCCs fabrication, high-modulus and strength oxidation resistant fibers with small diameter are ideal. Additionally, reinforcing fibers should be capable of retaining the structure, stiffness and strength under processing and service conditions. Keeping these in view, we have investigated the preparation of polycristalline boron nitride (BN) fibers from B-aminoborazine-derived polymers (= poly[(alkylamino)borazine]).[6-8] Indeed, BN fibers should be of a major interest in aerospace application as a reinforcing phase in a new generation of composites (BN/BN composites). As examples, in accordance to the good resistance of *h*-BN against oxidation and a graphite-like structure[9], these materials could favorably replace the traditional carbon/carbon composites which are oxidized around 400°C. Furthermore, in contrast to a large majority of non-oxide ceramic fibers, BN fibers offer the possibility to be used in radiation-transparent structures due to the low dielectric constant of *h*-BN. As the requirements to obtain high mechanical properties are rather stringent, an understanding of the polymer-to-ceramic conversion is of critical importance. In the present paper, the microtextural/microstructural changes occurring during the preparation of BN fibers were investigated by electronic microscopy and were correlated to XRD and tensile tests data.

EXPERIMENTAL SECTION
General comments
All reactions leading to the preparation of the starting polymer were carried out in a purified argon atmosphere using standard Schlenk techniques as previously reported.[8] Nitrogen and ammonia with 99.999% purity were used during the fiber preparation.

Tensile tests and diameter were obtained from single filaments with a gauge length of 10 mm. 50 single filaments were analyzed for each test. The diameter ϕ of each filament was measured by laser interferometry and an average diameter was deducted. Single filament tensile properties were determined using a standard tensile tester (Adamel Lhomargy DY 22). Young's modulus and failure strain were averaged from the 50 tests taking into account the system compliance. The strength distribution was described by Weibull statistics[10] and the average room temperature tensile strength σ was estimated for a failure probability P=0.632.

X-ray diffraction (XRD) measurements were performed using a Philips diffractometer (Cu$K\alpha$ radiation; λ = 1.5406 Å at 40 kV and 30 mA). Fibers were crushed prior characterization.

Scanning electron microscopy (SEM) (Hitachi S800) was used to investigate the cross-sectional microtexture of fibers. An Au/Pd film was deposed on fibers prior observation.

Transmission electron microscopy (TEM) was investigated using a Topcon 002B microscope. Samples were embedded in a resin and cut into thin foils with an ultramicrotome. Foils were then set on microgrids to observe the microtexture in the longitudinal sectional thin specimens.

Fiber preparation
The synthesis and characterization of the poly[(2,4,6-trimethylamino)borazine] (polyMAB) were previously described.[8]

Endless as-spun fibers, 15 μm in diameter, were prepared by the extrusion of the polyMAB around 180°C followed by the drawing of the emerging monofilament using a lab-scale spinning apparatus set up in a glove-box filled with nitrogen. The as-spun fiber was transferred into a furnace equipped with a silica glass tube directly connected to the glove-box in order to minimize contamination by oxygen (air) and moisture. Curing and pyrolysis were achieved at atmospheric pressure under the tension imposed by the important shrinking effects which occur in the fibers wound on the drum. The importance of the tension was previously reported to improve tensile strength and Young's modulus by producing straighter and stiffer fibers.[6] The as-spun fiber was cured in an ammonia atmosphere (25°C-400°C, 0.8°C/min), then the as-cured fiber was treated in an ammonia atmosphere to 1000°C (0.8°C/min) with a dwell time of 1h to remove the majority of carbon-based groups bearing by the polymer. After such treatments, the fiber was less sensitive to oxygen and moisture and could be handled and quickly transferred into a second furnace in the open air. An additional heat-treatment (10°C/min) was performed in a graphite furnace in a nitrogen atmosphere up to 1800°C for hold times of 1h. As-pyrolyzed BN fibers, 7.5 µm in diameter, were white colored and of flexible form. Their typical elemental composition was previously reported.[8]

RESULTS AND DISCUSSION
XRD investigations

Fig. 1. shows X-ray diffraction results taken on crushed fibers during the preparation of BN fibers through curing at 400°C (as-cured) then pyrolysis up to 1800°C (as-pyrolyzed).

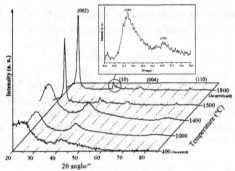

Fig. 1. X-ray diffraction patterns of polymer (polyMAB) fibers during their conversion into ceramic (BN) fibers up to 1800°C. All reflections can be assigned to h-BN.

As-cured fibers exhibited every broad X-ray reflections of h-BN with low relative peak intensity. XRD patterns did not change in appearance on further heating in an ammonia atmosphere at 1000°C and the amorphous nature of the BN phase was even preserved after pyrolysis to 1400°C. Only above 1400°C was crystallization detectable by XRD. An increase in the intensities of the (00l) reflections, which means an ordering of the structure in basal planes, accompanied by a sharpening of these reflections were detected at 1500°C. On further heating to 1800°C, fibers exhibited the majority of h-BN reflections. In particular, the diffraction from the (100) and (101) peaks was observed, which should indicate the formation of a three-dimensional crystal structure. However, the crystallization was not fully completed in as-pyrolyzed BN fibers as indicated by the lack of (h0l) type of reflection in the corresponding XRD pattern (1800°C). The results indicated a meso-hexagonal structure with an incomplete three-dimensional ordering.

In support of the amorphous-to-crystalline transition was the observations that the grain growth started from 1400°C to form a stabilized crystallization at 1800°C (Fig. 2).

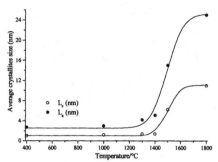

Fig. 2. Dependence of the average crystallite size of fibers during their conversion up to 1800°C.

The crystallites sizes at different temperatures were estimated on the basis of the FWHM of the (002) ($2\theta = 26.76°$) and (10) ($2\theta = 41.60°$) peaks using the Scherrer equation (Eq. 1)[14]:

$$B = K\lambda/(Lcos\theta) \hspace{2cm} (Eq. 1)$$

,where $K = 0.9$ (L_c) and 1.84 (L_a), λ, the wavelength of the X-rays (($\lambda = 0.15418$ nm), L represents the average stack height (L_c) and the average length (L_a) of the crystallites, θ is the diffraction angle of the corresponding reflections and B is the FWHM of the peaks.
For example, the average value of the particle size (L_a) increased from <6 nm at 1400°C to ~16 nm at 1500°C and reached ~25 nm after pyrolysis at 1800°C.

SEM observations
Fig. 3 shows the microtextural changes during the preparation of BN fibers. SEM observations are entirely consistent with XRD data. The cross-section of the as-spun fibers is common to that of organic polymer fibers. When fibers were cured at 400°C, then pyrolyzed at 1000°C and subsequently at 1400°C, their microtexture remained featureless reflecting the amorphous nature of the material below 1400°C. The amorphous-to-crystalline transition occurred through a featureless-to-granular microtexture transformation as fibers were heated from 1400 to 1500°C. The granular microtexture reflected the polycristalline nature of the BN fibers after such heat-treatments. Upon further heating at 1800°C, the coarsening of the grains through the cross-section of the as-pyrolyzed BN fibers was more pronounced in relation with the improvement of the structural ordering of the BN phase.

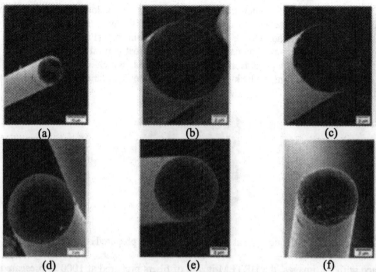

(a) (b) (c)

(d) (e) (f)

Fig. 3. SEM micrographs of the polymer fibers (a) and fibers prepared at 400°C (as-cured fibers) (b), 1000°C (c), 1400°C (d), 1500°C (e) and 1800°C (as-pyrolyzed fibers) (f).

TEM observations

In a previous work, TEM studies of BN fibers prepared at 1800°C indicated that the large crystallites were stacked in a mesohexagonal BN ordering and nearly oriented along the *fiber*-axis.[11] In order to complete TEM investigations, microstructures of intermediate fibers heated at 1000°C, 1400°C and 1500°C were observed in the longitudinal section.

Consistently with XRD results, the SAED of the fibers heated at 1000°C showed an amorphous halo imposed on the diffuse (002) arcs and poorly-resolved (10)/(100), (004) and (11)/(110) arcs which indicate the presence of a poorly-ordered structure (Fig. 4).

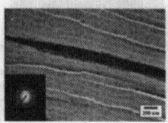

Fig. 4. BF TEM image of the longitudinal thin sectional fibers prepared at 1000°C.

In addition, the BF TEM image of such fibers was featureless and of low contrast and, in particular, it showed that the microstructure of the fibers prepared at 1000°C was damaged during the specimen preparation when the Leitz diamond knife cut the fibers into thin foils (Fig. 4). As an explanation, the knife moving along the *fiber*-axis damaged the poor microstructural cohesion of the fibers by separating the matter into elongated fibrils. This phenomena, which was also observed in fibers prepared at 1400°C, results from the lack of regular crystalline BN planes in the fibers heated below 1400°C as illustrated in Fig. 5.

Fig. 5. BF TEM image of the longitudinal thin sectional fibers prepared at 1400°C.

In accordance with BF images, the HRTEM image of fibers prepared at 1000°C presented in Fig. 6 exhibited a completely disordered microstructure with nanograins of varying size (less than 5 basal plane in thickness, d_{002} values around 0.350 nm) and orientation arising from an amorphous phase.

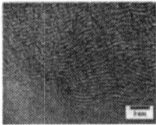

Fig. 6. HRTEM image of the longitudinal thin sectional fibers prepared at 1000°C.

It should be mentioned that a similar nanostructure was also detected in fibers prepared at 1400°C (d_{002} ~0.347 nm). As a consequence, we can postulate that the poor microstructural ordering and the low level of orientation which compose the fibers prepared below 1400°C result in poor mechanical properties (strength and modulus) as reported in Table 1. In contrast, in such poorly crystallized fibers, the slow crack propagation, due to the lack of well-defined crystalline planes in the structure, result in a high fiber flexibility and therefore, in a high strain (Table 1).

As observed in XRD and SEM analysis, the amorphous-to-crystalline transition was also reflected in TEM images when fibers were heated from 1400 to 1500°C. Figure 7 shows the BF TEM images of thin sectional fibers prepared at 1500°C.

Table I. Changes in the mechanical properties during the preparation of BN fibers.

As-pyrolyzed fibers (1800°C)	Temperature/°C	400 (as-cured)	1000	1400	1500
1.4	Average strength/GPa	0.2	0.3	0.4	0.7
360	Average modulus/GPa	17	23	40	150
0.35	Average strain/%	1.1	1.2	0.9	0.5

Fig. 7. BF TEM images of the longitudinal thin sectional fibers prepared at 1500°C.

Consistent with SEM observations, a granular microtexture is clearly seen in such fibers. No damage of fibers was observed which means that the microstructural cohesion of the BN network is stronger in accordance with the higher values of mechanical properties compared to those measured after thermal treatment at 1000 and 1400°C (Table 1). In contrast, as the microtexture becomes more grained, cracks proceed along the well-defined crystal planes, *i.e.*, along the basal (002) sheets. Such fibers are therefore accompanied by a higher brittleness and, therefore, a lower flexibility. SAED, which are composed of well-resolved (002), (10)/(100), (004) and (110) arcs, were also representative of a better ordered material. Furthermore, the HRTEM image presented in Fig. 8 well-reflected the BN phase evolution between the fibers prepared at 1000°C (Fig. 6), with a high proportion of a-BN and those heated at 1500°C exhibiting better ordered and extended (002) layers with $d_{002} = 3.40$ Å.

Fig. 8. HRTEM image of the longitudinal thin sectional fibers prepared at 1500°C.

The decrease in the d_{002} values from 1400 (d_{002} = 0.347 nm) to 1500°C (d_{002} = 0.340 nm) which were higher than the ideal value of h-BN (d_{002} = 0.333 nm) were assumed as an indicator in assigning the turbostratic structure to the fibers were prepared at 1500°C. On further heating to 1800°C, as-pyrolyzed fibers exhibited a well-ordered homogeneous nanostructure and basal layers nearly oriented along the *fiber*-axis in good agreement with their high mechanical properties as shown in a previous work.[11] d_{002} values decreased up to 3.35 nm indicating a meso-hexagonal structure.

CONCLUSION

The present study reported the microtextural/microstructural changes which occurred during the preparation of BN fibers from a B-aminoborazine-based polymer according to the Polymer-Derived Ceramic (PDC) route. In particular, it was shown that the microstructure of fibers prepared below 1400°C consisted of a mixture of disordered nanocrystals and amorphous regions, whereas the large crystallites tended to stack in a meso-hexagonal ordering and to nearly align along the *fiber*-axis in fibers prepared above 1400°C. The amorphous-to-crystalline transition occurred with the increase in the mechanical properties.

REFERENCES

[1]M. Peuckert, T. Vaahs, and M. Brück, "Ceramics from Organometallic Polymers," *Adv. Mater.*, **2**, 398-404 (1990).

[2]J. Bill, and F. Aldinger, "Precursor-Derived Covalent Ceramics," *Adv. Mater.*, **7**, 775-87 (1995).

[3]J. R. Strife, J. J. Brennan, and K. M. Prewo, "Status of Continuous Fiber-Reinforced Ceramic Matrix Composite Processing Technology," *Ceram. Eng. Sci. Proc.*, **11**, 871-919 (1990).

[4]S. Yajima, J. Hayashi, M. Omari, and K. Okamura, "Development of a SiC Fibre with High Tensile Strength," *Nature*, **261**, 683-85 (1976).

[5]E. Fitzer, and M. Heine, "Carbon Fiber Manufacture and Surface Treatment," In Fiber Reinforcements for Composite Materials, A. R. Bunsell (eds), Elsevier, New York, pp. 79-148 (1988).

[6]S. Bernard, K. Ayadi, J.-M. Létoffé, F. Chassagneux, M.-P. Berthet, D. Cornu, and P. Miele, "Evolution of Structural Features and Mechanical Properties During the Conversion of Poly[(methylamino)borazine] Fibers into Boron Nitride Fibers," *J. Sol. State. Chem.*, **177**, 1803-10 (2004).

[7]B. Toury, S. Bernard, D. Cornu, F. Chassagneux, J.-M. Létoffé, and P. Miele, "High-Performances Boron Nitride Fibers Obtained from Asymmetric Alkylaminoborazine," *J. Mater. Chem.*, **13**, 274-279 (2003).

[8]S. Bernard, D. Cornu, P. Miele, H. Vincent, and J. Bouix, "Pyrolysis of Poly[2,4,6-tri(methylamino)borazine] and its Conversion into BN Fibres," *J. Organomet. Chem.*, **657**, 91-97 (2002).

[9]R. S. Pease, "Crystal Structure of Boron Nitride," *Nature*, 5, 722-723 (1950).

[10]K. Goda, and H. Fukunaga, "The Evaluation of the Strength Distribution of Silicon Carbide and Alumina Fibers by a Multi-Modal Weibull Distribution," *J. Mater. Sci.*, **21**, 4475-80 (1986).

[11]S. Bernard, F. Chassagneux, M. P. Berthet, H. Vincent, and J. Bouix, "Structural and Mechanical Properties of a High-Performance BN Fibre," *J. Eur. Ceram. Soc.*, **22**, 2047-59 (2002).

INVESTIGATIONS ON GROWTH OF TEXTURED AND SINGLE CRYSTAL OXIDE FIBERS USING A QUADRUPOLE LAMP FURNACE

W. Yoon and W. M. Kriven
Department of Materials Science and Engineering, University of Illinois at Urbana-Champaign
Urbana IL 61801, USA

ABSTRACT

Fine ceramic oxide fibers are widely used as reinforcements in composites for high temperature applications. The primary goal of this research effort was to investigate the growth of single crystal or textured oxide fibers by heat treatment of polycrystalline or amorphous, extruded precursor fibers.

Mullite was selected for this study due to its excellent chemical stability, creep resistance and strength at high temperatures. Precursor polycrystalline fibers of mullite were prepared by (a) sol infiltration of silk or cotton threads and (b) extrusion. Green fibers of ~10 μm and ~150 μm could be made by sol infiltration (of single silk filament) and by extrusion, respectively. A quadrupole lamp furnace, with a small, disc-shaped, hot zone was used for the heat treatment. The effect of temperature and traverse rate through the hot zone, on the microstructure of the polycrystalline precursor fibers, was evaluated.

Mullite whiskers were synthesized and used as a template for introducing texturing in extruded mullite fibers. Textured growth of mullite fiber with elongated grains, ~400μm in length and aligned along the long-axis of the fibers, was achieved with heat treatment. The whisker templated fibers were further subjected to repeated cycles of heat treatment to form a transparent oxide. Preliminary investigations suggest that the transparent part of the heat treated fiber is a single crystal. However, rigorous optical microscopy, x-ray diffraction and TEM investigations are underway to confirm our finding.

INTRODUCTION

Small diameter oxide fibers are excellent reinforcements for composites for use in high temperature applications[1]. Mullite is a strong candidate material for advanced structural applications at high temperature, because of its good chemical stability, creep resistance and high temperature strength[2,3].

Single crystal or textured, oxide fibers usually exhibit better properties than do amorphous or polycrystalline oxide fibers. Crystallization and synthesis of the mullite fiber were studied with different methods[4,5,6,7]. The primary objective of this research was to investigate the potential to produce single crystal or textured fibers of mullite composition using polycrystalline or amorphous fibers as precursors. Extruded and mullite, sol-infiltrated, green fibers were the starting materials for these investigations.

In this research, the microstructure and properties of the resultant oxide ceramic fiber can be strongly influenced by the precursor fibers. Different precursor mullite fibers, both commercially available and those prepared using various methods were investigated in this study. These included (a) mullite sol infiltrated organic threads (single silk fiber and cotton thread), and (b) polycrystalline fibers extruded through a small orifice die. In order to produce textured fibers by extrusion, the use of mullite whiskers as a templating aid was also studied[8]. The precursor fibers were subjected to heat treatment in the quadrupole furnace. The effect of furnace temperature, traverse rate of the fiber through the furnace hot zone, and the number of heating cycles on microstucture modification of the fiber was studied.

51

EXPERIMENTAL PROCEDURE

A quadrupole lamp furnace was used for heat treatment of the polycrystalline ceramic oxide fibers to grow textured or single crystal fibers. Four halogen lamps (OSRAM 15V, 10 W, Osram, Germany) with ellipsoidal reflectors were mounted in the furnace and served as a heat source. The furnace hot zone shape and size was determined by measurement of both axial and radial distribution of temperature using an R-type thermocouple (Pt/Pt13%Rh). The axial temperature profile was measured as a function of position and furnace power.

As in any radiation furnace, the absorbance of the radiation (of emitted wavelength) is the primary mechanism of heating. Therefore, in order to enhance the absorption of the lamp radiation, an alumina (or zirconia) tube was placed in the furnace hot zone, and the fibers were passed through the tube. Commercial Nextel 550 Fibers (3M Inc., St. Paul, MN), used in this study, were obtained as fiber tows, and consisted of 100 single filament, 10μm diameter fibers covered with a protective polymer sizing. The fiber tows were heat treated at 55% of furnace power (at an estimated temperature of 1500°C) to synthesize mullite phase. Individual filaments were also extracted from fiber tows and heat treated. The fiber tows and single filaments were heat treated up to 75% of furnace power.

Infiltrated fibers: Small diameter mullite fibers were prepared by infiltration of organic thread with a mullite sol. Mullite sol was prepared from aluminum nitrate ($Al(NO_3)_3 \cdot 9H_2O$, Sigma-Aldrich, Milwaukee, WI) and colloidal silica (Ludox SK, Grace Davison, Brazil). The chemicals were dissolved in DI water and 5 wt% of PVA (polyvinyl alcohol, Airvol 540S, Celanese Chemicals, TX) solution was added. Different concentrations of mullite sol were used to infiltrate commercial cotton and single strands of natural silk thread, extracted from a silkworm cocoon. These fibers were first dried and were later heat treated with the quadrupole lamp furnace to induce grain growth. Subsequently, the microstructure and composition of heat treated fibers were characterized.

Extruded fibers: Polycrystalline mullite monofilament fibers were prepared by the extrusion through a fiber extruder (Marksman Fiber Drawing Machine, Chemat Technology Inc., Northridge, CA). Hydrothermally grown KM mullite powder (KM Mullite 101, Kyoritsu Ceramic Materials Co. LTD., Nagoya, Japan) was used as the starting powder. PVA (15 wt% solution in DI water) or 3 wt% solution of methyl cellulose (hydroxypropyl methyl cellulose, Sigma-Aldrich, Milwaukee, WI) was used as a binder for extruded fibers. Mullite powder was mixed with the binder to prepare a viscous paste which was extruded through a 200 mm orifice to produce the green polycrystalline fiber. The extruded fibers were dried in air and sintered at 1300°C for 2 hours in a $MoSi_2$ furnace. These pre-sintered fibers were later heat treated at 85% (approximately 1810°C) of power in quadrupole lamp furnace.

In this study, titania-doped mullite powders were used to synthesis extruded fibers. Titania doping was aimed at promoting anisotropic grain growth of mullite. The starting titania-doped mullite powder was synthesized by a sol-gel method. Mullite sol was prepared using aluminum nitrate nonahydrate ($Al(NO_3)_3 \cdot 9H_2O$, Sigma-Aldrich) and tetraethyl orthosilicate ($Si[OC_2H_5]_4$, Sigma-Aldrich) dissolved in ethanol. Titanium ethoxide ($Ti[OC_2H_5]_4$, Sigma-Aldrich) was used as a source for the titanium ions. After gellation with ammonium hydroxide, the gel was calcined at 1000°C for 1 hour and later crystallized at 1300°C for 2 hours in a $MoSi_2$ furnace. The synthesis of mullite phase was confirmed using X-ray diffractometry (Rigaku, Danvers, USA). The synthesized powder was mixed with 5 wt% PVA binder and extruded into fiber. The extruded fiber was pre-sintered and crystallized in a quadrupole lamp furnace. The microstructure of the heat treated thread was examined using scanning electron microscopy.

Templated fibers: Mullite whiskers were prepared for templating grain growth in the fibers. Mullite whiskers were abnormally grain grown mullite grains removed from a sintered mullite sample by HF leaching reaction. Titania-doped mullite powders were prepared by the sol-gel method, as explained before. The crystalline powder was pressed into a pellet, and sintered at 1600°C for 15 hours. The sintered mullite pellet was crushed and soaked in 49% HF solution, and the resulting cloudy dispersion was repeatedly rinsed with DI water to extract mullite whiskers. The morphology and composition of leached whiskers was examined by SEM and energy dispersive spectroscopy (EDS), respectively.

About 5 wt% mullite whiskers were mixed with KM mullite power and polymeric binder, and extruded as fibers. The templated green fiber was pre-sintered at 1300°C for 2 hours. The orientation of embedded whiskers was examined with SEM. The templated fiber was subjected to different heat treatment conditions in the QLF, up to 92% of furnace power, and several heating cycles, to enhance abnormal grain growth. The morphology of crystallized fibers was characterized with optical and electron microscopic techniques.

RESULTS AND DISCUSSION

Quadrupole Lamp Furnace: The temperature distribution in the QLF measured at different furnace powers is shown in Fig. 1. An alumina tube (of 6mm outer diameter), positioned in the center of the furnace, served as an auxiliary heater. Improved absorption of the lamp radiation by the alumina tube permitted more effective and uniform heating of the oxide fibers, which were positioned vertically inside the tube, to higher temperatures. The measurements were carried out up to 65% of the furnace power. As shown in Fig. 1, a hot zone was reproducibly created at the geometric center of the QLF. The measured temperature increased with the furnace power, but the rate of increase was reduced at higher powers. The small asymmetry observed in the temperature profiles is most likely due to the chimney effect above the hot zone, due to the rising hot air. When temperatures above 65% of power are extrapolated, it is expected that approximately 1825°C temperature can be attained at 90% of power. It should however be noted that, in the absence of an alumina tube, the actual temperature measured using a thermocouple will be different due to the difference in absorption of the lamp radiation by the thermocouple material. The temperature of the fiber might be different from the thermocouple temperature due to the different absorption of radiation.

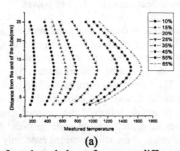

(a)

Fig. 1. Temperature profiles of quadrupole lamp furnace at different powers.

Nextel 550 fibers: The commercial Nextel 550 tows were heat treated in the quadrupole lamp furnace. The development of the mullite phase was examined by XRD. It was confirmed that the mullite phase belonged to the orthorhombic crystal system by the observation of splitting of the 120 and 210 X-ray peaks after heat treatment above 50% power (approximately 1425°C). The SEM micrograph of Fig. 2(b) shows the microstructure of fiber heat treated at 70% of power (approximately 1700°C) and a furnace traverse speed of 0.01 mm/s. As the fiber was heat treated at higher powers, elongated mullite grain growth was observed. The elongated grains were connected with adjacent filaments.

A single filament of the Nextel 550 fiber was withdrawn from the fiber tow and heat treated at 75% power (approximately 1775°C) with traverse rate of 0.01 mm/s. The original diameter was 14 μm and it decreased to 9μm after heat treatment (see Fig. 2(c)). It also exhibited elongated mullite grains along the length of the filament. These studies indicated the possibility of inducing textured grain growth along the length of the fiber with our quadrupole lamp furnace.

Infiltrated fibers: Single filaments of natural silk fiber were pulled out from a silk cocoon in order to obtain a thread without twists. The small diameter infiltrated fiber (~14 μm) retained flexibility even on heat treatment, and was expected to have smaller size and number of flaws. The infiltrated silk filament fibers were heat treated up to 44% of power (approximately 1320°C) with a furnace traverse rate of 0.05 mm/s. The diameter of the heat treated filament decreased as the heating power increased and grain growth was observed. However, no elongated mullite grain growth was observed, unlike in the case of the Nextel fibers, because heat treatment was performed at low power.

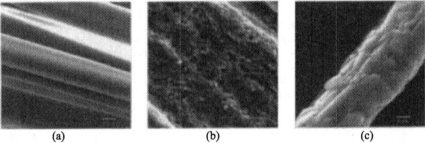

(a)   (b)   (c)

Fig. 2. SEM micrographs of Nextel 550 fiber: (a) as received, (b) heat treated at 70% of power, and (c) Nextel 550 single filament heat treated at 75% of power.

In the case of infiltrated cotton thread, the dried thread was first burnt with a cigarette lighter flame. The burnt fiber was then mounted in quadrupole lamp furnace and heat treated at 80% of power (approximately 1780°C) at 0.01 mm/s traverse rate. The micrographs in Fig. 3 show the crystallized mullite, sol infiltrated, cotton thread. The composition of the sintered thread was examined by EDS. The thread was composed of 66.3 mol% of Al_2O_3 and 33.7 mol% of SiO_2. The formation of mullite phase was confirmed with x-ray diffraction study. The morphology of the heat treated sample followed the original organic thread. As it was heat treated at high power, there was elongated mullite grain growth. While some elongated grains were aligned along the fiber direction, there still remained randomly oriented grains.

Fig. 3(c) shows the heat treated infiltrated cotton thread. The heat treatment was done at 87% of power (approximately 1815°C) with 1 μm/s transverse rate. All of the grains were turned to elongated grains and the average dimension was 3.2 μm in width and 22 μm in length. Although some of the elongated grains were aligned, most of the grains were randomly oriented.

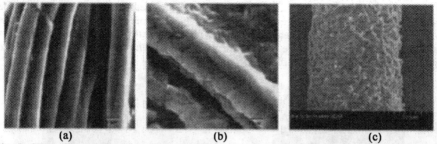

(a)   (b)   (c)

Fig. 3. SEM micrographs of mullite sol infiltrated cotton thread: (a) after infiltration, (b) heat treated at 80% of furnace power, at a transverse rate of 0.01 mm/s, (c) heat treated at 87% power.

Extruded fibers: Extrusion of mullite fiber was studied in order to obtain long, mullite green fibers having a small diameter. Different kinds of binders were mixed with synthesized mullite powder. In the case of PVA binder, the rapid drying of PVA binder made it hard to extrude long fibers and resulted in clogging of the extrusion die orifice. However, when methyl cellulose was used as a binder, longer and more flexible fibers (~ 150 μm diameter) could be extruded. The SEM micrograph in Fig. 4(a) shows the microstructure of the green fiber. The green fiber was pre-sintered at 1300°C for 2 hours to enable easier fiber handling.

(a)   (b)   (c)

Fig. 4. SEM micrographs of extruded mullite fiber: (a) green fiber, (b) heat treated at 1300 °C for 2 hours in $MoSi_2$ furnace, (c) heat treated at 85% of power, at a transverse rate of 0.01 mm/s.

As shown in Fig. 4(b), there was insignificant grain growth. The pre-sintered fibers were heat treated in QLF to obtain textured or single crystal fibers. As the radiative power of the lamp was increased, the surface morphology of the fiber changed. The morphology of the heat-treated, mullite fiber at 85% of power (approximately 1810°C), confirmed elongated grain growth as well as densification (Fig. 4(c)). Even though elongated mullite grain growth occurred during heat treatment, these grains were oriented randomly.

In order to enhance elongated grain growth, fiber was extruded with titania-doped-mullite powders. It is reported that doping of titania enhance the diffusivity due to a lower glass viscosity. A small amount of titania was added in order to promote elongated mullite grains. In the case of titania doping, several kinds of systems were evaluated in order to synthesize titania-doped mullite. The organic steric entrapment method could not be used as there was not a suitable cation source available which was compatible with an aqueous system. In the case of the ethylene glycol method, there was an immiscibility between tetraethyl orthosilicate and ethylene glycol. Furthermore, in the sol-gel system with aluminum isopropoxide, its solubility was minimal in the ethanol solvent. Therefore aluminum nitrate, tetraethyl orthosilicate, and titanium ethoxide dissolved in ethanol were selected for the synthesis. The formation of mullite phase powder was confirmed by X-ray diffraction.

(a)   (b)

Fig. 5. SEM micrographs of extruded 5 wt% titania added mullite fiber: (a) heat treated at 1400°C for 2 hours in MoSi$_2$ furnace, (b) heat treated at 65% of power (approximately 1650°C), at a transverse rate of 0.01 mm/s in quadrupole lamp furnace.

The heat-treated, titania-doped, mullite fiber is shown in Fig. 5(b). The heat treatment was carried out at up to 65% of power (approximately 1650°C) with a furnace traverse rate of 0.01mm/s. Elongated mullite grain growth occurred with increasing power. In order to enhance elongated grain growth the fiber was subjected to heat treatment cycles repeatedly. The micrograph shows evidence of textured grain growth of mullite grains along the furnace traverse direction. Mullite has a strong tendency to grow in needle-like shapes, due to the difference in activation energy for growth of difference faces. It has been reported that the activation energy for grain growth along the length is 690 kJ/mol while it is 790 kJ/mol for the thickness direction. The needle-like, mullite grains grew in the orthorhombic c-axis [001] orientation and were bound by {100} and/or {111} surfaces. This growth habit is related to the crystal structure of mullite.

Templated fibers: Mullite whiskers were prepared for use as seeds for textured grain growth of mullite. A sintered titania-doped, mullite body was crushed and leached by HF solution in order to obtain mullite whiskers. The glassy phase was removed and the individual grains were separated. The thickness of the mullite whiskers was about 3 μm. The aspect ratio between thickness and length was about 12. This aspect ratio increased with increasing titania doping. The formation of whisker-type mullite due to anisotropic grain growth was confirmed by SEM/EDS analysis. These mullite whiskers were mixed with KM mullite powder and methyl cellulose solution and extruded as fibers. The orientation of the whiskers in extruded fiber was aligned along fiber length direction, due to the high aspect ratio of whisker.

The SEM micrographs in Fig. 6 show the microstructure of heat treated 5 wt% whisker-added, mullite fiber. The heat treatment was done at 80% of power (approximately 1810°C) with 0.001 mm/s of transverse rate in Fig. 6(a). The large elongated mullite grains, along the fiber length direction, grew with heat treatment. As the power was increased to 92% (approximately 1830°C), elongated grains over 400 μm in length were also observed (see Fig. 6(b)). In this case the aspect ratio was over 40. During heat treatment cycles a considerably textured microstructure was achieved, although some randomly oriented grains were still present.

(a)   (b)   (c)

Fig. 6. SEM micrographs of extruded 5 wt% whisker-added, mullite fiber: (a) heat treated at 80% of power, at a transverse rate of 0.01 mm/s, (b) heat treated at 92% of power, at a transverse rate of 0.001 mm/s, and (c) heat treated at 85% of power, at a transverse rate of 0.001 mm/s.

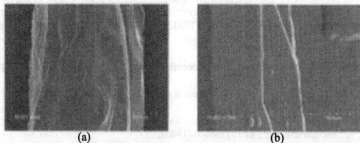

(a)   (b)

Fig. 7. (a) SEM micrographs of extruded 5 wt% whisker added mullite fiber heat treated 10 times at 89% of power, at a transverse rate of 0.001 mm/s , (b) high resolution of (a).

Therefore, the fiber was subjected to repeated heat treatments. The fiber heat treated at 89% of power (approximately 1820°C) for 10 heating cycles is displayed in the SEM micrograph of Fig. 7. As shown in SEM micrographs, the heat treated fiber has faceted surface with steps. The heat treated fiber was transparent, as verified using optical microscopy (Fig. 8(a)). Preliminary transmission electron microscopic evaluation suggests that the fiber is a single crystal.

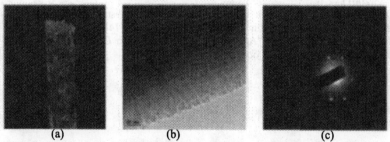

(a)   (b)   (c)

Fig. 8. (a) Optical microscope image, (b) TEM micrograph, and (c) diffraction image of whisker added mullite fiber heat treated 10 times at 89% of power, at a transverse rate of 0.001 mm/s.

CONCLUSIONS

The feasibility of growing textured and possibly single crystal mullite fibers by heat treatment of polycrystalline fibers was demonstrated using the QLF. The shape and temperature distribution of the QLF was determined as a function of furnace power. The experimental design was modified, by placement of an alumina tube through the furnace center, to ensure uniform and effective heating of the precursor mullite fibers. The effects of temperature and the duration of heat treatment, on the microstructural evolution of different kinds of polycrystalline precursor mullite fibers, were studied. Heat treatment of polycrystalline, extruded and sol-infiltrated fibers resulted in randomly oriented, elongated grains. Elongated mullite grain growth could be enhanced by the addition of titania. Texturing in the mullite fiber was significantly improved by templating the grains grown using mullite whiskers. Repeated heat treatment of templated fibers resulted in a transparent oxide fiber, which is expected to be a single crystal.

ACKNOWLEDGEMENTS

This work was supported by the AFOSR under grant F49620-03-1-0082. Use of the facilities in the Center for Microanalysis of Materials, University of Illinois, which is partially supported by the U.S. Department of Energy under grant DEFG02-91-ER45439 is gratefully acknowledged.

REFERENCES

1   A. R. Bunsell and M. –H. Berger, "Fine Ceramic Fibres," *J. Eur. Ceram. Soc.*, **20** [13] 2249-60(2000),

2   H. Schneider, K. Okada and J.A. Pask, "Mullite and Mullite ceramics," John Wiley, New York, NY, 1994,

3   W. M. Kriven and J. A. Pask, "Solid Solution Range and Microstructures of Melt-Grown Mullite," *J. Am. Ceram. Soc.*, **66** [9] 649-54(1983),

4   M. D. Petry and T. Mah, "Effect of Thermal Exposures on the Strengths of Nextel550 and 720 Filaments," *J. Am. Ceram. Soc.*, **82** [10] 2801-807(1999),

5   A. Sayir and S. Farmer, "Directionally Solidified Mullite Fibers," pp. 11-21 in *Ceramic Matrix Composites-Advanced High Temperature Structural Materials*, Edited by R. A. Lowden, M. K. Ferber, J. R. Hellmann, K. K. Chawla, and S. G. DiPietro, Mat. Res. Proc., [365] (1995),

6   S.T. Mileiko, V.M. Kiiko, M. Yu. Starostin, A.A. Kolchin and L.S. Kozhevnikov, "Fabrication and Some Properties of Single Crystalling Mullite Fibers," *Scri. Mat.*, **44** [2] 249-55(2001),

7   B. R. Johnson, W. M. Kriven and J. Schneider "Crystal structure development during devitrification of quenched mullite," *J. Eur. Ceram. Soc.*, **21** [14] 2541-62(2001),

8   S. Hong and G. L. Messing, "Anisotropic Grain Growth in Diphasic-Gel-Drived Titania-Doped Mullite," *J. Am. Ceram. Soc.* **81** [5] 1269-77 (1998).

THERMO-MECHANICAL PROPERTIES OF SUPER SYLRAMIC SiC FIBERS

HeeMann Yun, Donald Wheeler, Yuan Chen, and James DiCarlo, NASA Glenn Research Center, 21000 Brookpark Road, Cleveland, OH 44135

ABSTRACT

Ceramic matrix composites (CMC) reinforced by SiC fibers, such as SiC/SiC, are being targeted for application in hot-section components of advanced propulsion and power generation engines and in first walls of advanced nuclear systems. Two "Super Sylramic" SiC fiber types, recently developed at NASA using the Sylramic fiber from COI Ceramics, are candidates for providing these components with improved thermal capability and improved performance. This paper reports on the ability of these new fiber types, Super Sylramic-iBN and Super Sylramic-SiC, to meet the key fiber requirements of these applications: high strength, high creep-rupture resistance, and high thermal conductivity. For example, creep-rupture tests performed at 1300 to 1450°C show that the creep resistance of these fibers is ~20 and ~7 greater than the current Sylramic and Sylramic-iBN fiber types, respectively, that have already been used to demonstrate state-of-the-art SiC/SiC composites. TEM and AES microscopic observations are presented to indicate that these improvements can be correlated with the replacement of weak grain boundary phases with stronger phases that hinder grain boundary sliding more effectively. Preliminary SiC/SiC composite results are also provided for the Super Sylramic fiber types.

INTRODUCTION

SiC/SiC ceramic matrix composites are being developed for a variety of high-temperature structural components in advanced gas turbine engines [1-2] and in future nuclear systems [3]. Hot-section applications that are often exposed to through-thickness thermal gradients will typically require ceramic composite materials with high mechanical performance (high cracking strength, high ultimate strength and strain, and high creep-rupture resistance), as well as high thermal conductivity in all directions in which the component will experience significant service-related stresses. Thus SiC/SiC constituent microstructures, fiber pre-form architectures, and component wall thicknesses need to be optimized to maximize these properties. NASA has demonstrated a variety of high-performance SiC/SiC systems [4-5] by choosing (1) fiber types that retain as much of their as-produced strength as possible during the various SiC/SiC processing steps, and (2) SiC fiber architectures that enhance uniform interphase coating and easy matrix infiltration with small residual porosity, thereby increasing composite maximum use temperature, thermal stress resistance, through-thickness thermal conductivity, and inter-laminar strength.

For reinforcement of the latest NASA SiC/SiC composite systems, NASA recently developed the Sylramic-iBN SiC fiber which displays various processing advantages: high weave-ability during textile pre-forming and robustness with CVI-BN, CVI-SiC, and slurry plus molten metal processing during typical NASA SiC/SiC densification steps [4]. In addition, in comparison to other near stoichiometric SiC fibers, such as Sylramic from COIC (formerly produced by Dow Corning), Hi-Nicalon Type S from Nippon Carbon, and Tyranno SA from Ube Industries, the Sylramic-iBN SiC fiber not only displays superior creep and rupture behavior, but also more chemical stability with comparable room temperature tensile strength. The improved creep-rupture behavior is achieved by subjecting the Sylramic fiber to specific heat treatment process conditions that remove boron sintering aids in the grain boundaries of the as-produced fiber. The process conditions also allow the in-situ growth of thin BN-based surface coatings that enhance the fiber's environmental resistance while maintaining high fiber tensile strength.

For this study, commercial Sylramic SiC fibers were heat treated under two enhanced process conditions different than those employed for the Sylramic-iBN fiber in order to further improve fiber properties, such as creep-rupture resistance. Like the Sylramic-iBN fiber, the two new fiber types, called Super Sylramic-iBN (SSBN) and Super Sylramic-iC (SSC), also contain in-situ grown surface layers that are based on BN and carbon, respectively. The objectives here are focused on evaluating the key mechanical and chemical properties of these newly formed stoichiometric Super-Sylramic SiC fibers and on comparing these properties with other currently available SiC fibers under potential SiC/SiC composite fabrication and service conditions.

EXPERIMENTAL

For tensile strength evaluation, the four Sylramic-based SiC fibers, Sylramic, Sylramic-iBN, Super-Sylramic-iBN, and Super-Sylramic-iC, were tested after various treatment conditions as single fibers, multifilament tow, and 2D 0/90 5-harness satin woven fabric. The specimens were tensile tested to fracture at room temperature using a 25mm gauge length and a constant displacement rate of 1.27 mm/min. In addition, the woven fabrics of each fiber type were exposed to air at 800°C for 100 hours, and then tensile tested at room temperature to assess environmental resistance under oxygen attack. Further, the creep-rupture behavior at 1300 to 1450°C of the single fiber specimens were determined using the fiber creep and rupture test facilities at NASA and the same procedures reported in detail elsewhere [6]. Creep deformation versus time was recorded under a constant deadweight load at a constant temperature using 25mm hot zone lengths in both air and argon environments.

For surface chemistry evaluation, Auger electron spectroscopy (AES) depth-profile analyses were conducted on single fibers at every 10 nm from the fiber surface to a depth of ~400 nm. In addition, thermo–gravimetric analyses (TGA) were conducted in air from room temperature to 1500°C on bundled tows whose initial weight was ~30 to 50mg. Also TEM analyses were performed to evaluate SiC grain size, in-situ grown surface layer crystalline morphology, and overall chemistry in the cross sections of the fibers. The TEM studies were conducted on longitudinal sections of single fibers.

One set of woven fabrics out of Super Sylramic-BN and Super Sylramic-C was then used by a CVI-BN or CVI-C interphase coating and CVI-SiC densification vendor (GE Power Systems Composites) to fabricate 8-ply SiC/SiC pre-forms with partial content of SiC matrix. Final panels from these pre-forms were produced by filling the open porosity with SiC slurry plus melt-infiltrated (MI) Si by an MI vendor (GE Power Systems Composites). Rectangular shaped flexural specimens and square-shaped thermal diffusivity specimens were prepared from the CMC panel. The inner span length of the four point flexural specimen was 40.6mm, and the width 5.1mm. Four-point flexural tests were conducted at room temperature using conventional test procedures [5]. Thermal diffusivity values were measured by Thermo-physics Inc. using the laser-flash at room temperature up to 1500°C in a controlled inert environment. The thru-thickness thermal conductivity could then be determined by multiplying thermal diffusivity, specific heat, and the average panel density.

RESULTS AND DISCUSSION
Super Sylramic Fiber Development. For the development of the Sylramic-iBN, Super Sylramic-iBN, and super Sylramic-iC fibers, a variety of heat-treatment process conditions had to be optimized, including treatment temperature, time, gas composition, gas pressure, and gas flow

60

rate. In comparison to the precursor Sylramic fiber, key process goals were (1) little or no loss in fiber strength, (2) significant improvement in fiber creep-rupture resistance and temperature capability by alteration of the fiber microstructure, and (3) the in-situ development of fiber surface coating that has the potential of improving fiber environmental resistance and also acting as functional interface coating (interphase) in high-performance structural ceramic composites. At the time of development for the Sylramic-iBN fiber, all process conditions had not yet been optimized towards these goals. Thus the Super fibers represent a more complete examination of the Sylramic fiber process conditions that has resulted in higher attainment of the fiber goal properties. The following sections detail these enhanced property results.

Fiber Tensile Strength. The average room-temperature tensile strengths of the four as-produced Sylramic-based fiber types in various fiber forms are shown in Fig. 1a. Also shown in Fig. 1a are the average strengths of other high performance SiC fibers in their as-produced condition. The strength values are the average of at least 12 tests. It is clear that the optimized process conditions for the SSBN and SSC fibers produced little loss in strength relative to Sylramic and Sylramic-iBN and that as a group the Sylramic-based fibers displayed comparable or higher as-produced strength than the other SiC fiber types.

Another important fiber property goal is strength retention during SiC/SiC composite

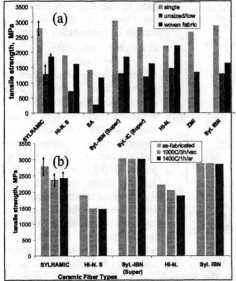

processing. To evaluate this property, all the SiC fiber types were treated in vacuum and in argon at 1000 and 1400°C and then strength tested at room temperature. Fig. 1b shows that all the Sylramic fibers as well as Tyranno SA retained a high percentage of their as-produced strength, but the Hi-Nicalon types lost measurable strength. These results can be well correlated with the maximum process temperatures of the various fiber types [7].

Fiber Creep-Rupture Resistance. For each SiC fiber type, creep strain versus time curves generally displayed a short transient primary stage followed by a pseudo steady-state creep stage. Because total creep is important for understanding residual stress development in SiC/SiC components with through-thickness temperature gradients, a convenient method for evaluating fiber creep resistance is by measuring the total creep

Fig. 1. Room temperature fiber tensile strength of (a) the as-fabricated Super Sylramic SiC fibers and (b) after exposure in a possible processing condition.

strain value after a given time for a constant stress, temperature, and test environment. For example, Fig. 2 shows, as a function of applied stress, the 10 hr-creep strain at 1400°C in air for the Super-Sylramic SiC fibers and several other SiC fiber types [8]. With increasing stress, the creep strain increased by a power dependence of ~ 2 for the Super-Sylramic fibers like most of the other SiC fibers. This power dependency was similarly observed for creep data measured in argon (not shown). For nearly all creep strains, the 10-hr creep-strengths of the Super Sylramic-iBN and Super Sylramic-iC fibers were raised by a factor of ~2 and ~8 in comparison to the Sylramic-iBN and the Sylramic fiber, respectively. This creep strength increase means that at one stress, such as 250 MPa, the creep strains of the Super-Sylramic fibers are ~20 times and ~7 times lower than those of the Sylramic and Sylramic-iBN, respectively. In argon (not shown), the Super-Sylramic fibers showed a similar superiority against the other SiC fiber types. As will be discussed, there was little change in grain size between the four Sylramic-based fiber types, so that the creep resistance improvement can be attributed to changes in composition of phases at the grain boundaries.

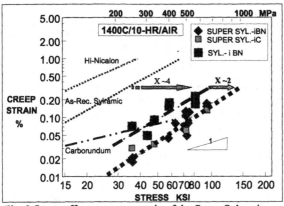

Fig. 2 Stress-effects on creep strain of the Super Sylramic SiC fibers.

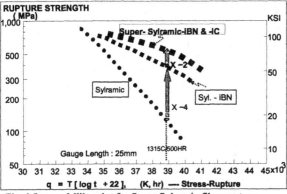

Fig. 3 Larson-Miller plot fro Super Sylramic fiber rupture strength in air.

The creep-rupture lives of the Sylramic-based fibers in single fiber form were determined from 1300 to 1450°C at various stress levels in air, and results used to construct Larson-Miller (LM) creep-rupture master curves using the q-parameter [8]. As shown in Fig. 3, these master curves describe the stress or rupture strength of each fiber as a function of time and temperature as contained in the parameter $q = T (\log t_r + 22)$, where rupture time t_r is in hours and test temperature T is in degrees Kelvin. Clearly, at the higher q-values the Super-Sylramic fibers showed improved rupture strength behavior than the other Sylramic-based fibers. For instance at a q-value of 1315°C-500hr, rupture strength increased by a factor of ~8 and ~2 increase against the Sylramic and the Sylramic-iBN fibers, respectively. This means that at one stress, such as at 400 MPa, the 1315°C rupture lives of the Super-Sylramic fibers improved by ~1000 and 20 times against these same fibers.

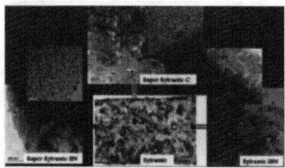

Fig. 4 Typical TEM microstructures of the Super Sylramic fibers, as-fabricated

Fiber Microstructures.

Typical TEM microstructures near the fiber surface zone are shown in Fig. 4 for the Super-Sylramic fibers in comparison to the Sylramic-iBN and Sylramic fibers. Also included in Fig. 4 are high-resolution images of the surface layers of the SSBN and SSC fibers. All microstructures showed equi-axed and stacking-faulted β-SiC grains with various sizes, from ~ 70 to ~ 170 nm. Comparing the various fiber types, several observations can be drawn: (1) the treated fibers have less stacking faults, meaning less atomic-scale defects, (2) the treated fibers have less boron-containing phases, meaning cleaner grain boundaries, (3) the Super-Sylramic fibers contain slightly larger grains on the average than the Sylramic-iBN fiber, (4) the Super Sylramic-iBN fiber has a more crystalline and denser BN-based layer than the Sylramic-iBN fiber, and (5) the Super Sylramic-iC fiber has a clearly definable crystalline C-rich layer.

Fiber Surface Coatings.   Nearly all the Sylramic-based fibers were quantitatively analyzed by AES techniques from the fiber surface toward the fiber center. The key elements were C, Si, B, N, and O, and the concentration of each element was calculated from known standard materials [9]. In general, the Sylramic-iBN and Super Sylramic-iBN fibers displayed B-N-containing layers with thicknesses of ~120 and ~160nm, respectively. These values are close to theoretical assuming that all available boron in the as-produced Sylramic fiber was converted to a stoichiometric BN surface layer. On the other hand, the carbon-rich layer on the Super Sylramic-iC fiber could be varied by controlling the time or temperature of the heat treatment. For the SSC fiber data presented here, the C-rich layer thickness was ~200nm Hence, the Super-Sylramic fibers represent two new SiC fiber types that display high as-produced tensile strengths with uniformly thick surface protection layers that are selectable in the two compositions that are currently being used as interphases in structural ceramic composites. Furthermore, the in-situ grown C-rich layer of the SSC fiber is not only uniformly thick, but controllable in thickness by varying the heat treatment time conditions.

Fiber Environmental Resistance.   The change in initial weight of bundled tows of the Sylramic-based fibers was measured by TGA in air at a warm-up rate of ~60°C/min. Weight changes at 500, 800, and 1500°C are summarized in Table 1.   Typically, a positive change is related to formation of silica (silicon dioxide)

Table 1. TGA weight change in air tested on the wrapped tows, 60C/min.

FIBER TYPE	500C	800C	1500C
	Weight Change in Air		
SYLRAMIC	+0.6	+1.2	>3
SYLRAMIC-iBN	-0.2	+0.3	>4
SUPER SYLRAMIC-iBN	-0.2	+0.1	>4
SUPER SYLRAMIC-iC	-0.3	-0.8	+1.8

and/or other impurity oxides on the fiber surfaces; while a negative change is related to removal of carbon-containing phases, such as those contained in manufacturer-applied polymer sizing or in the in-situ grown carbon coatings. At 500°C, the as-produced (and sized) Sylramic fiber showed a weight gain, while the treated (and un-sized) fibers showed a weight reduction with the SSC fiber displaying a slightly greater loss. It is currently concluded that the weight increase for Sylramic is mainly due to B_2O_3 formation, while the weight losses for the other fiber are related to removal surface contaminants for the iBN fibers and to partial removal of the iC layer. At 800°C, the Sylramic fibers gained more weight, probably caused by enhanced silica growth due to the B_2O_3, while the iBN fibers began to oxidize after loss of their BN layers and the iC fibers continued to lose their carbon layers. At this temperature, the Super Sylramic-iBN fibers showed the lowest weight gain, suggesting a denser and thicker iBN layer than the Sylramic-iBN fiber. At 1500°C, all the boron-containing or boron-coated fibers displayed large weight increases, suggesting again enhanced silica growth due to the B_2O_3 formation. Interestingly, the Super Sylramic-iC fiber showed a lower weight gain at 1500°C, perhaps due to the absence of boron.

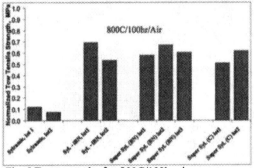

Fig. 5 Tow strength after 800C/100hr air exposure

To evaluate the mechanical effects of these environmental results, single tows of the various Sylramic-based fiber types were exposed to air for 100 hours at 800°C, and then their tensile strengths were measured at room temperature. The strength results in show that the Sylramic fibers displayed a significant strength loss, which can be attributed to fiber-fiber bonding related to its enhanced oxidation. On the other hand, the treated fiber types, which showed the lowest weight gains, were more capable of retaining their original strengths.

BENEFITS OF SUPER-SYLRAMIC FIBER REINFORCEMENT. Based on the state-of-the-art performance observed for SiC/SiC composites reinforced by the Sylramic-iBN fiber [4], it is expected that the Super Sylramic-iBN fiber should provide SiC/SiC composites with similar or even better performance since it displays significantly better creep-rupture resistance and possesses a slightly higher fiber strength with a thicker and denser BN-rich layer. The enhanced creep-rupture resistance should allow composite structural performance beyond 1450°C, an upper use temperature already achieved by SiC/SiC reinforced with Sylramic-iBN [4]. In addition, the enhanced strength and environmental resistance of the Super Sylramic-iBN fiber should provide stronger and tougher composites with improved capability to withstand adverse environmental conditions during composite fabrication and service. Although the Super Sylramic-iC fiber may not provide the oxidation resistance of the iBN fibers, particularly at intermediate temperatures, it should be well suited for application in SiC/SiC components that operate under low oxygen conditions, such as those that exist in the first wall of future fusion reactors [3]. In contrast to the iBN fibers, this fiber type has the advantage of having an in-situ grown carbon layer that is controllable in thickness, which may in turn obviate the need and cost of extrinsically applied interphase coatings. Furthermore for nuclear applications, the carbon-rich coating can be expected to be more dimensionally stable than BN coatings.

Currently, four types of SiC/SiC panels have been successfully fabricated for thermo-structural evaluation: SSBN fiber with CVI-slurry-MI SiC-Si matrix; SSBN fiber with CVI-PIP SiC matrix; SSC fiber with CVI-slurry-MI SiC-Si matrix; and a baseline of Sylramic-iBN fiber and CVI-slurry-MI SiC-Si matrix with panel densities of 2.75, 2.70, 2.80, and 2.83 g/cc, respectively. Densities > 2.70g/cc indicate small amount of residual porosity (< 3v/o). The calculated fiber and CVI-SiC volume fractions were in the range of 35 to 39% and 33 to 39%, respectively. Dog-bone and disk specimen machining is in progress, and the as-fabricated tensile stress-strain and interlaminar behavior will be measured shortly. Room-temperature flexural strength and thru-thickness thermal conductivity values for the SSBN/MI, SSC/MI, and SBN/MI composites are 410, 510, 480 MPa, and 31, 33, 28 W/m.K, respectively. These values are comparable to state-of-the-art SiC/SiC composite values reported elsewhere [4].

SUMMARY AND CONCLUSIONS

The properties of two new types of developmental SiC fibers, Super Sylramic-iBN and Super Sylramic-iC, were determined and compared to those of commercially available Sylramic SiC fiber and the previously developed Sylramic-iBN SiC fiber. Key test results show:

a) The as-produced tensile strength and strength retention of the Super-Sylramic fibers were comparable to the Sylramic and Sylramic-iBN fibers.

b) The Super-Sylramic fibers possessed optimum microstructures for significantly improved creep and rupture resistance over the other Sylramic-based fibers, plus in-situ grown BN-rich and C-rich surface layers for enhanced environmental resistance over the Sylramic fiber.

c) Preliminary results on SiC/SiC composites with the two new SiC fiber types are comparable to state-of-the-art values for flexural strength and thermal conductivity with creep-rupture measurements still to be performed.

d) Super Sylramic-iC fiber should be well suited for application in SiC/SiC components that operate under low oxygen conditions, and also has the advantage of having an in-situ grown carbon layer that is controllable in thickness for cost-effective, uniform interphase coatings.

REFERENCES

[1] D. Brewer, "HSR/EPM Combustor Materials Development Program," Mat. Sci. and Eng., A261, 284-291 (1999).

[2] J.D. Kiser, M.R. Effinger, D.E. Glass, et al, "CMC Component Development Supporting the NASA-Led 2nd and 3rd Generation RLV Investment Areas: Propulsion Applications", in the proceedings of 25th Annual Conference on Composites, Materials, and Structures, Cocoa Beach, FL.

[3] A.R. Raffray, R. Jones, G. Aiello, M. Billone, L. Giancarli, H. Golfier, A. Hasegawa, Y. Katoh, A. Kohyama, S. Nishio, B. Riccardi, and M. Tillack, "Design and Materials Issues for High Performance SiC/SiC-Based Fusion Power Cores," Fusion Eng. Des. 55, 1, 55 (2001).

[4] J.A. DiCarlo, H-M. Yun, G.N. Morscher, and R.T. Bhatt, "SiC/SiC Composites for 1200C and Above", Chapter for *Handbook of Ceramics and Glasses*, 77-98, 2004.

[5] H.M. Yun, J.Z. Gyekenyesi, Y.L. Chen, D.R. Wheeler, and J.A. DiCarlo, "Tensile Behavior of SiC/SiC Composites Reinforced By Treated Sylramic SiC Fibers", Cer. Eng. Sci. Proc., 22 [3], (2001), pp. 521-531

[6] H.M. Yun, J.C. Goldsby, and J.A. DiCarlo, "Tensile Creep and Stress - Rupture Behavior of Polymer Derived SiC Fibers", NASA TM 106692, 1994

[7] J.A. DiCarlo and H-M. Yun, "Non-Oxide (Silicon Carbide) Ceramic Fibers", Chapter for *Handbook of Ceramics and Glasses*, 33-52, 2004.

[8] H.M. Yun and J.A. DiCarlo, "Tensile Fracture and Creep-Rupture Behavior of SiC Fibers", *Advances in Fracture Research, Proceedings of ICF10*, Elsevier Applied Science, London, 2001, paper ICF100582OR

[9] Private communication: D. Wheeler, AES operational manual, 2000.

HIGH TEMPERATURE TENSILE TESTING METHOD FOR MONOFILAMENT CERAMIC FIBERS

B-M. Yee, P. Sarin and W.M. Kriven
University of Illinois at Urbana-Champaign
1304 West Green Street
Urbana, IL 61801

ABSTRACT

An alternative method for testing tensile strength of ceramic fibers at high temperatures has been developed, whereby the entire sample is in the hot zone of the furnace. This eliminates any effect that a thermal gradient might have on the mechanical properties of the fiber. The method proposed here was calibrated with Nextel™ 720 fibers. Tensile strengths comparable to those published by 3M were obtained.

INTRODUCTION

Many high temperature tensile testing methods have limitations for testing fibers of shorter lengths. The method that is most commonly used employs a form of grip that necessitates keeping the fiber ends outside of the furnace. This results in a temperature gradient along the length of the fiber, as well as a long gauge length.[1] These two issues motivated the development of a method to test fibers at elevated temperatures. Moreover, this test method was also designed to observe the ferroelastic behavior of $DyNbO_4$ at high temperatures. $DyNbO_4$ has been identified as a potential smart material for applications such as a large force actuator. Ferroelastic materials have domains or twins that can be rearranged by the application of stress.[2] Since ferroelastic behavior is facilitated by elevated temperature, it is important that the entire fiber is exposed to a uniform temperature distribution. The objective of this work was to develop and calibrate a method for testing tensile strengths of monofilament ceramic fibers at high temperatures, where the entire gauge length is within the hot zone of the furnace.

Another high temperature tensile test method had been proposed by Unal, et al.,[3] and involved a lengthy preparation. It was limited for testing the number of samples necessary for reasonable statistical analysis and only had a 60% success rate.[3] The method proposed here has a success rate of over 80% and has a relatively short sample preparation time. After the development of this method, it was calibrated by testing Nextel™ 720 fibers and the results were compared with tensile strengths published by 3M.

EXPERIMENTAL PROCEDURE

Nextel™ 720 fibers were obtained from 3M (3M, St. Paul, MN) in tows of 3000 denier (grams/9000 m). A section, approximately 200 mm (~8 in) long, was cut from the tow and heated to 700°C for five minutes to remove the protective organic sizing. Individual fibers were carefully separated from the tow and prepared for testing. Fiber test sample preparation followed the procedure outlined by ASTM C 1557-03 for testing at ambient temperature with a gauge length of 25.4 mm (1 in).[4] For reproducibility, this standard suggested using paper tabs with a preprinted line down the center for the grip and a suitable adhesive, most commonly an epoxy. Since epoxy and paper fails at elevated temperatures, high temperature ZrO_2 cement, functional up to 2200°C, was used to attach the fiber to the tab (904 ZrO_2, Cotronics Corp., Brooklyn, NY) and extruded Al_2O_3 tubes (99% pure, Vesuvius McDanel, Beaver Falls, PA), onto which the grips would be applied, were affixed to each end of the fiber. The preprinted line on the tab ensured alignment of the fiber and tubes. A schematic of this setup can be seen in Fig. 1. The cement was allowed to cure at room temperature for 24 hours before being tested. The setup of the test method as mounted on an Instron 4500 Series Universal Testing Machine (Instron, Canton, MA) is shown in Fig. ?

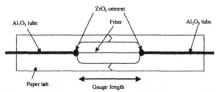

Fig. 1. Schematic diagram of fiber and tube assembly.

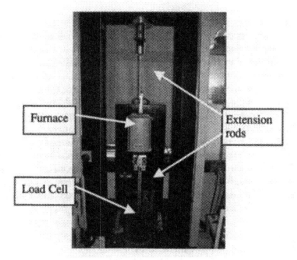

Fig. 2. Instron 4502 Universal Testing Machine used for high temperature tensile tests. A 10N load cell was placed at the bottom of the instrument to avoid drift from temperature effects.

A 1600°C DTA tube furnace was used for heating the fiber samples. Its temperature profile was measured using a thermocouple by moving it in small increments inside the furnace along the length of the tube. A 10N load cell was positioned at the base of the Instron. Aluminum extension rods were mounted on the load cell and on the crosshead to isolate the furnace from the Instron components, particularly the load cell. Two different connectors, shown in Fig. 3, were designed to join the test sample to the aluminum extension rods.

A commercially available universal joint, with a collet inserted in the lower segment to grip the tube, was the top connector. The bottom connector was a specially designed, hollow aluminum cylinder, into which the lower alumina tube was inserted and held in place with glue. Both the top and bottom connectors were attached to the extension rods by pins. Once the fiber sample was mounted, the cardboard tab supporting the fiber was burned away with a wire heater.

68

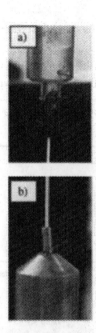

Fig. 3.(a). Upper connector that joins the alumina tube to the aluminum rod with a universal joint fashioned with a collet in the lower portion. (b). Lower connector that attaches the alumina tube to the aluminum rod. Motion is limited in one plane.

Nextel™ 720 fibers were tested at a constant crosshead speed of 0.2 mm/min at three temperatures: room temperature, 800°C, and 1000°C. Room temperature tests were executed without the furnace in place. For high temperature tests, the furnace was raised to a level so that the entire gauge length of the fiber was within the hot zone. Test temperature was reached at a rate of 40°C/min and the fibers were allowed to equilibrate at that temperature for at least 20 seconds before a load was applied until fracture. Fiber ends were not retrievable after fracture, and so could not be used for individual diameter determination for precise strength calculations. Instead, a theoretically calculated value of fiber diameter, using the same method as adopted by 3M, was used in fracture strength calculations.[5] The results were compared to published data for room temperature and high temperature tests on Nextel™ 720 fibers.

For materials that fail by the weakest link theory, the Weibull modulus is typically reported to describe the flaw distribution. As a function of volume, the probability of failure, F, is given by

$$F = 1 - \exp [- (V/V_o)(\sigma/\sigma_o)^m] \qquad (1)$$

where V = tested volume, σ = failure strength, m = Weibull modulus, and V_o, σ_o are constants. Since the gauge length remained constant, Equation 1 can be reduced to

$$\ln \{ \ln [1/(1-F)] \} = m \ln \sigma + k \qquad (2)$$

where k is a constant. Sufficient data was gathered at room temperature and 800°C for Weibull analysis.

RESULTS
    The temperature profile for the tube furnace is shown in Fig. 4.

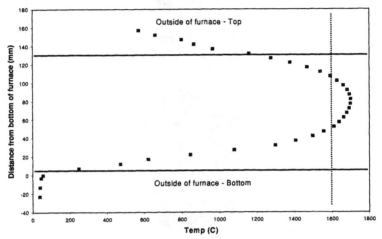

Fig. 4. Temperature profile of the 1600°C tube furnace showing that the hot zone was ~51mm (~2in) in length.

For this particular lot of fibers, the diameter was calculated to be 12.82 μm. Figures 5(a) and (b) show representative stress-strain plots obtained at room temperature and 800°C, respectively.

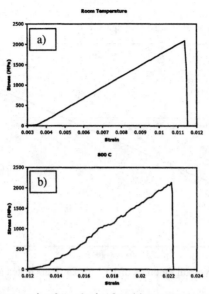

Fig. 5. Representative stress-strain plots obtained at (a) room temperature and (b) 800°C using the proposed method.

The corresponding tensile strengths for Nextel™ 720 fibers at each test temperature are listed in Table I, along with the values reported by 3M.

Table I. The tensile strengths obtained with the method proposed here compared to those reported by 3M at four temperatures.

	Proposed Method	3M
Room Temperature	1820 ± 170 MPa	2100 (1700) MPa
800°C	1780 ± 310 MPa	1700 MPa
1000°C	1530 ± 100 MPa	1700 MPa
1200°C	TBD	1500 MPa

71

Figure 6 shows the Weibull modulus plot at room temperature. The Weibull moduli for room temperature and 800°C were 10.8 and 5.9, respectively.

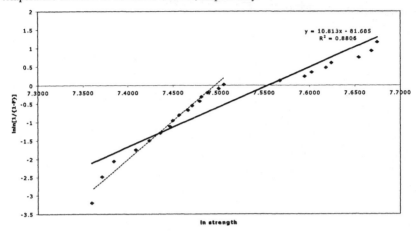

Fig. 6. Plot to obtain the Weibull modulus of 10.8 for 24 fibers tested at room temperature.

DISCUSSION

Fig. 4 indicates that the hot zone of the furnace was ~51 mm (~2in) long. In addition, it indicates that the location of this hot zone was such as to ensure that the test fiber was positioned within it. As shown, the measured temperature approximately one inch outside the furnace at the top was still almost 600°C. Since the load cell was temperature sensitive, it was placed at the base of the Instron to keep it away from the heat from the furnace due to the "chimney effect". An extension rod was attached to the load cell to further isolate it from the heat of the furnace. The bottom connector, (Fig. 3(b)) permitted motion in one plane by pivoting around the pin. A matching rod on the other end extended from the crosshead, onto which the top connector (Fig. 3(a)) was attached. Alignment of the rods had to be ensured prior to testing. The top connector, a universal joint, was capable of pivoting 360° at a maximum working angle of 35°. Together, these two connectors allowed the alumina tube-fiber assembly to automatically align itself along the load axis to minimize the bending moment when a small tensile load was applied.

Fig. 5(a) depicts the typical elastic deformation and brittle failure expected of ceramics. At 800°C, the plot was not as linear, but it still followed the same elastic trend. The roughness was possibly due to convection currents inherent in the tube furnace that the load cell sensed. Another reason could be the elongation of the grips at higher temperatures, which is part of the system compliance. System compliance has not been accounted for at the time of this paper.

Nextel™ 720 fibers are reported as having room temperature tensile filament strengths of 2.1 GPa for a 25.4 mm (1 in) gauge, determined by standard methods, and having high creep resistance up to 1000°C. In Table I, the strength in parentheses, 1700 GPa, was obtained by 3M's own method for high temperature tensile testing described elsewhere[6], as well as at elevated temperatures. The fibers retained their strength up to at least 1000°C, before decreasing by ~15% to 1500 MPa.[5,6] Using the proposed method, a general trend of decreasing strength with increasing temperature was observed. The average tensile strength at 1000°C, however, was determined from only five samples, whereas at least 20 tensile strengths were averaged at the other temperatures.

These tensile strengths were calculated using theoretically determined diameters based on the geometry, roving denier, and density for each lot of fibers produced. Diameter variation across the fibers was not taken into account, which could increase or decrease the strengths. Individual diameter determination at the point of fracture was not possible in most cases because the fiber fractures were not retrievable from the furnace. The strengths reported by 3M were also determined by a similarly calculated diameter. Since these tests were performed for calibration of the proposed method by comparison to other data as a standard, the analysis was kept constant. At room temperature and 800°C, the strengths were comparable to the published data when error was taken into account.

Weibull statistics give a better idea of the scatter in data than do standard deviation when a material fails by a critical flaw. The slope of the line of best fit in Fig. 6 yields the Weibull modulus, where a higher number indicates less scatter in data. At room temperature, the Weibull modulus was 10.8, which is comparable to 9.7, the value reported by 3M.[6] The lower data points, indicated by the dashed line, appear to have a steeper slope. When Weibull analysis was performed on these 16 data points, the modulus increased significantly to 23.3. This data set corresponded to an average strength of 1700 MPa, which was the same strength reported by 3M. At 800°C, the Weibull modulus was 5.9, but similarly to room temperature, the plot produced regions of different slopes. The lower region had an average strength of 1700 MPa and a Weibull modulus of 11.1.

When this method will be applied to the $DyNbO_4$ fibers, it is planned that the sample preparation method be modified. Instead of placing the alumina tubes on top of the fibers, as done with the Nextel 720 fibers, the $DyNbO_4$ fibers will be inserted inside the tubes and then secured with ZrO_2 cement.

CONCLUSION

A method for testing ceramic fibers at high temperature with the entire length of the sample in the hot zone of the furnace was developed and calibrated up to testing temperatures of 1000°C. The method described here can be applied to ceramic fibers of ~10 μm in diameter, or modified slightly to accommodate thicker fibers.

Future work includes continuing calibration at higher temperatures, obtaining the system compliance for accurate determination of the tensile modulus, and applying this method to extruded $DyNbO_4$ fibers.

ACKNOWLEDGEMENTS

Consultations with Dr. D.M. Wilson of 3M, Dr. P. Kurath, Mr. J. Grindley, Dr. D.K. Kim, and Mr. W. Yoon, are greatly appreciated. This work was funded by AFOSR account number F49620-03-1-0082. The work was partially carried out in the Center for Microanalysis of Materials, University of Illinois, which is partially supported by the US Department of Energy under Grant DEFG02-91-ER45439.

REFERENCES
[1] M-H. Berger, "Fine Ceramic Fibers: From Microstructure to High Temperature Mechanical Behavior," Advances in Ceramic Matrix Composites IX, pp. 3-26 (2003).

[2] R.J. Harrison, S.A.T. Redfern, and E.K.H. Salje, "Dynamical Excitation and Anelastic Relaxation of Ferroelastic Domain Walls in LaAlO$_3$," Physical Review B **69**, 144101 (2004).

[3] O. Unal and K.P.D. Lagerlof, "Hot-Grip Tensile Method for Alumina-Fiber Testing," *J. Am. Ceram. Soc.*, **76**[12] 3167-69 (1993).

[4] ASTM C 1557-03, "Tensile Strength and Young's Modulus of Fibers."

[5] D.M. Wilson and L.R. Visser, "High Performance Oxide Fibers for Metal and Ceramic Composites," Composites: Part A, 32, pp. 1143-1153 (2001).

[6] D.M. Wilson, S.L. Lieder, and D.C. Lueneburg, "Microstructure and High Temperature Properties of Nextel 720 Fibers," Cer. Eng. Sci. Proc. **16** (5) 1005-1014 (1995).

Fracture and Damage of Ceramics and Composites

COMPUTER MODELING OF CRACK PROPAGATION USING FRACTAL GEOMETRY

George W. Quinn
National Institute of Standards and Technology STOP 8940
Gaithersburg, MD, 20899-8940

Janet B. Quinn
ADAF Paffenbarger Research Center
National Institute of Standards and Technology STOP 8546
Gaithersburg, MD, 20899-8546

John J. Mecholsky, Jr.
Dept. of Materials Science and Engineering
University of Florida
Gainesville, FL, 32611

George D. Quinn
National Institute of Standards and Technology STOP 8520
Gaithersburg, MD, 20899-8520

ABSTRACT
    A model of the propagation of two-dimension traveling cracks in a biaxially-stressed glass disk was derived using concepts of fractal geometry. A computer program employing this model was written using known fractographic equations, empirical observations and generic algorithms for generating fractals. Inputs include material property data, fracture load, disk size and the sizes of the load-bearing rings. The outputs include fracture stress, initial crack size before branching, and a prediction of the expected number of radial cracks. A typical image of a fractured disk with the input conditions is also produced. The image incorporates empirically-determined randomness and a degree of crack curvature dependent on the stress state. The program is user-friendly, and may be easily adapted for other materials and conditions.

INTRODUCTION
    Fractography involves the interpretation of features observed on fracture surfaces and overall breakage patterns. Quantitative analyses of features on fracture surfaces can reveal significant information concerning failure conditions, such as the critical stress (breaking point) and environmental effects, as well as properties of the material. The aircraft industry makes considerable use of fractography for design and failure analysis[1] and the automotive field uses fractographic techniques to evaluate parts such as valves, crankshafts and steering columns.[2,3] This project utilizes fractographic theory in developing a computer model of cracks in brittle materials using fractal geometry.
    Fractal geometry is a branch of mathematics devised by Mandelbrot[4] to characterize various non-smooth natural phenomena. Fractal analysis describes the shapes of these phenomena as containing a degree of self-similarity (scale symmetry) that can broadly be described by a non-integer fractal dimension. Mecholsky et al.[5] first summarized how fracture in brittle materials can be described using fractal geometry. These fractal analyses were first applied to the roughness of the fracture surfaces. Mecholsky also showed that gross crack

branching is distinct from the fractal dimension obtained from surface roughness.[6] While continuing work on relating fractal geometry to fracture characteristics is not extensive, several other authors have published on the topic.[7,8,9] None of these authors, however, has yet made practical models that predict gross fracture patterns in brittle materials. A primary reason for this lack of modeling is that the current fractographic theory is insufficient to produce a complete model of a topological fracture surface. For a complete model, it was necessary to perform original research and gather empirical evidence. This process description is included within the Methodology and Results sections along with a description of the programming aspects.

METHODOLOGY

Because of the vastness of the fractographic field, it was decided to restrict this project to model topological fracture patterns in 76 mm diameter, 5.3 mm thick glass disks, broken in biaxial tension (ring-on-ring tests, Fig. 1). The inner and outer load rings were 31.8 mm and 63.6 mm, respectively. Biaxial disk fracture patterns were chosen because much information relating to such disks exists in published literature and we had access to a large number of broken disks donated by NIST (National Institute of Standards and Technology). The formulae for calculating stress conditions for this load configuration have been well documented.[10] It was hoped that the techniques and approaches used here, in a simple case, can be expanded to encompass other materials and stress configurations.

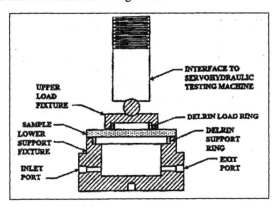

Figure 1. A schematic diagram of the biaxial stress test. A load is applied *via* the Delrin rings to the disk until fracture. The lower support fixture has ports to enable testing in different environments.

Model Basis

In order to model the crack patterns in glass disks, theoretical calculations, empirical data and physically measured data were used. Cracks initiated in the highly stressed central portions of the disks, propagated, branched, and continued to grow and branch until they intersected the disk edges. Some of the variables, such as the distance the crack grows until the first branch, are well known, and can be calculated by simple formulae. Other variables, such as whether the cracks branched into two, three or more cracks, are not well documented. This section outlines the ideas and concepts used in this model.

First, it was necessary to confirm that the crack patterns in the disks were approximately scale invariant and could be modeled using fractals. Initial research involved examination of broken disks at different magnifications under a Wild stereo optical microscope (Model M10, Leitz, Ltd., Heersbrugg, Switzerland). These examinations did show approximate scale invariance, which is illustrated by Figs. 2a and 2b. Fig. 2a shows each of 4 failures at different loads at the same magnification. Fig. 2b shows the same disks, but magnified by different amounts. Comparing the photos, the higher stress failures appear similar to lower stress failures when magnified to an appropriate degree, demonstrating their approximate scale invariance.

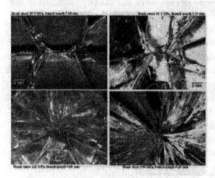 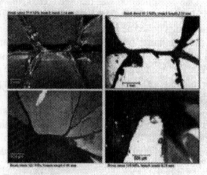

Figure 2a. Fracture origins of four failed disks. The stresses at failure are, clockwise from upper left: 55.0 MPa, 81.3 MPa, 198 MPa, and 121 MPa.

Figure 2b. The same four disks with the origins shown at different magnifications to illustrate the shape similarity among the fractures.

The next consideration was the frequency of bifurcation to trifurcation. There has been limited research in this area. To derive the model, empirical data were collected on 25 amenable disks. Empirical observation of the failed disks reveals that sometimes a branch would bifurcate, and sometimes it would trifurcate. The frequency of trifurcations seemed to occur more often at higher stresses. Furthermore, the frequency of bifurcation as opposed to trifurcation appeared to increase with subsequent branching. For example, trifurcation was observed to occur more often at secondary branch points than at ternary branch points. Statistical variation was simulated in the model by incorporating a degree of randomness in the branching frequency consistent with the empirical observations.

Preliminary research indicates a great deal of controversy regarding predictions of final branching angles in specimens such as those shown in the figures. It was necessary to choose the model that seemed the most accurate. Sakai et al.[11] used a model that suggested multiple cracks emanate from one single central location and then bifurcate at constant distances from this location at constant branching angles (10 degrees in the example in the reference). None of the twenty five fractured disks in this project seemed to follow this pattern, however, and this model may be better suited for other fracture conditions such as that for tempered or ion-exchanged glasses. Frechette suggested an initial branching angle of 180 degrees for the equibiaxial stress state, and smaller subsequent angles.[3] This, too, did not match our observations and subjective judgment, as branching angles did not seem to occur as high as 180 degrees, although the cracks seemed to approach this angle as stresses became high. The model we chose assumes branching occurs at angles that would allow cracks to travel and branch to most evenly divide the specimen surface. This model is based on the idea that the disk strain energy is initially evenly distributed within the inner load ring and is released in the immediate vicinity of cracks as fracture proceeds. Disk areas without cracks contain the highest amount of stored strain energy, so the nearest cracks travel into those areas. This idea was vaguely implied by Rice in his observations on disk fracture[12] and the resulting model best corresponded to our subjective observations, although it was the most difficult to incorporate mathematically.

One complication was that the cracks, branched or not, were not exactly linear from the source. Initial experimentation with preliminary versions of the program showed that very realistic results could be obtained from random curvature, but in this case, it was possible to calculate a curvature from a study of the known stress fields. In mode I fracture (opening tensile mode), cracks propagate perpendicular to the direction of highest tensile stresses. By calculating the stresses at each segment of a growing crack in our model, a shift in crack direction was made proportional to the magnitude and in the direction perpendicular to the highest tensile stress. Boundary conditions were important. The cracks were assumed to be perpendicular to the disk edge when they reached it, because at that location there are no stresses in the radial direction. The stress ratio (S_R), which defines how sharply the traveling crack curves towards perpendicularity to the greatest stress, was determined by dividing the radial stress equation by the circumferential stress equation. The stress equations are from Fessler and Fricker.[10]

$$S_R = \frac{[(1+v)\ln R_s/R_2 + (1-v)(R_s^2-R_l^2)/2R_o^2 - (1-v)(R_2^2-R_l^2)/2R_2^2]}{[(1+v)\ln R_s/R_2 + (1-v)(R_s^2-R_l^2)/2R_o^2 + (1-v)(R_2^2-R_l^2)/2R_2^2]} \qquad (1)$$

where S_R is the stress ratio, v is Poisson's ratio (0.26 for the glass in this study), R_l is the inner ring radius, R_s is the outer ring radius, R_o is the whole disk radius and R_2 is the location of the line segment from the disk center. As the line segment nears the outer ring ($R_2 \rightarrow R_s$), the radial stress decreases faster than the circumferential stress, and the ratio decreases. The program is written such that smaller ratios result in proportionally smaller deviations in a crack path perpendicular to the circumferential direction.

One final set of data incorporated into the model was the number of cracks, N, that reached the disc edge as a function of failure stress. This data was collected for a fractography course at Alfred University, NY, in July of 2003 by one of the co-authors of this report, and is included as Fig. 3. For the disc and ring sizes of this study and stress measured in ksi:

$$N = 0.74 \text{ failure stress}^{1.43} \qquad (2)$$

Programming

User inputs include the material constants Poisson's ratio, v, and the branching constant, A_b, both found in material engineering tables. (The branching constant is defined in Eq. 4 in a subsequent section.) Inputs also include the disk size: t and R_o, the disk thickness and radius, respectively. Finally, the last three inputs are the inner loading ring radius, R_l, the outer loading ring radius, R_s, and the break load.

80

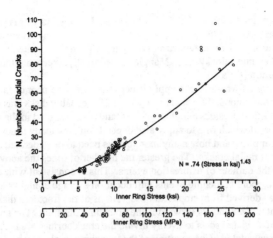

Figure 3: Graph showing the direct relationship between the inner ring (failure) stress and the number of radial cracks.

The first step is to parse the inputs and use them to calculate the stress in MPa. The failure stress was calculated using the equation for stresses within the inner load ring:[10]

$$\text{Failure stress} = 3(\text{load})[(1+v)\ln R_s/R_l + (1-v)(R_s^2 - R_l^2)/2R_o^2]/\pi t^3 \quad (3)$$

Once the maximum stress is calculated, the predicted number of radial cracks can be calculated using Eq. 2 (Fig. 3).

The equation to calculate the initial branching distance, R_b, (distance a crack travels before it first branches) has been well documented:[3,12,13]

$$R_b = (A_b / \text{failure stress})^2 \quad (4)$$

where A_b is the crack branching constant, a material property that can be found in fractography tables for different brittle materials.[14]

The previous variables are used to determine an initial branching distance, and then the Lindenmayer model is used to simulate crack propagation. It begins with an initial generating string and recursively applies a context-free text substitution scheme. Geometric interpretation of the resulting string after n iterations produces a fractal image within a range. The number of iterations, n, is dependant on the stress. A crack can either bifurcate, trifurcate, or not branch at all. These events are represented respectively by the following grammar rules, which were used in the model:

F<a> = F<d>-<angle>[F<a>]+<2*angle>[F<a>]
F<a> = F<d>-<angle>[F<a>]+<angle>[F<a>]+<angle>[F<a>]
F<a> = F<d>F<a>

81

where F<d> is a terminal symbol. The first rule can be described in words as follows: Move the crack forward by an amount ("F<d>"), rotate the crack tip direction clockwise by "angle" degrees ("-<angle>"), create a crack tip in the current direction ("[F<a>]"), rotate the original crack tip by 2*"angle" degrees counter-clockwise (<2*angle>), and finally, create another crack tip pointing in the current direction ([F<a>]).

Determination of which rule to apply is dependant on a number of factors that affect actual fracture configurations. Statistical analysis of 25 amenable disks was used to aid in calculating probabilities. Besides a high degree of randomness, noted in the actual fractured disks, determination of which rule to apply also depends on how many times the above rules have already been applied, and how many more cracks need to be generated in order to create the necessary number of radial cracks. The greater the number of times the above rules have been applied, the fewer the number of trifurcation events. This is consistent with empirical data on the disks. Furthermore, more branching events will occur on average at higher loads than at lower ones. This also was derived from empirical observations of the fractured disks.

The amount a crack moves forward is the branching distance. The angle for each branching event is calculated so as to evenly divide the crack surface area. At each branching event, this value is calculated. Computation of the branching angle involves some basic geometric calculations such as calculating the intersection between a line and a circle, and calculating the intersection point and angle of intersection between two lines.

Curvature of the branch segments is calculated after each iteration over the grammar string. It is calculated at draw-time and is not visible in the grammar string. In the L-System model, F<a> and F<b> or other terminal and non-terminals are interpreted as line segments. As the program interprets the grammar string into a fractal image, it breaks up each segment into a sequence of smaller connected line segments. These connected segments simulate curvature. The curvature amount is determined by Eq. 1 in the Basis for Model section.

RESULTS

The program outputs several fracture-related properties as well as a graphical image representing the fracture surfaces of glass disks. The image accurately portrays the number of cracks for the calculated stress, as well as branching properties, such as angle and location. Examples of images are included in Fig. 4. Fig. 4a compares a computer-generated image with a photographed disk that fractured at a low stress of 41 MPa. Fig. 4b is a corresponding high stress fracture of 116 MPa. The circular cracks in the middle of the photographed disk are due to deformation around the inner ring after initial fracture, and are not included in the model. Crack curvature is apparent in the low stress fracture, and is incorporated in the model.

User friendliness is reflected in the ease of inputting variables, and includes reasonable default values. This might be especially helpful in the case where Poisson's ratio is unknown. This property is sometimes not easily available, but is almost always between 0.2 and 0.3 for brittle materials. The stress that caused fracture, the initial calculated branch length and predicted number of radial cracks are also outputs.

The program could be easily applied to other materials and conditions. The stress state would be calculated the same way as it depends only on the load, ring and disk dimensions, and the material constants of Poisson's ratio and branch length are already inputs. Values for the incidence of bifurcation vs. trifurcation and for number of cracks vs. failure stress are not published for many materials, but different values could be easily substituted, and pictures in the literature suggest similar relationships to the glass of this study.

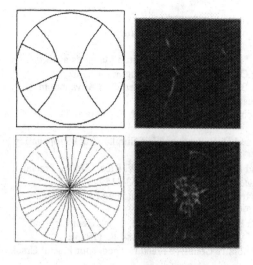

Figure 4(a). A computer-generated image of a low stress (41 MPa) fracture left, and a photograph of a disk that fractured at the same stress, right.

Figure 4(b). A computer-generated image of a higher stress (116 MPa) fracture left, and a photograph of a disk that fractured at the same stress, right. The circular crack pattern in the photographed disk formed after the radial cracks and is not included in the model.

DISCUSSION AND CONCLUSION

There is currently no easily available program for simulating fracture patterns for biaxial flexure. This program attempts to fill that gap, and produces an image that is subjectively similar in appearance to actual fracture surfaces. Helpful in achieving a reasonably realistic appearance is the use of randomness and crack curvature, notions that are seldom included in any fracture computer models. Parts of these concepts are also innovative in the field of fractography, which tends to generally focus on other aspects, such as initial crack sizes and fracture surface roughness.

A difficulty often encountered in producing the model was the combination of well-known relationships in some areas and lack of published research in other areas. This necessitated melding empirical data with published observations. The subjectivity was also frustrating. For example, whether to classify a crack as bifurcating or trifurcating when the crack appeared to branch and rebranch only a very small distance apart. Some cracks appeared to start and stop and some went only partially through the disk. Future efforts might include developing "classification rules" to assist in consistent crack property measurements.

Other future efforts could involve improving the model's accuracy, expanding its applicability to other materials and conditions, and utilizing the program in design and forensic analysis. This project also opens doors to innovative areas of research in fractography, such as the degree of crack curvature as functions of stress and material properties.

Contact the corresponding author, George W. Quinn, at gwquinn@nist.gov for a copy of the modeling program described here.

ACKNOWLEDGEMENTS

This work was supported by ADAF and NIH DE14534. It is an official contribution of the National Institute of Standards and Technology; not subject to copyright in the United States.

REFERENCES

[1]C. J. Peel and A. Jones, "Analysis of Failures in Aircraft Structures," *Metals and Materials*, **6** 496-502 (1990).

[2]R. Danzer, M. Hangl and R. Paar, "How to Design with Brittle Materials against Edge Flaking," 6th International Symposium on Ceramic Materials for Engines, 658-62 (1997).

[3]V. Frechette, "Failure Analysis of Brittle Materials," Chapter 28 in Advances in Ceramics. American Ceramic Society, Westerville, OH, 1990.

[4]B. B. Mandelbrot, "The Fractal Geometry of Nature," W. H. Freeman and Co., N.Y., NY (1982).

[5]J. J. Mecholsky, Jr., T. J. Mackin and D. E. Passoja, "Self-Similar Crack Propagation in Brittle Materials," *Adv. in Ceram.*, **22** 127-38 (1988).

[6]J. J. Mecholsky, Jr., R. Linhart and B. D. Kwitkin and R. W. Rice "On the Fractal Nature of Crack Branching In MgF_2," *J. Materials Res. 13*, **11**, 3153-3159 (1998).

[7]X. Heping, "The Fractal Effect of Irregularity of Crack Branching on the Fracture Toughness of Brittle Materials," *Int. J. of Fracture*, **41** 267-74 (1989).

[8]M. Tanaka, "Relationship between Fracture Surface Pattern, Crack Geometry and Fracture Toughness in Brittle Materials," *J. Mater. Sci. Lttrs.*, **15** 1184-85 (1996).

[9]A Yavari, "Generalization of Barenblatt's Cohesive Fracture Theory for Fractal Cracks," Fractals, **10** [2] 189-198 (2002).

[10]H. Fessler and D. Fricker. "A Theoretical Analysis of the Ring-on-Ring Loading Disk Test," *J. Am. Ceram. Soc.*, **67** [9] 582-588 (1984).

[11]T. Sakai, M. Ramulu, A. Ghosh and R. C. Bradt, "A Fractal Approach to Crack Branching (Bifurcation) in Glass," Fractography of Glasses and Ceramics II, *Ceram. Trans.* **17** 131-146 (1991).

[12]R. W. Rice, "Ceramic Fracture Features, Observations, Mechanisms, and Uses," Fractography of Ceramic and Metal Failures, ASTM STP 827, J. J. Mecholsky, Jr., and S. R. Powell, Jr., eds., ASTM, 5-103 (1984).

[13]J. B. Quinn, "Extrapolation of Fracture Mirror and Crack-Branch Sizes to Large Dimensions in Biaxial Strength Tests of Glass," *J. Am. Ceram. Soc.*, **82** [8] 2126-32 (1999).

[14]ASTM C1322. Fractography and characterization of fracture origins in advanced ceramics. Annual Book of ASTM Standards, Vol. 15.01, ASTM, West Conshohocken, PA, (2004).

GEOMETRY OF EDGE CHIPS FORMED AT DIFFERENT ANGLES

Janet Quinn and Ram Mohan
American Dental Association Foundation
National Institute of Standards and Technology, 100 Bureau Drive, STOP 8546
Gaithersburg, MD 20899-8546

ABSTRACT
In the edge-chip test, an increasing force is applied near the edge of a specimen until a chip is formed. At greater distances from the specimen edge, higher forces are required for chip formation. A plot can be constructed by graphing the force necessary to form a chip against the distance from the specimen edge where the force is applied. The slope of the line resulting from such a plot constitutes the edge toughness. Studies have shown that chip geometries are self-similar, such that the chip width, depth and height ratios are independent of material or total chip size. Such previous studies, however, refer to edge chips made with a force perpendicular to the specimen surface. The current work addresses the issues of applied force direction and subsequent changes in edge toughness and chip geometry when chips are formed from forces that are not perpendicular. Greater force is required to produce a chip when the force is angled away from the edge, for example, and some aspects of the resulting chip geometry resemble a flattened cone. Quantitative analyses of such geometric changes can enable back calculation of force and direction in performing edge chip failure analyses, and would aid in designing components for edge integrity when the forces are not perpendicular.

INTRODUCTION
The edge chip test has been used in a variety of applications, including valve design,[1] K_{Ic} measurement,[2] efficacy of flint knapping conditions[3] and insight into material property and machining techniques.[4] In this test, an increasing force is applied near the edge of a specimen until a chip forms. The force to form the chip is graphed against the distance from the edge at which the force was applied. The slope of the resulting line has been defined as the edge toughness.[5]

Edge toughness tests utilize a force perpendicular to the specimen surface, but edge chips in failed components are unlikely to have been formed by forces applied at exactly 90°. A study of edge chipping at non-ideal conditions[6] notes that asymmetrical flakes form at different angles of applied force, but it would be helpful to quantize such results. In the current study, geometric measurements and edge toughness values were determined for flakes formed at different force angles.

EXPERIMENTAL PROCEDURE
Annealed soda lime glass was chosen as the model material for this study. Glass is plentiful, inexpensive, and various types of glass, such as tableware, mirrors and windows, are commonly subject to edge failures. Since it had been shown that chip geometry was independent of material,[7] it was also believed that the glass results would be applicable to other brittle materials.

The soda lime plate glass (Corning 0080, NY)* used in the experiments has hardness, KNH_{100} of 465, density of 2.47 g/cc, and elastic modulus of 10.2 GPa according to the company data sheets. The glass thickness was 0.6 cm, about five times the measurements of the largest chips. Lateral plate dimensions of 3.5 cm to 13 cm were easily accommodated in the edge-chip machine platform. The most difficult preparation step was achieving an initial clean, unchipped edge in the glass.

After several attempts, a procedure was established to produce a 90° edge with minimal damage and rounding. Two pieces of plate glass were sandwiched together with melted machinists' wax on a hot plate, with the wax softened at about 100 °C. The layered specimens were allowed to cool slowly, and then were sliced through simultaneously with a diamond wheel-cutting blade. The resulting sandwiched cut surfaces were polished parallel to the layers with successively finer grit sandpaper, from #300 to #4000. The layered specimens were then separated by remelting the wax, allowing several hours for the specimens to cool sufficiently for handling. The slow cooling rates were used to avoid inadvertent glass tempering. The residual wax on the specimen surfaces was removed with acetone. This method resulted in fairly clean edges along the surfaces that had been waxed together.

Figure 1. Edge chip machine with 15° wedge (left) and diamond scribe (above).

Edge chips were made using an Engineering Systems Model CK 10 edge-chip machine (Nottingham, UK), fitted with a conical 120° diamond scribe indenter, as shown in Fig. 1. The indenter was frequently viewed under an optical microscope to confirm the integrity of the diamond tip between series of edge chips. The machine normally tests specimens at 90°, but an aluminum block was machined to form a 15° wedge in order to test the glass specimens at different angles. The specimens were mounted with machinist's wax to the block, and the block was attached to the machine platform with double-sided tape. Test force directions were 75°, 90° and 105° as shown in the schematics in Figs. 2 - 4. Chips were also made with an angled force parallel to the specimen sides.

---

* Equipment and materials are identified in order to adequately specify the experimental details. This does not imply recommendation by NIST or ADAF, nor does it imply the materials are necessarily the best for the purpose.

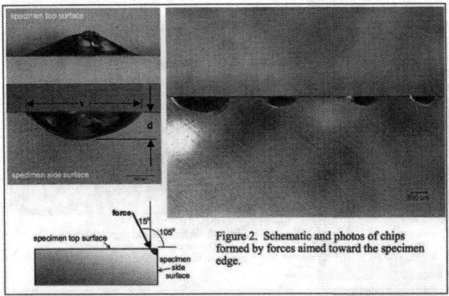

Figure 2. Schematic and photos of chips formed by forces aimed toward the specimen edge.

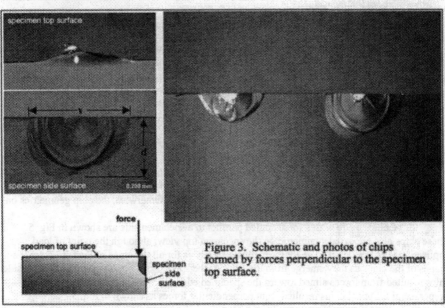

Figure 3. Schematic and photos of chips formed by forces perpendicular to the specimen top surface.

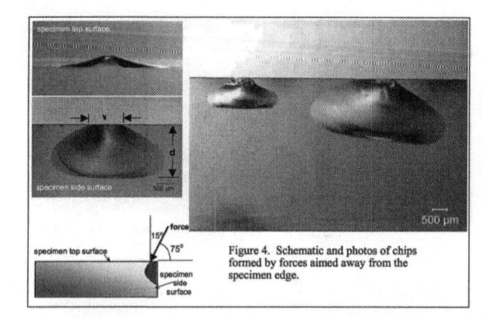

Figure 4. Schematic and photos of chips formed by forces aimed away from the specimen edge.

The samples were loaded until the edge chipped, indicated by a sudden drop in recorded force and frequently accompanied by a popping sound. The chipped edges were then examined and chip sizes measured with a Leica MZ16 optical stereomicroscope system (Heerbrugg, Switzerland) with high depth of field, adjustable oblique light sources and traveling stage accurate to the nearest micrometer. These microscope features were helpful in producing images of the chips, as the translucency and reflectivity of the glass made this a particularly difficult material to photograph.

Line plots by least squares analyses were made of the edge chip distance *vs.* force for each angle, where the slopes constitute the edge toughness.

RESULTS

Three distinct types of chips were formed at the three different angles shown in Figs. 2 - 4. In comparing the top surfaces of the figures, more contact damage is evident in the immediate vicinity of the indenter for forces aimed away from the edge. Otherwise, the chip geometries on the specimen top surfaces appear similar.

Edge chips made with a force angled parallel to a specimen side are shown in Fig. 5. These chips also appeared symmetric on the specimen top view, although the point of load application is not centered. The most obvious differences for all the chips in Figs. 2 - 5 are the shapes of the chips as they emerge from the sides of the specimens. Shallow, semi-elliptical side chips resulted from forces aimed toward the specimen edge (105°) as shown in Fig. 2, while deeper, semi-elliptical chips resulted from perpendicular forces as shown in Fig. 3.

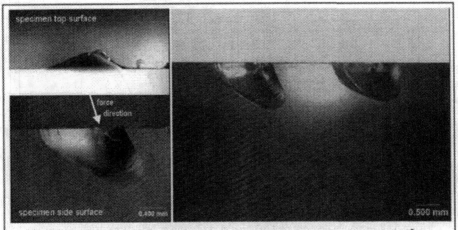

Figure 5. Chips formed by a force in the plane of the specimen side, but angled 15° as shown by the arrow.

The chips resulting from forces aimed away from the edge toward the bulk of the specimen (75°) have a distinctive geometry as shown in Fig. 4. These flakes are quite large viewed at the specimen sides compared with the top surface chip size, spreading out in almost a triangular shape before rounding into a flat bottom.

Measurements were made of the ratios of the linear dimensions of the chips as they emerged from the specimen sides for the force directions in Figs. 2 – 4. The shallow 105° chips resulting from the indenter aimed toward the specimen edge have width/depth ratios ± standard deviation of 3.43 ± 0.22. The chip widths are thus approximately three and one half times as wide as deep. The chip width/depth ratio for a perpendicular force is 1.66 ± 0.22. These chips are thus about one and one half times as wide as deep. The chip width/depth ratios measured at the edges of the specimens indented with 75° forces aimed away from the specimen edges continue to decrease to 0.62 ± 0.05. These chips are only slightly more than half as wide as deep at the specimen edges. However, these chips gradually become wider on the specimen side surfaces as shown in Fig. 4, with a maximum chip width/depth ratio of 2.55 ± 0.61 near the base of the "triangle".

The edge chipping plots are shown in Figure 6, with least squares slope ± standard deviation of (206 ± 13) N/mm and (67.0 ± 2.5) N/mm, for the chips formed at 90° and 105° respectively. The edge distances were not difficult to estimate for these chips; the points of load application are discernable on the top surfaces of the specimens in Figures 2 and 3.

It was not possible to obtain accurate values for the edge distances for the chips formed at 75°. For these specimens, the edge distance was measured from the tangent of the circular damage region to the specimen edge, as shown in the graph inset. The resulting slope ± standard deviation is 175 ± 14, but the actual force vs. edge distance slope associated with the 75° angled force should be much steeper, as the graphed edge distances are much larger than the distances of load application.

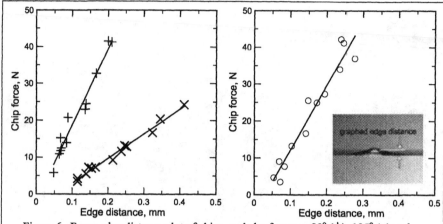

Figure 6. Force-edge distance plot of chips made by forces at 90° (+), 105° (X) and 75° (O); the lines are the least squares slopes given in the text. The edge distances for chips formed by 75° forces were measured differently, as shown in the inset. The slopes for the actual point of force application would be much higher.

SOMMARY AND DISCUSSION:

Analyses of edge chip geometry in terms of contact mechanics are complex and generally attempted for the initial crack path angle towards the specimen edge.[3,10,11] It is the final emergence of the crack at the specimen side, however, that was shown to be the most obvious indicator of the applied force direction, and the most useful in terms of fractographic failure analysis. This report quantitatively characterizes the shallow width/depth ratios for forces aimed toward the specimen edge and the distinctive triangular shape of the chip for forces aimed away from the specimen edge. For perpendicular applied forces, the chip width/depth ratio of approximately 1.5 measured by Almond and McCormick[7] in tungsten carbide/cobalt is within the range measured in the present study. The higher degree of contact surface damage when the applied force is aimed away from the edge is another fractographic indication of force direction. Presumably, the chip/width ratios increase or decrease in relation to the force angle. The standard deviations in the chip/width ratio measurements indicate that it might be possible to forensically determine force angles to within 5° or 10° degrees for edge chip failures.

Theoretically, the size of the chips should scale with the forces needed to form them for a brittle material. This presumes an energy balance where all the energy used to chip the material results in the formation of new surface area. Thus, the surface areas of the chips in this study should proportionally scale with the forces that formed them, regardless of angle. This is roughly the case here, in spite of the different geometries. The 105° chips which easily formed at low loads are more shallow than the chips formed at higher loads, while the 75° chips are fanned out in a triangular shape with more surface area than chips formed at lower loads.

While geometry alone appears to be sufficient to estimate applied force angles, the edge chip plots are necessary to forensically estimate the magnitude of a force that caused a chip to form. Once the force angle is estimated, the distance of force application on the specimen top surface can be used to extrapolate the force magnitude from an appropriate plot.

Not all the chips in this study were of the "ideal" shapes shown in the representative figures. There was some variability of chip shape as noted by Morrell and Gant.[6] This may be due to inhomogeneities in material as well as slight differences in loading conditions. Fragmentation was occasionally noted, resulting in partial chips, either undetached or partially detached. Examples are shown in Fig. 7, where the side view of a 90° partial chip is quite similar to the side views of chips formed by angled forces in Fig. 4. In such cases, the two types of chips can be distinguished by the top views. An example of a partial chip formed by a 75° angled force that was removed prior to chip formation is also shown in the figure. The black areas are reflections from subsurface crack growth. Had the load continue to increase, the contact damage would have formed a circle around the point of load application, and the subsurface crack would have grown to the translucent arch before the chip detached, resulting in a chip similar to those pictured in Fig. 5.

The "ripple effect" noted in soda lime glass by Morrell and Gant[6] is also evident in this study, especially in Fig. 3. The authors attribute this to small interruptions in crack growth due to local fragmentation under the indentation. This is likely the case, as small changes in force measurements were noted as the indenter in this study steadily descended into the specimen. This behavior may be peculiar to the material, however.

In hindsight, glass might not have been the most judicious choice of model material for this study. There were difficulties in obtaining a clean edge and photographing the highly transparent, reflective surfaces. Also, the small zero offset in the edge toughness plots in Fig. 6 might be due to the relative plasticity and compressibility of glass compared with other ceramics. There is a small plastic region at the indenter tip that may affect data very close to the specimen edge, and glasses can densify under compressive loads because of the nature of the lack of molecular structure. This could also explain why edge chip plots of some glassy porcelains, unlike those of polycrystalline materials, become non-linear as the edge is approached.[8]

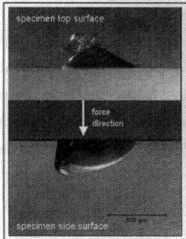

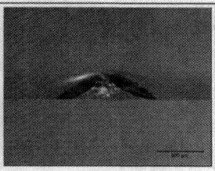

Fig. 7. Partial chips. The partial chip on the left has a side view similar to the side view of the Fig. 5 chips, despite the perpendicular force. Above, partial formation of a chip under the Fig. 4 loading conditions.

91

An important point is that the forces to form chips in this study, as in all the referenced studies, were measured while the forces were slowly increasing. In the glass tests, however, it was observed that a chip could form from a lower applied force if the force were removed prior to observed cracking. This is because residual stresses result from compressed material beneath the indenter, and these stresses cause cracking when the applied load is reduced. This problem may be exacerbated in a material that is more subject to local densification. An analogy is the formation of crack systems beneath Vickers indentations during load removal as observed and recorded by Cook and Pharr.[9] The authors note that the cracking sequence in soda-lime glasses may be atypical for Vickers hardness tests, and there is evidence in this study that chip formation in soda-lime glasses might also be atypical. Several times during this study a test was deliberately stopped prior to observed cracking, and a chip subsequently formed and popped out during load removal. This data was not included in the edge toughness calculations.

In conclusion, the edge-chip test can be very useful in forensic estimation of failure force direction as well as magnitude, but is somewhat limited by lack of precision that may be due to material characteristics such as densification, residual stresses and inhomogeneity. However, the ease of the test and the small amount of required material still render it attractive for material evaluation, particularly for applications where edge failures are frequent or problematic.

ACKNOWLEDGEMENTS
This work was supported by ADAF, NIST, NIH DE14534 and a Fellowship from Colgate-Palmolive Co. to R. Mohan.

REFERENCES
[1] R. Danzer, M. Hangl and R. Paar, "Edge Chipping of Brittle Materials"; pp. 43-56 in Ceramic Transactions, Vol. 122, *Fractography of Glasses and Ceramics I.* Edited by J. R. Varner and G. D. Quinn. American Ceramic Society, Westerville, OH, 2001.
[2] F. Petit, P. Descamps, J. P. Erauw and F. Cambier, "Toughness (K_{Ic}) Measurement by a Sliding Indentation Method," *Key Eng. Mater.*, 206-213, 629-632 (2002).
[3] B. Cotterell and J. Kamminga, "The Formation of Flakes," *Amer. Antiquity*, 52[4], 675-708 (1987).
[4] J. B. Quinn and I. K. Lloyd, "Flake and Scratch Size Ratios in Ceramics"; pp. 57-72 in Ceramic Transactions, *ibid.*
[5] N. J. McCormick, "Edge Flaking as a Measure of Material Performance," *Metals and Mater.*, **8**, 154-57 (1982).
[6] R. Morrell and A. J. Gant, "Edge Chipping – What Does it Tell Us?"; pp. 23-42 in Ceramic Transactions, *ibid.*
[7] E. A. Almond and N. J. McCormick, "Constant-Geometry Edge Flaking of Brittle Materials," *Nature* **321** [6065], 53-55 (1986).
[8] J. Quinn, L. Su, L. Flanders and I. Lloyd, "'Edge Toughness' and Material Properties Related to the Machining of Dental Ceramics," *Mach Sci and Tech,* **4** [2], 291-304 (2000).
[9] R. F. Cook and G. M. Pharr, "Direct Observation and Analysis of Indentation Cracking in Glasses and Ceramics," *J. Am. Ceram. Soc.,* **73** [4], 787-817 (1990).
[10] T. J. Lardner, J. E. Ritter, M. L. Shiao and M. R. Lin, "Behavior of Indentation Cracks Near Free Surfaces and Interfaces," *Int. J. Fract.*, **44**, 133-143 (1990).
[11] J. D. Speth, "Miscellaneous Studies in Hard-Hammer Percussion Flaking: The Effects of Oblique Impact," *Amer. Antiquity*, **40**[2], 203-207 (1975).

THRESHOLD STRESS DURING CRACK-HEALING TREATMENT OF STRUCTURAL CERAMICS HAVING THE CRACK-HEALING ABILITY

Wataru Nakao, Koji Takahashi and Kotoji Ando
Yokohama National University
79-5, Tokiwadai, Hodogaya-ku
Yokohama, Kanagawa, 240-8501

ABSTRACT

Alumina/ 15 vol% SiC particles, alumina/ 20vol% SiC whiskers, mullite/ 15vol% SiC particles and mullite/ 15 vol% SiC whiskers composites having excellent crack-healing ability are subjected to crack-healing under elevated static and cyclic stresses at elevated temperatures from 1000 °C to 1200 °C. The bending tests were done after the specimens had been subjected to the crack-healing temperature. From the obtained result, the threshold stresses during crack-healing, which is upper limit stress been safely able to apply during crack-healing, are determined. The determined threshold stresses have been 64 % fracture strengths of the as-cracked specimens, respectively.

INTRODUCTION

Structural ceramics such as alumina, mullite and silicon nitride have excellent heat, corrosion and wear resistance. However, fracture toughness is low. This low reliability has, thus, limited their applications. Many investigators tried to improve the fracture toughness of structural ceramics by admixing with whisker and fiber. On the other hand, the present authors[1-10] have proposed to guarantee the reliability by endowing with the crack-healing ability. When ceramics admixed with SiC are kept in air at high temperature, SiC located on the crack surface reacts with O_2 in air and then crack is completely restored by the products and the heat of the reaction. Moreover, the restored part is mechanically stronger than the other parts. If the crack-healing mechanism is used on structural components in engineering use, great benefits can be anticipated not only an improvement in reliability but also a decrease in machining and polishing costs of ceramics elements. For example, mullite admixed with 15 vol% SiC particles[1-3] can completely recover the strength reduced by cracking so that this has adequate self-crack-healing ability.

In the previous study[4, 10, 11], it had been found that crack-healing also occurs though the crack is applied tensile stress. Thus, fatigue strength was held almost equal to fracture strength at the same temperature so that the crack-healing prevents fatigue crack growth. A prolongation of the lifetime of ceramics can be anticipated by using this behavior. For applying fully crack-healing under stress, it is necessary to determine the threshold stress during crack-healing, which is upper limit stress been safely able to apply during crack-healing.

In this study, some ceramics composites having crack-healing ability are subjected to crack-healing under elevated static and cyclic stresses at 1200 °C. The bending strengths of the crack-healed specimens were investigated at the crack-healing temperature. From the result, the threshold stresses during crack-healing were determined.

EXPERIMENTAL

Alumina/ 20 vol% SiC whiskers and alumina/15 vol% SiC particle composites, abbreviated AS20W and AS15P, respectively, were prepared for this study. Also mullite/15 vol% SiC whiskers and mullite/15 vol% SiC particles composites, abbreviated MS15W and MS15P, respectively, were prepared. The details of the raw materials and sample preparations were written elsewhere[1, 4, 6, 7].

A semi-elliptical surface crack was made at the center of the tensile surface of specimens with a Vickers indenter, using a load of 19.6 N. The surface length of the pre-crack was 100 μm. The ratio (a/c) of depth (a) to half the surface length (c) of the crack (aspect ratio) was 0.9. The pre-crack was healed in air at elevated temperatures from 1000 °C to 1200 °C under the some arbitrary tensile stress. The static stress, $\sigma_{ap,s}$, and cyclic stress, $\sigma_{ap,c}$, were applied by the three-point bending system shown in fig. 1. The applied cyclic stress was sinusoidal wave with a stress ratio, maximum stress / minimum stress, of 0.2 and frequency of 5 Hz. The value of $\sigma_{ap,c}$ was indicated that of the maximum stress. The pre-crack was subjected to tensile stress before the crack-healing was started by heating the furnace to avoid the crack-healing under no-stress. Then, the specimen was kept at the above condition for 2 - 80 h to finish completely the crack-healing process.

All fracture tests of the crack-healed specimens were performed on a three-point loading system with a span of 16 mm at the crack-healed temperature or room temperature. The cross-head speed in the monotonic test was 0.5 mm/min.

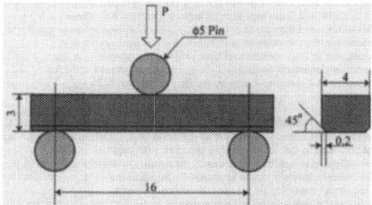

Figure 1 Dimensions of the specimen and the three-point loading system used for this investigation

RESULTS AND DISCUSSIONS

Figure 2 shows the fracture surface of the MS15P crack-healed at 1000 °C for 10 h under $\sigma_{ap,c}$ = 88 MPa. The specimens fractured from the crack-healed part because the crack-healing is not finished. Thus, the bright area shows the crack-healed part. The surface length of the crack-healed part increased from 100 μm to 240 μm. The result revealed that the crack-healing occurs although the pre-crack is grown by the applied stress.

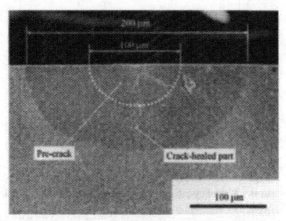

Figure 2 SEM image of fracture surface of MS15P crack-healed at 1000 °C for 10 h under cyclic stress of 88 MPa.

Figure 3 shows the bending strength at the crack-healing temperature of AS15P having the pre-crack crack-healed at 1200 °C in air under stress. The open and the closed triangle indicate the bending strength of the specimen crack-healed under static stress and cyclic stress, respectively. The bending strength of 0 MPa indicates the specimen fractured during crack-healing. The specimens crack-healed under static stresses below 150 MPa were never fractured during crack-healing treatment, and had the same bending strength as the specimens crack-healed under no-stress at 1200 °C. A few specimens crack-healed under a static stress above 180 MPa were fractured during crack-healing. Therefore, the threshold static stress during crack-healing of AS15P having the pre-crack, $\sigma^C_{ap,S}$, was found to be 150 MPa. Also, the threshold cyclic stress, $\sigma^C_{ap,S}$, was found to be 180 MPa. The same experiments were performed for MS15P, MS15W and AS20W. The obtained threshold stresses during crack-healing are listed in Table I. The $\sigma^C_{ap,S}$ were smaller than $\sigma^C_{ap,C}$ for all specimens. From a comparison with the values of $\sigma^C_{ap,S}$ and $\sigma^C_{ap,C}$, it may be confirmed that the crack growth behaviors of these specimens are time dependent rather than cyclic dependent. It is, therefore, concluded that applying static stress is the easiest condition to fracture during crack-healing under stress. The threshold stresses of every condition during crack-healing, σ^C_{HS}, has been defined as the value equal to $\sigma^C_{ap,S}$.

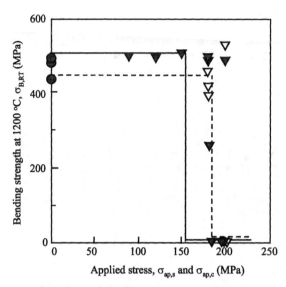

Figure 3 Bending strength at the crack-healing temperature of AS15P crack-healed at 1200 °C for 24 h under static stress (▼) and cyclic stress (▽), with that of AS15P crack-healed under no-stress(●).

Table I Threshold stresses during crack-healing of AS15P, AS20W, MS15P and MS15W

Specimen	$\sigma^c_{ap,s}$ (MPa)	$\sigma^c_{ap,c}$ (MPa)	σ_c (MPa)
MS15P	115	120	175
MS15W	100	150	250
AS15P	150	180	225
AS20W	250	300	400

Figure 4 shows the relation between the threshold stress during crack-healing and the fracture strength of the cracked specimens, σ_C. All data except that of MS15W are satisfied the proportional relation. The proportional constant of the relation between σ^C_{HS} and σ_C was found to be 64 %. The ratio of σ^C_{HS} to σ_C for MS15W is smaller than the other ceramic, because MS15W did not have the adequate crack-healing ability[4]. Therefore, it is found that ceramics components having adequate crack-healing ability can heal the pre-crack even if the pre-crack was applied tensile stress below 64 % σ_C.

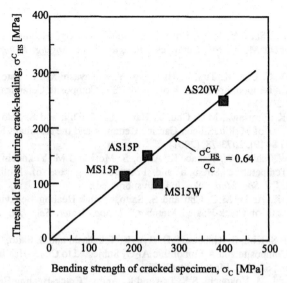

Figure 4 Threshold static and cyclic stresses during crack-healing as a function of bending strength of the cracked specimens.

CONCLUSION

AS15P, AS20W, MS15P and MS15W having excellent crack-healing ability were subjected to crack-healing under elevated static and cyclic stresses at elevated temperatures from 1000 °C to 1200 °C. The bending strengths of the specimens crack-healed under stress are investigated at the crack-healing temperature. From the obtained result, the crack-healing behavior under stress was discussed and the threshold stresses during crack-healing are determined. Following conclusions were derived.

It is found that the crack-healing occurs although the pre-crack is grown by the applied stress. The crack-healed under stress had the same bending strength as the specimens crack-healed under no-stress. Moreover, it is found that ceramics components having adequate crack-healing ability can heal the pre-crack even if the pre-crack was applied tensile stress below 64 % σ_C.

REFERENCES

[1]M.C. Chu, S. Sato, Y. Kobayashi and K. Ando, "Damage Healing and Strengthealing Behavior in Intelligent Mullite/SiC Ceramics," *Fatigue Fract. Enging. Mater. Struct.*, **18** [9], 1019-1029, (1995).

[2]K. Ando, M.C. Chu, K. Tuji, T. Hirasawa, Y. Kobayashi and S. Sato, "Crack Healing Behaviour and High-Temperature Strength of Mullite/SiC Composite Ceramics," *J. Eur. Ceram. Soc.*, **22**, 1313-19, (2002).

[3]K. Ando, K. Furusawa, M.C. Chu, T. Hanagata, K. Tuji, and S. Sato, "Crack Healing Behavior Under Stress of Mullite/Silicon Carbide Ceramics and the Resultant Fatigue Strength," *J. Am. Ceram. Soc.*, **84** [9], 2073-78, (2001).

[4]M. Ono, W. Ishida, W. Nakao, K. Ando, S. Mori and M. Yokouchi, "Crack-Healing Behavior, High Temperature Strength and Fracture Toughness of Mullite/SiC Whisker Composite Ceramic," *J. Soc. Mater. Sci. Jpn.*, (in submitted).

[5]B.S. Kim, K. Ando, M.C. Chu and S. Saito, "Crack-Healing Behavior of Monolithic Alumina and Strength of Crack-Healed Member," *J. Soc. Mater. Sci. Jpn.*, **52** [6], 667-673, (2003).

[6]K. Ando, B.S. Kim, S. Kodama, S.H. Ryu, K. Takahashi and S. Saito, "Fatigue Strength of An Al_2O_3/ SiC Composite and A Monolithic Al_2O_3 Subjected to Crack-Healing Treatment," *J. Soc. Mater. Sci. Jpn.*, **52** [11], 1464-1470, (2003).

[7]K. Takahashi, M. Yokouchi, S.K. Lee and K. Ando, "Crack-Healing Behavior of Al2O3 Toughened by SiC Whiskers," *J. Am. Ceram. Soc.*, **86** [12], 2143-47, (2003).

[8]K. Ando, M.C. Chu, Y. Kobayashi, F. Yao and S. Sato, "The Study on Crack Healing Behavior of Silicon Nitride Ceramics," *Jpn. Soc. Mech. Engng. Intl. J.*, **64 A**, 1936-1942, (1998).

[9]F. Yao, K. Ando, M.C. Chu and S. Sato, "Static and Cyclic Fatigue Behaviour of Crack-Healed Si_3N_4/SiC Composite Ceramics," *J. Eur. Ceram. Soc.*, **21**, 991-997, (2001).

[10]K. Ando, K. Takahashi, S. Nakayama and S. Sato, "Crack-Healing Behavior of Si_3N_4/SiC Ceramics Under Cyclic Stress and Resultant Strength at The Crack-Healing Temperature," *J. Am. Ceram. Soc.*, **85** [9], 2268-72, (2002).

[11]W. Nakao, M. Ono, S.K. Lee, K. Takahashi and K. Ando, "Critical Crack-Healing Condition under Stress of SiC Whisker Reinforced Alumina," *J. Eur. Ceram. Soc.*, (accepted).

ABLATION OF CARBON/CARBON COMPOSITES : DIRECT NUMERICAL SIMULATION AND EFFECTIVE BEHAVIOR

Yvan Aspa, Michel Quintard,
Institut de Mécanique des Fluides de Toulouse (IMFT),
1, Allée du Professeur Camille Soula,
F 31000 Toulouse, France

Frédéric Plazanet, Cédric Descamps,
Snecma Propulsion Solide
Les Cinq Chemins – Le Haillan
F 33187 Le Haillan Cedex, France

Gerard L. Vignoles,
Lab. des Composites ThermoStructuraux (LCTS),
Université Bordeaux 1 – 3, Allée La Boëtie,
F 33600 Pessac, France

ABSTRACT

In this paper, we are interested in the ablation of carbon/carbon composites, as found, for instance, in rocket nozzle applications, where aggressive gases cause the composite surface recession. Global recession velocity and the appearing roughness are function of the subscale description. We model ablation in terms of surface oxidation of a heterogeneous material by a binary gas mixture. We assume the gas transport to be purely diffusive and temperature almost uniform.

A finite volume code has been developed based on a VOF method using complex representation of interface and elementary surfaces, and sequential solving of the coupled equations for fluid transport and surface recession. The code was used to perform many numerical experiments, which allowed us to identify different recession regimes as a function of the characteristic dimensionless numbers.

A one-dimensional model was built using the concept of effective surface reaction. This approach replaces the non-uniform reactivity of a geometrically complex surface by an effective reactivity of a plane homogeneous interface that gives macro-scale fluxes similar to the average micro-scale fluxes.

INTRODUCTION

In the last decade, with the development of new solid-propellant generations, gases in rocket nozzles have reached higher and higher temperature. Carbon/Carbon composites (C/C) are among the few materials that are able to withstand such conditions. Indeed, they combine a good ratio high temperature mechanical properties to density and a high thermal conductivity.

When exposed to flow, C/C are attacked by the solid-propellant gaseous products. This attack causes a surface recession and a morphological deformation of the throat. This phenomenon which regroups several causes (mechanical erosion, gasification,...) is called *abla-*

tion. The recession modifies the mass transports in gas phase as heat flow in both phases introducing a strong coupling between the evolution of the solid and the dynamical behavior of the fluid.

Ablation has two negative effects :

- During motor operation, the throat section increases, causing a loss of performance;

- The thickness of the C/C part reduces, increasing the heat flow received by weak parts.

However, as the phenomenon of ablation is globally endothermal, gasification of C/C consumes a non-negligible amount of heat.

Considering all these effects, we understand that nozzle design has to take ablation into account. Unfortunately, several difficulties prevent us from a full understanding of the phenomenon. On one hand, there are theoretical limitations because of the strong coupling between a turbulent flow and a rough surface that recedes and the roughness of which may evolve. On the other hand, there is a lack of experimental knowledge, for example on local reactivities of C/C components, due to extreme conditions of ablation.

Studies have been done on the evolution of a heterogeneous surface, but without coupling it with the recession source[1] ; other works have been made on the effective behavior of heterogeneous catalytic surface[2] . The aim of this work is to blend these approaches and set-up a tri-dimensional time-efficient numerical model for coupled surface ablation and to use it in order to investigate the impact of heterogeneity on effective behavior material.

MODEL SETUP

Phenomenology

By design, at throat level, the Mach number of gases equals unity. In typical conditions, the pressure is close to 5 MPa and the temperature reaches 3000 K[3] . There are various boundary layers close to the solid/gas interface. One can define a thin diffusive layer included in the dynamic boundary layer (*i.e.* with respect to fluid velocity). In this diffusive layer, mass transport can be seen as predominantly diffusive. Its thickness is a function of the distance from the throat and of the wall roughness. Inside the diffusive layer, species diffuse to the wall and cause the surface ablation, mainly by oxidation according to the two following heterogeneous reaction balances[3,4] :

$$C_{(s)} + H_2O \rightarrow CO + H_2 \qquad (1)$$
$$C_{(s)} + CO_2 \rightarrow 2\,CO \qquad (2)$$

Accordingly, there exists a backward diffusion of the oxidation products : CO et H_2. As a matter of fact, the chemistry is somewhat more complicated, due to the existence of radical species, but eqs. (1-2) are sufficient to describe what happens.

In typical solid propellants, the combustion products also contain aluminium particles. However, in the studied nozzles, most of them are rejected out of the boundary layers and do not reach appreciably the surface. Therefore, in the following study, focused on wall effects, the multiphasic character of the flow is ignored.

Figure 1: Example of the elementary pattern

Composite surface model and conditions

One of the simplest models for a receding composite is a periodic pattern, as represented in 1 containing two materials (the convex phase, noted 1, may be though of as a fiber or a bundle, and the other one, noted 2, as a filling matrix) and a volume of surrounding gas. Let us denote ϕ the volume fraction of material 1 ; note that since the model is created by extrusion, the ratio of the vertically projected areas of phases 1 and 2 is equal to the volume fractions ratio $\phi/(1-\phi)$. A law-of-mixture average of any quantity ψ is defined by $\langle\psi\rangle = \phi\psi_1 + (1-\phi)\psi_2$.

Simplifying assumptions are made in order to produce a rapidly tractable model. They are : i) the system is assumed isobaric and isothermal, in order to keep focus on mass transfer only ; ii) the oxidation chemistry is restricted to eq. (2) with an assumed first order kinetic law ; iii) CO and CO_2 are the only two considered species. The transport reduces then to simple binary diffusion, and the CO partial pressure is simply calculated from the CO_2 partial pressure by pressure conservation ; iv) it is assumed that, during surface recession, the mean boundary layer thickness is conserved.

With such assumptions, only two equations are necessary : they are mass balances for the solid (described by its interface $z = h(x, y, t)$) and for CO_2 molar concentration C :

$$
\begin{aligned}
\partial_t C + \nabla\cdot(-D\nabla C) &= 0 & \text{in the fluid} & \quad (3)\\
C &= C_0 & \text{on boundary layer upper limit} & \quad (4)\\
(-D\nabla C)\cdot\mathbf{n} &= -k_i C & \text{at the fluid-solid interface } (z = h) & \quad (5)\\
\partial_t h_i &= -v_s k_i C\mathbf{n} - \langle\partial_t h\rangle & \text{at the fluid-solid interface } (z = h) & \quad (6)
\end{aligned}
$$

where the following quantities have been introduced : D is the diffusion coefficient, i is the phase index, k_i are the heterogeneous reaction constants, and h_i the local boundary layer heights. Maintaining the interface at a constant altitude implies the use of the second term on the right-hand side of eq. (6), which is an average recession velocity.

Numerical implementation

Mathematically, this problem can be seen as the resolution of two equations in different domains with moving interface. Such a problem can be easily solved with a Finite-Element method, but one has to pay the cost of a mesh update of the interface at each time step. To prevent this cost, the conservation equations are recast into a continuous form, defined on the whole domain (fluid + solid) : it is a VOF (Volume of Fluid) method[5-7] . To do this, one uses a phase indicator, for instance the local solid fraction $\varepsilon_s = \frac{V_{solid}}{V}$. The normal to the interface is obtained with the gradient of the phase indicator. The balances are now :

$$\partial_t C + \nabla \cdot (\mathbf{J}) = 0 \qquad \text{in the fluid} \tag{7}$$

$$\partial_t \varepsilon_s + v_s \nabla \cdot (-\mathbf{J}) = 0 \qquad \text{in the solid} \tag{8}$$

$$C = C_0 \qquad \text{on boundary layer upper limit} \tag{9}$$

$$\mathbf{J} = \begin{cases} -(1 - \varepsilon_s) D \nabla C & \text{in the fluid and solid} \\ -kC \|\nabla \varepsilon_s\|^{-1} \nabla \varepsilon_s & \text{at the fluid-solid interface} \end{cases} \tag{10}$$

Application of the finite volume method to this equation set is an integration on an elementary volume Ω containing possibly sub-volumes Ω_f of fluid and Ω_s of solid. Then, using Green-Ostrogradski theorem, one obtain two equations of the following kind:

$$-\int_{\Omega} \frac{\partial}{\partial t}\left(\frac{C}{v_s^{-1}\varepsilon_s}\right) d\Omega = \int_{\partial \Omega_i}\binom{\mathbf{n} \cdot \mathbf{J}}{-\mathbf{n} \cdot \mathbf{J}} d\Sigma = \underbrace{\int_{\partial \Omega_{f/f}}\binom{\mathbf{n} \cdot \mathbf{J}}{-\mathbf{n} \cdot \mathbf{J}} d\Sigma}_{\text{Diffusion}} + \underbrace{\int_{\partial \Omega_{f/s}}\binom{\mathbf{n} \cdot \mathbf{J}}{-\mathbf{n} \cdot \mathbf{J}} d\Sigma}_{\text{Reaction}} \tag{11}$$

One can observe in 11, that the evaluation of the integrals requires the knowledge of the interface area. Let us assume that the interface is planar at elementary volume scale. Following ref. [6] , analytical equations give for each cell the area values for the outer fluid/fluid interfaces and outer and inner fluid/solid interfaces.

Moreover, the two equations will be solved sequentially over a time step with the following scheme :

$$\begin{cases} \varepsilon_s(x,y,z,t+dt) = f(C(x,y,z,t),\varepsilon_s(x,y,z,t)) \\ C(x,y,z,t+dt) = g(C(x,y,z,t+dt),\varepsilon_s(x,y,z,t+dt)) \end{cases} \tag{12}$$

A FORTRAN 90 implementation called Diabl3D of the model has been done, using the SPARSKIT[8] library solvers.

DIRECT NUMERICAL SIMULATION RESULTS

A series of test cases has been run in Diabl3D. In all of them, the fiber is less reactive than the matrix. Under typical conditions, the diffusion coefficient D is approximately[9] $D \sim 10^{-5} m^2 s^{-1}$. In the following simulations, the diffusive layer thickness is $h = 12\mu m$ and the fiber diameter is $7\mu m$, a typical value for individual carbon fibers.

The varied parameters are :

• The size of the diffusive layer ;

• ϕ, the fiber fraction ;

- $\tilde{k} = k_2/k_1$, the reactivity ratio ;

- $Da_i = \frac{h_0 k_i}{D}$, the Damköhler numbers.

The reactivity contrast is known[10] to be as high as 1000 between different forms of pyrocarbon. Accordingly, $\tilde{k} \in [1, 10^3]$. The Da_i numbers have been varied between 10^{-3} and 10^3.

All the simulations done have lead to a stationary profile. This means that, after a transient period of deformation, the surface reaches a steady state for which the velocity is the same at all surface points. As we can see in 2a), the fiber which is less reactive remains closer to the oxidation source and has a larger surface area. The steady morphology depends on $\tilde{k}$ as illustrated at figures 2a) and 2c). However there is also a role played by the absolute average reactivity as seen by comparing figures 2b) and 2c) for which $\tilde{k} = 100$ in both cases.

ABLATION RATE AND EFFECTIVE BEHAVIOR

In addition to surface morphology evolution, Diabl3D also provides the global ablation rate $\langle \partial_t h \rangle$. Figure 3 is a plot of the scaled ablation rate $vs.$ time. The time units are $\tau = h_0^2/(DC_0 v_s)$, and the parameters were chosen as in figure 2b). Here it appears that the ablation rate in the first 2 time units is larger than in the stationary rate. This is explained by the fact that in the transient regime, the surface is relaxing towards its steady morphology by excess consumption of the most reactive phase.

An effective model gives the average behavior of the wall in terms of roughness and reactivity. It gives effective properties that are extrinsic to the C/C because they also depend on the flow conditions. It has been shown[11] that for an homogeneous plane surface, the ablation rate is given by :

$$R_a = \frac{-1}{S_0 h_0} \partial_t V_s = \frac{1}{\tau}(1 + 1/Da)^{-1} \tag{13}$$

with $\tau = \frac{h_0^2}{DC v_s}$. In the case of a rough heterogeneous surface, we naturally define, Da_{eff} such as :

$$R_a = \frac{1}{\tau}(1 + 1/Da_{eff})^{-1} \tag{14}$$

with $Da_{eff} = \frac{k_{eff}\langle h \rangle}{D}$ which gives the definition of k_{eff}, the effective reactivity of our system. Using the fact that all points have the same recession velocity, it is shown that a prediction of k_{eff} if diffusion were strictly vertical for both phases is a harmonic average :

$$k^\star = \langle k^{-1} \rangle^{-1} \tag{15}$$

This prediction is lower than a previous estimation[2] , which is the arithmetic average $< k >$. Nevertheless, because of transverse diffusional effects, k_{eff} is not equal to the predicted $k^\star$. For example, in the given case, inverting relation 14 gives $Da_{eff} = 0.41$, which lies between the two predictions $< Da >= 5.05$ and $Da^\star = 0.198$. This is an encouraging result, showing that it is possible to provide interesting numerical estimates as well as analytical bounds for the recession of a composite surface.

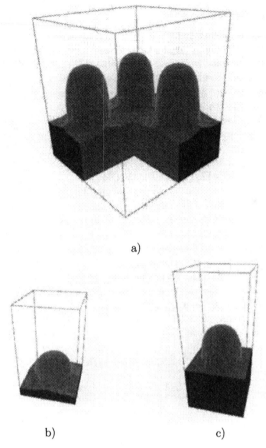

a)

b)   c)

Figure 2: Stationary profiles for : a) $[Da_1 = 1, Da_2 = 10^3]$, b) $[Da_1 = 0.1, Da_2 = 10]$, c) $[Da_1 = 10, Da_2 = 1000]$

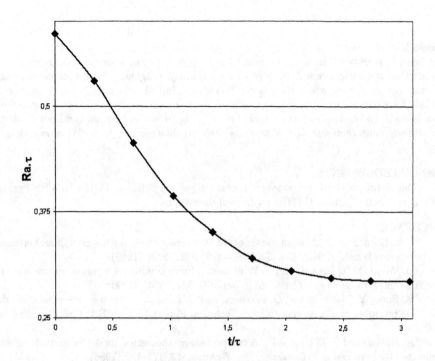

Figure 3: Ablation rate with $Da_1 = 0.1$ and $Da_2 = 10$

105

CONCLUSION AND OUTLOOK

The problem of surface recession of a composite material under mass transfer conditions, featuring competition between diffusion and heterogeneous reaction has been approached : a VOF-based numerical method has been designed and first results are given. It appears that the surface morphology (and, consequently, roughness) is conditioned by various parameters, among which the reactivity contrast, but also some transverse diffusional effect that are difficult to capture analytically. A more comprehensive study of the parameter space has still to be done in future work, before going to non-isothermal conditions, which are certainly more representative of true macroscopic ablation of C/C rocket nozzle throat parts.

ACKNOWLEDGEMENTS

The authors wish to acknowledge Snecma Propulsion Solide and DGA for a PhD grant to Y. A. and Jean Lachaud (LCTS) for fruitful discussions.

REFERENCES

[1] I. V. Katardjiev, "A kinematic model of surface evolution during growth and erosion : Numerical analysis", *J. Vac. Sci. Technol. A*, **7**, 3222–3232 (1989).

[2] B. Wood, M. Quintard, and S. Whitaker. "Jump conditions at non-uniform boundaries: The catalytic surface." *Chem. Eng. Sci.*, **55**, 5231–5245 (2000).

[3] V. Borie, Y. Maisonneuve, D. Lambert, and G. Lengellé. "Ablation des matériaux de tuyères de propulseurs à propergol solide." Technical Report 13, ONERA, Châtillon, France (1990).

[4] K. K. Kuo and S. T. Keswani. "A comprehensive theoretical model for carbon-carbon composite nozzle recession." *Combust. Sci. Technol.*, **42**, 177–192 (1986).

[5] S. W. J. Welch and J. Wilson. "A volume-of-fluid based method for fluid flows with phase change." *J. Comput. Phys.*, **160**, 662–682 (2000).

[6] D. Gueyffier, J. Li, A. Nadim, R. Scardovelli, and S. Zaleski. "Volume-of-fluid interface tracking with smoothed surface stress methods for three-dimensional flows." *J. Comput. Phys.*, **152**, 423–456 (1999).

[7] R. Scardovelli and S. Zaleski. "Analytical relations connecting linear interfaces and volume fractions in rectangular grids." *J. Comput. Phys.*, **164**, 228–237 (2000).

[8] E. C. Anderson and Y. Saad. "Solving sparse triangular system on parallel computers." Technical report 794, U. of Illinois, CSRD, Urbana, IL (1988).

[9] R. C. Reid, J. M. Prausnitz, and B. E. Poling. *The properties of gases and liquids.* McGraw Hill Book Company, 4th edition (1987).

[10] J. Nagle and R. F. Strickland-Constable. "Oxidation of carbon between 1000-2000 °C." In *Proceedings of 5th Conference on Carbon*, Pergamon Press, 154–164 (1962).

[11] G. Duffa, G. L. Vignoles, J.-M. Goyhénèche, and Y. Aspa. "Ablation of carbonbased materials : investigation of roughness set-up from heterogeneous reactions." submitted to *Int. J. of Heat and Mass Transfer* (2005)

MODELING OF DEFORMATION AND DAMAGE EVOLUTION OF CMC WITH STRONGLY ANISOTROPIC PROPERTIES

Dietmar Koch, Kamen Tushtev, Meinhard Kuntz, Ralf Knoche, Juergen Horvath, Georg Grathwohl
University of Bremen, Ceramic Materials and Components, IW3 / Am Biologischen Garten 2, D-28359 Bremen, Germany

ABSTRACT
    A new model was elaborated describing the mechanical performance of various advanced ceramic matrix composites with weak matrices as oxide/oxide composites, C/SiC with polymer derived pyrolized matrix, or C/C. The model is based on the different response of the composites in dependence on the angle between the load direction and the fiber orientation. If loaded in fiber orientation the weak matrix composites show an almost linear elastic behavior but in off-axis loading the composites behave strongly inelastic. Experimental results from tensile, shear, and mixed mode loading tests as well as from experiments with notched specimens (DEN) are used as basis for the model which is implemented in a finite element code. In addition, the damaging and failure mechanisms are investigated by acoustic emission analysis and by evaluation of loading-unloading cycles. The model allows then the prediction of the elastic/inelastic stress-strain behavior and the strength under complex loading conditions which helps to optimize the design of components with complex shape and variable fiber orientation.

INTRODUCTION

Ceramic matrix composites (CMCs) are mainly used for applications in the aerospace industry but are also developed in order to increase the efficiency of gas turbines. Beside an increased damage tolerance the CMCs are designed for high temperature stability under static and cyclic conditions and for stability under oxidative atmospheres. The classical concept for enhanced fracture toughness is based on the concept of a weak fiber matrix interface [1-3]. If the interface is adjusted in an optimal way a propagating matrix crack will not proceed through the fiber but will be deflected at the interface. The fibers are then able to bridge the matrix crack and the well described processes as debonding and pullout take place. In the cracked area the fibers carry the total load which will be transferred back to the matrix according to the stiffness ratio and the volume fraction of fiber and matrix. The interface is generally adjusted by using various interphases as e.g. pyrolytic carbon. This weak interface concept is studied quite extensively and several models describe the mechanical behavior of these composites. The models are mainly based on the elementary processes in the microscale dealing with fiber matrix toughness and fiber matrix friction [4-6].

In this paper a different concept to reach damage tolerant CMCs will be considered [7-8]. Here the fiber matrix interface does not play this important role as the matrix itself is characterized by low stiffness and strength compared to the fibers. Typical composites are oxide/oxide CMCs where the matrix is produced by slurry impregnation technique and nonoxide CMCs with carbon or SiC matrix derived from pyrolysis processes. The mechanical properties of these composites are mainly dominated by the stiff and strong fibers. The occurring cracks propagating through the matrix cover a larger volume in the material compared to the weak interface composites and crack deflection occurs within the matrix close to the reinforcing fibers. The failure behavior is not dominated by the microstructural interaction between fiber and matrix [9-11]. In contrast to the weak interface composites these materials are named as weak matrix composites. The weak matrix composites are on focus of recent developments as the processing costs are believed to be cheaper and – especially in case of oxide/oxide composites – the applications demand inherent stable composites in oxidative atmosphere. The failure mechanisms, however, are still not understood and described completely [12-13]. This article will propose a model to explain the mechanical behavior of weak matrix composites described on a pyrolytic derived C/C-composite. The model accounts for occurring damage and inelastic deformation under quasistatic loading conditions. Resulting characteristic values of the material will be used for numerical calculations in a finite element code. The strong anisotropic behavior of CMCs with weak matrix will be explained by this model.

MATERIALS AND EXPERIMENTALS

As representative material for CMCs with weak matrix a 2D continuous fiber reinforced Carbon/Carbon-composite (SGL Carbon) is investigated. The matrix is produced via polymer pyrolysis and features porous and cracked characteristics. As seen in figure 1, the microstructure features pores in between fiber bundles as well as within fiber bundles. In addition matrix material is accumulated in some areas between fiber tows.

As the stiffness and strength of the carbon fibers are much higher than of the matrix, it has to be expected that the material behaves strongly anisotropic dependent on loading direction. Therefore tensile tests with various angles between loading direction and fiber orientation are performed. These tests lead to a superposition of tensile and shear stresses while pure shear mode is measured in a Iosipescu testing rig. Additional DEN-tests show the notch insensitivity of the C/C composite. The geometries of the specimens are shown in figure 2.

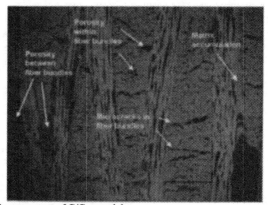

Figure 1: Microstructure of C/C material

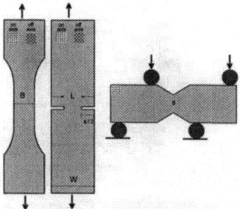

Figure 2: Geometries of the specimens for tensile testing, DEN-tests, and Iosipescu tests (from left to right).

EXPERIMENTAL RESULTS

Under pure tensile mode, i.e. when loading direction and fiber orientation coincide, an almost linear elastic behavior is observed in the stress-strain diagram. The fibers dominate the mechanical response and are responsible for the high strength values of about 400 MPa (see figure 3). The failure strain with 0.4% indicates that the matrix will be cracked dramatically even if it is not reflected in a stiffness reduction of the composite. Since the matrix is porous and weak its contribution to stiffness and strength in pure tensile mode seems to be negligible. However, recording of acoustic emission shows that already at stresses far below strength acoustic signals are observed which is evidence for damage due to enhanced matrix cracking (figure 4). Furthermore only slight inelastic strain values are measured when the specimen is unloaded prior to failure.

If specimens are loaded off axis, i.e. loading direction and fiber orientation differ, then a totally different stress-strain-behavior is observed. Strength is reduced significantly to values below 150 MPa at a slight angle of 15° down to 65 MPa at 45°. In these cases the matrix properties become more important as superimposed shear loading leads to damage and multiplied crack formation in the matrix starting already at low loading levels. Under pure shear mode, measured in the Iosipescu rig, a strong nonlinearity of the mechanical response combined with a high strain-to-failure is observed (figure 3). Following loading-unloading cycles show increasing inelastic strain values which are induced by imperfectly closed matrix cracks as debris blocks the cracks (figure 5). The elastic properties decrease only due to damage processes in the matrix while the fibers remain intact up to composite failure. In pure shear loading the acoustic emission signals indicate qualitatively the damage of the matrix while the reduction of stiffness can be interpreted quantitatively as a measure of damage.

Combined tests where the specimens are preloaded under pure tensile mode and then are stressed under pure shear mode show that tensile preloading already induces matrix damage which leads to reduced stiffness in the following shear tests [14]. It can be concluded that CMCs with weak matrices behave strongly anisotropic. Depending on loading direction either shear deformation becomes relevant for failure behavior or tensile failure is the dominating deformation process.

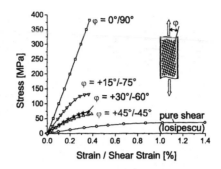

Figure 3: Typical stress strain curves of C/C under tensile, mixed, and shear testing mode.

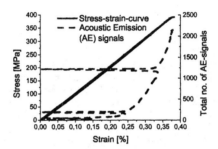

Figure 4: Stress-strain loading-unloading curve of C/C under pure tensile mode and additional acoustic emission signal demonstrating matrix cracking and damage accumulation independent on macroscopic stress-strain behavior.

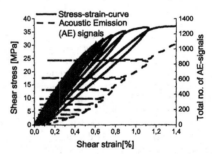

Figure 5: Stress-strain loading-unloading curve of C/C under pure shear mode and additional acoustic emission signal demonstrating matrix cracking and damage accumulation which coincides well with macroscopic stress-strain behavior.

MODELING OF THE MECHANICAL BEHAVIOR

The mechanical properties of the C/C CMC with weak matrix is constituted by reinfiltration cycles and succeeding pyrolysis. Therefore a definition of properties of single layers with specific elastic and inelastic values is not feasible. Even if there are local inhomogeneities the model will treat the material as homogenous in macroscopic manner. Necessary experimental input data as elastic constants, yield stress, inelastic work, and change of stiffness are derived from mechanical tests under pure tensile and under pure shear mode.

The planar strain state is described vectorial and is broken down into elastic and inelastic strains. To describe inelastic deformations a plasticity model is used which is based on incremental plasticity theory [15]. For a multiaxial stress state the quasi-plastic yield is described by Hills function with tensile and shear yield stresses measured in a uniaxial test [16]. The actual size of the yield surface is defined by isotropic hardening rule using the inelastic work as hardening parameter. The equivalent stress – equivalent strain relation (figure 6), characterizing the hardening of the material, is experimentally determined using a shear test (figure 5) and is reduced to a tensile stress – plastic strain curve for the uniaxial stress test. The plastic strain increment is expressed by an associated flow rule.

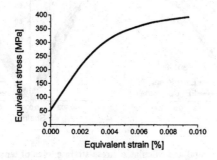

Figure 6: Equivalent stress-strain curve for C/C derived from experimental data.

The elastic strains are modeled with Hooke's law where the damage is considered via a damage parameter describing stiffness reduction in shear mode. As both, inelastic strains and stiffness reduction result from the same physical process namely the damage accumulation in the matrix. It is reasonable to link the damage variable with the inelastic strain using inelastic work. The dependency of the damage variable on the inelastic work is determined by a shear test. It is assumed that this relationship is independent from the stress state.

The implementation of the model in the finite element program MARC allows a precise calculation of stress-strain behavior as well as the prediction of strength and strain to failure. In figure 7 the results of the calculations are compared with the experimental data. Assuming that damage and inelastic strain are based on matrix damage accumulation only it becomes possible to interpret the mechanical behavior of CMCs with weak matrix.

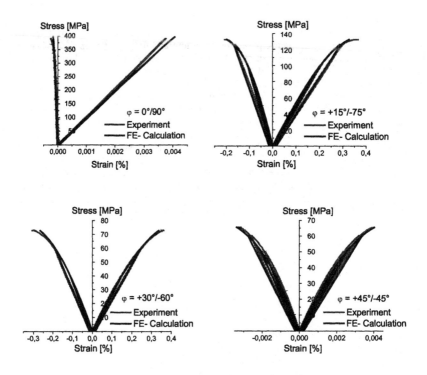

Figure 7: Experimental and calculated stress-strain curves of tensile tests on C/C with different angles between fiber orientation and loading direction.

RESULTS FROM DEN-TESTS

When the specimens are notched in DEN mode a significant change in stress-strain behavior can be observed. While the tensile specimens behave linear elastic up to maximum applicable stress the DEN-tests show a pronounced nonlinear behavior although fibers are oriented in loading direction (figure 8). When the tensile specimens fail finally they spontaneously splice within a large volume which results in large damage effects at maximum stress level. In case of DEN configuration the damage zone is first concentrated around the notched area but this damage zone can grow in a controlled manner into wider specimen region thus leading to a nonlinear stress-strain curve. Finite element calculations based on the above described model show that shear bands, which are induced by the notches, and additional normal inelastic strains occur and lead to deformation and high strain-to-failure. The nominal strength which is defined as the maximum load based on the ligament cross section increases only slightly in case of 0°/90° (axial) reinforcement whereas specimens loaded in +45°/-45° (diagonal) orientation rises quite strongly (figure 9). In this case the total length of the intact fibers running through the cross section is the strength defining factor and not the nominal cross section.

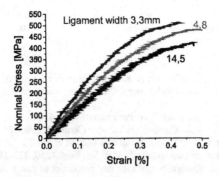

Figure 8: DEN tensile test with 0/90° fiber orientation with varying residual cross section (ligament width) showing enhanced nonlinear behavior compared to tensile test from figure 7.

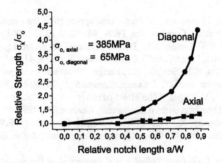

Figure 9: DEN tensile test with 0/90° (axial) and +45°/-45° (diagonal) fiber orientation with varying residual cross section (relative notch length) showing strong increase in strength with reduced cross section.

CONCLUSIONS

Ceramic matrix composites with weak matrices behave strongly anisotropic compared to CMCs with weak interfaces as the elastic mismatch between fibers and matrix is very high. In this article a model is presented which allows to predict this anisotropic behavior dependent on the angle between fiber orientation and loading direction. The model is based on damage and inelastic deformation of the composites which are both a result of fracture of the matrix. Depending on loading direction the failure of the composite is either induced by tensile stresses or by shear stresses. Once shear loading occurs the composites fail at quite low stress levels as the low mechanical properties of the matrix play the determining role. The model takes the different shear and tensile behavior into account and predicts the damage of the matrix resulting in inelastic deformation and stiffness reduction. The crack insensitivity of this material could be shown using DEN-tests where a tensile specimen is notched from two sides. The nominal strength values increase as the ligament cross section is not the decisive area since the specimens fail in a volume defined manner. In case of on-axis tests the damaged volume grows slowly leading to enhanced nonlinearity in the stress-strain diagram while in case of off axis tests the embedding of the crack bridging fibers are decisive and not the cross section of the notched specimen. As the finite element model takes into account the massive volume damage, observed experimentally, the stress-strain curves of the DEN-tests can be predicted correctly.

LITERATURE

[1] J. Aveston, G.A. Cooper, A. Kelly, "The Properties of Fiber Composites", In: *Conf. Proc. National Physical Laboratory. IPC Scientic and Technical Press*, Guildford 15-26 (1971).

[2] R.J. Kerans, R.S. Hay, T.A. Parthasarathy, M.K. Cinibulk, "Interface Design for Oxidation-Resistant Ceramic Composites", *J. Am. Ceram. Soc.* **85**, 2599-2632 (2002).

[3] M. Kuntz, G. Grathwohl, "Coulomb friction controlled bridging stresses and crack resistance of ceramic matrix composites", *Mat. Sci. Eng. A250*, 313-319 (1998).

[4] R. Naslain, "Design, preparation and properties of non-oxide CMCs for application in engines and nuclear reactors: an overview", *Compos. Sci. Tech.*, **64**, 155-170 (2004).

[5] W.A. Curtin, B.K. Ahn, N. Takeda, "Modeling brittle and tough stress–strain behavior in unidirectional ceramic matrix composites", *Acta Materialia* **46**, 3409-3420 (1998).

[6] J. Lamon, "A micromechanics-based approach to the mechanical behavior of brittle-matrix composites", *Compos. Sci. Tech.*, **61**, 2259-2272 (2001).

[7] W.C. Tu, F.F. Lange, A.G. Evans, "Concept for a Damage-Tolerant Ceramic Composite with "Strong" Interfaces", *J. Am. Ceram. Soc.*, **79**, 417-424 (1996).

[8] J.J. Haslam, K.E. Berroth, F.F. Lange, "Processing and properties of an all-oxide composite with a porous matrix", *J. Europ. Ceram. Soc.*, **20**, 607-618 (2000).

[9] K. Anand, V. Gupta, "The effect of processing conditions on the compressive and shear strength of 2-D carbon-carbon laminates", *Carbon*, **33**, 739-748 (1995).

[10] J.A. Heathcote, X.Y. Gong, J.Y. Yang, U. Ramamurty, F.W. Zok, "In-Plane Mechanical Properties of an All-Oxide Ceramic Composite", *J. Am. Ceram. Soc.*, **82**, 2721-2751 (1999).

[11] F.W. Zok, C.G. Levi, "Mechanical Properties of Porous-Matrix Ceramic Composites", *Adv. Eng. Mater.* **3**, 15-23 (2001).

[12] G.M. Genin, J.W. Hutchinson, "Composite Laminates in Plane Stress: Constitutive Modeling and Stress Redistribution due to Matrix Cracking", *J. Am. Ceram. Soc.*, **80**, 1245-1255 (1997).

[13] K. Tushtev, J. Horvath, D. Koch, G. Grathwohl, „Versagensverhalten keramischer Faserverbundwerkstoffe mit poröser Matrix - experimentelle Untersuchungen und Modellierung", *Materialwissenschaft und Werkstofftechnik*, **35**, 143-150 (2004).

[14] D. Koch, K. Tushtev, G. Grathwohl, "Shear Properties of Carbon Fiber Reinforced Ceramic Composites". 28th International Cocoa Beach Conference and Exposition on Advanced Ceramics & Composites, January 25-30, 2004, Cocoa Beach, Florida (2004).

[15] W.F. Chen, D.J. Han, "Plasticity for structural engineers", Springer, New York (1988).

[16] R. Hill, "The mathematical theory of plasticity", Oxford University Press, New York (1950).

114

Materials Characterization, NDE and Novel Techniques

A NOVEL TEST METHOD FOR MEASURING MECHANICAL PROPERTIES AT THE SMALL-SCALE: THE THETA SPECIMEN

George David Quinn, Edwin Fuller, Dan Xiang,[a] Ajit Jillavenkatesa, Li Ma, Douglas Smith
National Institute of Standards and Technology
100 Bureau Drive
Gaithersburg, MD, 20899-8529

James Beall
National Institute of Standards and Technology
325 Broadway Road
Boulder, CO, 80305-3328

ABSTRACT

A test method has been developed for measuring mechanical properties of material structures at the small-scale. Round or hexagonal rings are compressed vertically on their ends thereby creating a uniform tension stress in a horizontal crossbar that serves as the gauge section. The compression loading scheme is simple and eliminates the need for special grips. A conventional nanoindentation hardness machine serves as a small-scale universal testing machine that applies load and monitors displacement. Prototype miniature silicon specimens were fabricated by deep reactive ion etching (DRIE) of a single crystal wafer and were tested to fracture. Finite element analysis confirmed that the stress distribution was very uniform in the web portion of the specimen. The theta specimen is a versatile configuration and has great potential for use with a variety of materials and for testing extremely small structures.

INTRODUCTION

Miniaturization technologies developed by the semiconductor industry such as thin-film deposition, photolithography, etching, and micromachining are used to make microelectro-mechanical systems (MEMS) and nano-electromechanical systems (NEMS) structures. Mechanical properties at the small scale may play a vital role in design, fabrication, and application of those structures and systems. Testing methodologies must keep pace with these technologies. Ideally, measurements should be taken on specimens at roughly the same size as the micro-devices themselves and manufactured in the same manner. Test procedures for evaluating the mechanical properties of very small test coupons have been devised [1,2,3,4,5,6,7], but most are scaled down traditional configurations such as microtensile specimens that are pulled apart by grips or electrostatic devises. Sometimes cantilever beams are flexed to fracture. Nevertheless, it is challenging to manipulate, install, and apply load to such tiny specimens. There are few commercial instruments designed to do tensile testing of materials at such a small scales. Many of the testing schemes are prone to the normal misalignments and loading problems associated with testing rigid ceramics. Serious experimental errors can occur. For example, a small interlaboratory comparison (round robin) study on miniature polysilicon specimens generated elastic moduli that varied by as much as ± 12 % and average strengths that varied by a factor of two.[6]

---

[a]   Now with XD Technologies, Gaithersburg, MD 20878.

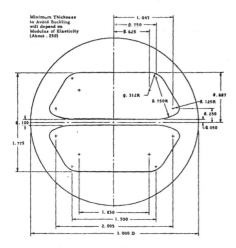
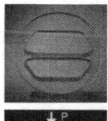
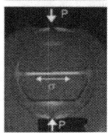

Figure 1 The Durelli and Parks theta specimen. The drawing is the original 1962 configuration with dimensions in inches. The specimen was 6.35 mm (0.25 in) thick. The photos on the right show a plexiglass specimen prepared at NIST to these dimensions. The bottom photo shows it illuminated between crossed polarizers while compressed on its rim to show the stress state.

A clever approach for testing brittle materials, shown in Figure 1, was devised by Professor Augusto Durelli and Dr. V. J. Parks in the early 1960's.[8,9,10,11] Durelli was a famous experimental stress analyst[b] known for his Moiré and photoelasticity work on ring and disk structures.[11,12] The theta specimen was an outgrowth of this work and was designed to be loaded between two parallel platens with the central web parallel to the platens. Compressing the ring on its rim by a force P causes the sides to expand outwards thereby creating a tensile stress σ in the web. They evaluated 60 to 70 different configurations and settled on the one shown in Figure 1 since it had a uniform web stress and usually fractured only in the web. Microflash photography confirmed that fractures elsewhere in the specimen were usually secondary. They tested Plexiglass, graphite, and a plastic[c] and compared strength results to data from direct tension, bending, single and two hole diametrally compressed rings. They observed a very clear trend of decreasing strength with increasing specimen size and were one of the first groups to show that strengths scaled in accordance with Weibull effective volumes.[d][10]

Nearly all of their theta specimen work was with 76 mm (3 inch) diameter, 2.5 mm (0.1 inch) web thickness specimens, but they did vary the web dimensions somewhat to study the sensitivity of the stresses to dimensional variations. They used both photoelasticity and Moiré analyses to determine the web stresses and strains in a specimen with a web height (w) to

---

b   The Society for Experimental Mechanics has an award that is named in his honor.

c   Columbia Resin-39, which is a hard ADC plastic (Allyl Diglycol Carbonate) from PPG, Pittsburgh. It is still in use today for "regular plastic lenses."

d   Unfortunately, they were unable to compute the Weibull effective volumes for all twelve testing configurations, since closed form solutions only existed for tension and pure bending in 1962.

118

diameter (D) ratio of 1/30. Stress (σ) is independent of the elastic modulus and depends only on the load (P), the specimen diameter, and the overall specimen thickness (t):

$$\sigma = K\frac{P}{Dt} = 13.8\frac{P}{Dt} \qquad (1)$$

where K is a dimensionless constant that depends upon the w to D ratio. The strain (ε) is:

$$\varepsilon = -0.585\frac{\delta_v}{D} = 1.293\frac{\delta_h}{D} \qquad (2)$$

where δ_v and δ_h are the change in length of the vertical and horizontal diameters.

There is some doubt about the accuracy of their K = 13.8 in equation 1. That value appeared in all of Durelli and Park's publications where the formula was shown.[8,10-11] On the other hand, some of their work is included in a 1962 government report [13] that listed a K factor of 16.4. Otherwise the reported work and data was identical. Their developmental work was supported by the United States Air Force in conjunction with the Illinois Institute of Technology. This early discrepancy in K's was noted elsewhere.[14][e] A clue may be found in Ref. 10 wherein the formula 1 is listed, but with two additional factors K_1 and K_2 which were not explained or defined. It is possible these two other terms are correction factors of some sort. Ref. 10 stated that the isochromatic fringe count (~ 7 shown in photos in [8,11,13]) was used with the stress-optic law for plane stress loading to estimate the primary axial principal stress, with the further observation that the other principal stresses were zero.

A literature search revealed that the theta specimen was almost never used in subsequent years for ceramics for obvious reasons.[f] It requires an extraordinary amount of machining and specimens would be very sensitive to machining and edge damage. It is also very wasteful of material since only a tiny fraction is exposed to the uniform tensile stresses. On the other hand, the latter limitation is of no concern for miniature specimens that may be made by lithographic or focused ion beam methodologies. Indeed, the fact that only a small portion of the specimen is tested is *advantageous* since extremely tiny web sizes ("specimens") may be tested as part of a much more manageable ring structure.

In this study, we have fabricated and tested miniature silicon theta specimens with both classic round and new hexagonal theta shapes as shown in Figures 2 and 3.

 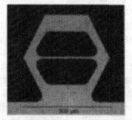

Figure 2 Miniature round and hexagonal shaped theta specimens etched from a single crystal silicon wafer. Both are 300 μm wide and have web heights of approximately 7.5 μm.

---

[e]   The provenance of the theta specimen is confirmed by Weil in page 168 of the 1962 report [13]: "The concept and development of the theta specimen .... is due to A. J. Durelli and V. J. Parks."
[f]   Guckel et al. [15] recently used tiny silicon rings with webs, but they were pulled in tension via elaborate end tabs. Web bucking was used to study residual stresses in thin films.

Figure 3 Strips of 10 specimens on a 150 μm tall base were prepared by standard lithographic techniques. The plane of the strips is (001). The hexagonal specimens on the right are clamped on their base by glass microscope slides in a rudimentary holder.

MATERIAL

The 300 μm diameter specimens were fabricated from 100 μm thick single crystal (100) silicon wafers. Ten specimens on a single strip were prepared as shown in Figure 3. All round specimens were designed to have nominal 10 μm webs to match Durelli's original shape, but at 1/250th its size. The wafers were aligned and the strips etched such that the long axis of the strip was parallel to a <110> direction as shown in Figure 3. The specimens were micromachined by deep reactive ion etching (DRIE) at about 3.3 μm/min with a standard photoresist type mask. Hexagonal specimens were made with 10 μm, 20 μm, and 50 μm nominal web heights.

EXPERIMENTAL PROCEDURE

Specimens were placed in a holder and loaded to fracture in a nanoindentation machine with a flat tipped indenter.[g] Load and displacement were monitored throughout the loading cycle and the instrument was usually operated in load control with a total ramp time to fracture of about 15 s. Break loads were of the order of 500 mN to 4 N depending upon the configuration. Initially strips were mounted in a rudimentary holder that held the 150 μm tall base strip by two glass slides lying flat as shown in Figure 3, but this proved unsatisfactory for two reasons. First, pieces flew about and were difficult to retrieve after fracture. Second, the indenter itself was too complaint and its follow-through tended to pulverize any standing fragments or specimen remnants after the initial web fracture.

A new aluminum holder shown in Figure 4 solved both problems. Two common glass microscope slides standing on their edges sandwiched a 102 μm (0.004 in) wide steel shim which created a slot into which the 100 μm thick specimen strip was inserted and held erect, but with no constraint. The specimen tops protruded slightly above the glass slides. The slides held the fragments in place after fracture and also served as end stops to limit indenter travel.

The flat indenter contacted the top of the round specimens automatically centering the load. The quality of the loading on the hexagonal specimens is less certain since we could not ensure the specimen top surface was exactly parallel to the flat indenter face.

Specimen dimensions were measured on a compound optical microscope at magnifycations up to 1600 power. Images were captured with a digital camera and software used to measure the overall dimensions to within 1 μm and the web height to 0.1 μm. All fractures were examined in a stereo binocular microscope at up to 200 power and selected fragments examined in a field emission scanning electron microscope.

---

[g]   Nanoindenter XP, MTS, Oak Ridge, TN.

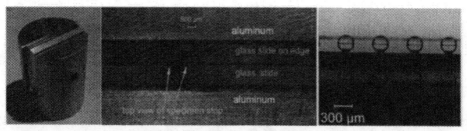

a   b   c

Figure 4 A holder for theta specimen strips. (a) shows the holder. (b) shows a top view. (c) shows a front view looking through the glass slides. The specimens protrude slightly above the glass slides so that the indenter can load them without interference.

STRESS ANALYSIS

  The K factors for our specimens were estimated. For the ring specimens, we started with a simple modification of Durelli and Parks' formula. Our ring diameters, D, were very close to specified (298 µm – 300 µm), but the web height, w, was only 7.5 µm and not the intended 10 µm after DRIE preparation. Durelli and Parks [8] noted that a ± 5% variation in the w/D ratio from the baseline ratio of 1:30 led to only a 4 % variation in their nominal K of 13.8. The 1962 government report [13] has more information and shows K varies linearly with the ratio 30w/D from 0.8 to 1.4. That report lists a different K for the baseline configuration (16.4), but their trend for the change in K with the w/D ratio matches that of Durelli and Parks. On the basis of these trends and Durelli's K = 13.8 for his baseline configuration, an approximation for K for free standing rings of homogenous and isotropic materials is:

$$K = 25.0 - 336\frac{w}{D} \tag{3}$$

Our ring specimens had w = 7.5 µm and D = 300 µm for a K of 16.6, but also have base strips so this is only an approximation for our rings.

  Two finite element analyses also were used for comparison. Both incorporated the strip base. One of our group (Ma) used Abaqus software to analyze 300 µm sized round and hexagonal specimens with 7.5 µm webs. The analysis took into account the specimen's crystallographic orientation and used single crystal elastic properties from Reference.[16] An alterative analysis was also done for comparison that arbitrarily assumed the material was isotropic and homogeneous. For the latter, a polysilicon elastic modulus of 169 GPa and a Poisson's ratio of 0.22 were used. The K factor was almost identical for the two analyses and was 18.0 ± 0.3 for the rings and 10.5 ± 0.3 for the hexagonal specimens. K also was not particularly sensitive to whether the stress state was assumed to be plane strain or plane stress and it was relatively insensitive to the loading conditions.

  A second analysis was done with the OOF program (Object Oriented Finite Element) by one of our group (Fuller) for the 7.5 µm web round and hexagonal specimens. Single crystal elastic constants were used and the orientation of the specimen taken into account. The K factors for the round and hexagonal specimens were estimated to be 17.7 ± 0.3 and 10.1 ± 0.3, respectively. Additional analysis for isotropic and homogeneous specimens showed that K is independent of elastic modulus and may be weakly dependent upon Poisson's ratio. Additional details have been presented elsewhere.[17]

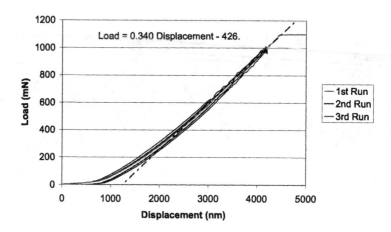

Figure 5. Repeatability test of load-displacement curves from a theta ring specimen using a nanoindentation instrument. The 1st and 2nd loading/unloading runs were repeated with a maximum load of 1.0 N, while the 3rd run loaded the specimen to fracture at 1.1 N.

Table I. Strength test results for hexagonal and round silicon specimens with 7.5 μm tall webs. The uncertainties are one standard deviation. K factors are in brackets [ ].

Specimen type	# of tests	Stiffness (N/μm)	Fracture Force (N)	Strength (MPa) Durelli Formula Eq. 3 [K] **	Strength (MPa) Finite Element OOF [K] *	Strength (MPa) Finite Element Abaqus [K] *
Hexagonal	4	0.98 ± 0.18	1.77 ± 0.12	-	596 ± 40 [10.1]	620 ± 42 [10.5]
Hexagonal	7	0.79 ± 0.15	1.63 ± 0.17	-	549 ± 57 [10.1]	570 ± 60 [10.5]
Round	8	0.44 ± 0.13	1.02 ± 0.21	564 ± 116 [16.6]	602 ± 124 [17.7]	612 ± 126 [18.0]
Round	8	0.38 ± 0.03	1.01 ± 0.18	559 ± 100 [16.6]	596 ± 106 [17.7]	606 ± 108 [18.0]
Round – 3 strips	12	0.36 ± 0.04	0.86 ± 0.12	476 ± 66 [16.6]	507 ± 71 [17.7]	516 ± 72 [18.0]
Round – strip with webs	4	0.34 ± 0.04	0.83 ± 0.06	459 ± 33 [16.6]	490 ± 35 [17.7]	498 ± 36 [18.0]
Round – strip without webs	4	0.20 ± 0.01	0.78 ± 0.04	-	-	-

-    Not applicable
*   Directionally dependent single crystal properties were used.
**   Assumes the material was homogenous and isotropic and ignores the effect of the base strip.

(a)   (b)

Figure 6. Fractured specimens. (a) shows ring fragments with the front cover glass slide removed. (b) shows a broken hexagonal specimen after a test. The glass cover sides help retain the fragments.

RESULTS

Figure 5 shows a load-displacement record for a ring specimen. The stiffness (0.34 mN/nm) was obtained from linear curve fitting the last twenty points before the sample broke. Table 1 shows the stiffness and break loads for several strips of specimens. Fracture strengths are listed in three columns corresponding to the different estimates of the K factors discussed above. Usually fewer than the full ten on a strip were tested. Occasionally a specimen was broken in handling. Our initial experiments had greater scatter until we realized that the strip often would shift when a specimen broke. Sometimes a shard or tiny fragment would land under or on top of an intact specimen thereby disturbing subsequent tests. Shifting was eliminated by application of a tiny drop of super glue to the strip ends as a tack bond.

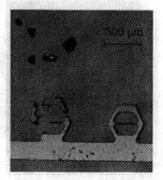

Figure 7 Broken Hexagon

The last series of experiments on one strip were with four rings with and four rings without webs. Webs were manually broken by gentle application of a lateral force to the web with a sewing needle. The data show a clear compliance difference between rings with and without webs, but the break loads were comparable.

The holder retained most fragments as shown in Figure 6b. Figure 7 shows the same specimen in Figure 6b after removal from the holder. Secondary fractures occurred in every case. The finite element stress analyses showed the maximum ring tensile stresses indeed were in the web, but a moderately high stress region (with stresses only 13% less) occurred in a small region inside the ring directly opposite the loading point. The web section had much greater Weibull effective surface and volume, however. The maximum secondary stress in the hexagons was located on the inside top corners, but was less than half the web stresses.

The nanoindenter is designed to apply and measure small loads and monitor tiny displacements, but it is not stiff enough to immediately stop the indenter head after first web

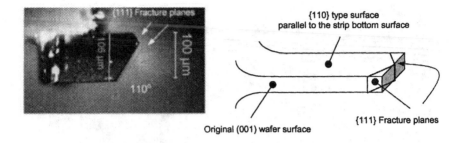

Figure 8 Top view of a fractured web from a 20 μm web hexagonal specimen. The fractures occurred on {111} planes that were at an angle to the web axis as shown on the right.

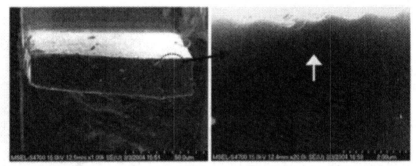

Figure 9 FESEM view of the same piece in Figure 8 showing the fracture origin. The close-up shows a cathedral fracture mirror centered on a small surface fracture origin (white arrow). The origin is a flaw associated with the grooves created by the DRIE process.

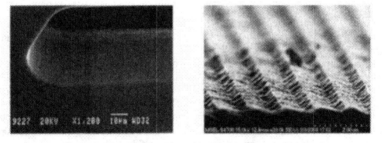

Figure 10 The DRIE process created 0.2 to 0.4 μm deep surface grooves and pits in the side wall surfaces.

124

fracture. It can continue to apply load that could break the ring and hexagon frames. Single crystal silicon is very brittle, so preventing a second break would be difficult. It is also possible that once a web breaks, the release of stored elastic energy in the specimen causes secondary fractures. Irregular breakage patterns that might have arisen from uneven loading of the top surface of the hexagons were not detected, but that possibility cannot be ruled out. The rings are less susceptible to this problem.

The specimens broke into characteristic faceted shapes. The webs almost always broke on {111} planes creating bent fracture surfaces as shown in Figures 8 and 9. The {111} is a preferred cleavage plane in single crystal silicon.[18] Figure 9 shows the fracture planes and a close-up of the tell tale cathedral shaped fracture mirror. A simple estimate of the critical flaw size can be made with fracture mechanics. Given a fracture toughness on the {111} of 0.8 MPa√m [17], and using a shape factor Y of 1.13 for a semicircular surface flaw and a stress of 550 MPa (and ignoring that the fracture plane is at an angle to the stress), then the critical flaw should be about 1.3 μm in size. The DRIE process involves alternating etching-passivating steps and grooving is normal. Figures 10 shows the surface of a DRIE prepared hexagonal specimen. The strength-limiting flaw is deeper than the 0.2 to 0.4 μm deep groves.

DISCUSSION

The strengths are on the low end of the range commonly reported for miniature single crystal silicon structures.[2] The surface condition controlled the strengths in our hexagonal and round theta specimens and we believe the strength values are those that actual components prepared by the same processing would have. Refined DRIE processing could improve strength. DRIE surface faults were recently identified as fracture origins in chemical vapor deposited miniature silicon carbide specimens.[19,20]

This project is a work in progress and the method and testing procedures will be refined. The stress solutions will be confirmed by additional finite element analyses. The specimen shapes will be optimized and other derivative shapes investigated. Ring specimens with thinner webs may be used to guarantee that initial breakage is in the web. The loading conditions will be refined to ensure correct load application and deflection measurement. A stiffer loading system will be developed to control the loading conditions more precisely and to limit secondary breakage. Testing with specimens having indented webs has already commenced. There is no reason that the webs cannot be notched or tapered by lithography or by focused ion beam micro machining. In a sense, the outer ring may be considered the loading frame or "universal testing machine" and the web is the test coupon.

We have plans to test some new advanced materials with specimens as much as ten times smaller than those in this paper and with web heights of only 1 μm. We see no inherent limit to scaling the theta specimen down in size. As long as a small specimen can be made in a ring or hexagon shape, it can be loaded in compression on its rims to put the web in tension. In situ testing in electron microscopes is feasible.

The theta specimen is an inherently simple and practical configuration. Very thin specimens may not be suitable lest they buckle during compression loading, but even this could be solved by affixing a thin film specimen to the circular face of a thicker hollow ring specimen with a matching diameter. The combination could then be loaded on the rim. The symmetry and harmony of the theta specimen along with the benefits of St. Venant's principle suggest that that minor irregularities in the end loading conditions will have negligible effect upon the stress distribution in the web.

CONCLUSIONS

Durelli's theta specimen and a hexagonal variant were used to measure the strength and compliance of miniature test specimens. Data for single crystal silicon accurately reflected the surface condition of the micro component. This versatile method can be applied to a variety of other materials and is ideal for testing ultra miniature test specimens.

REFERENCES

[1] S. M. Allameh, "An Introduction to Mechanical-Properties-Related Issues in MEMS Structures," *J. Mat. Sci.*, **38**, 4115-23 (2003).

[2] O. M. Jadaan, N. N. Nemeth, J. Bagdahn, and W. N. Sharpe, Jr., "Probabilistic Weibull Behavior and Mechanical Properties of MEMS Brittle Materials," *J. Mat. Sci.*, **38**, 4087-113 (2003).

[3] T. E. Buchheit, S. J. Glass, J. R. Sullivan, S. S. Mani, D. A. Lavan, T. A. Friedmann, and R. Janek, "Micromechanical Testing of MEMS Materials," *J. Mat. Sci.*, **38**, 4081-86 (2003).

[4] W.N. Sharpe, Jr., D.A. LaVan, and R.L. Edwards, "Mechanical Properties of LIGA-Deposited Nickel for MEMS Transducers," in *Proc. Transducers '97*, Chicago, IL, 607-10 (1997).

[5] J. Bagdahn, W.N. Sharpe, Jr., and O. Jadaan: "Fracture Strength of Polysilicon at Stress Concentrations," *J. Microelectromechanical Systems*, **12**, 302-12 (2003).

[6] W. N. Sharpe, Jr., S. Brown, G. C. Johnson, and W. Knauss, "Round-Robin Test of Modulus and Strength of Polysilicon," *Proc. Mat. Res. Soc.*, **518**, 57-65 (1998).

[7] A. M. Fitzgerald, R. S. Iyer, R. H. Dauskardt, and T. W. Kenny, "Subcritical Crack Growth in Single Crystal Silicon Using Micromachined Specimens, *J. Mater. Res.*, **17** [3] 683-91 (2002).

[8] A. J. Durelli, S. Morse, and V. Parks, "The Theta Specimen for Determining Tensile Strength of Brittle Materials," *Mat. Res. and Stand.*, **2**, 114-7 (1962).

[9] A. J. Durelli and V. J. Parks, "Influence of Size and Shape on the Tensile Strength of Brittle Materials," *Brit. J. Appl. Phys*, **18**, 387-8 (1967).

[10] A. J. Durelli and V. Parks, "Relationship of Size and Stress Gradient to Brittle Failure Stress," pp. 931 – 8 in *Proc. 4th U.S. National Congress on Applied Mechanics*, ASME, New York, 1962.

[11] A. J. Durelli, *Applied Stress Analysis*, Prentice-Hall, New Jersey, 1967.

[12] A. J. Durelli and V. J. Parks, *Moire Analysis of Strain*, Prentice-Hall, New Jersey, 1970.

[13] N. A. Weil, "Studies of the Brittle Behavior of Ceramic Materials," Technical Report ASD-TR-61-628, April 1962, prepared by the Illinois Institute of Technology, Chicago, IL, for the Air Force Directorate, Wright Patterson Air Force Base, OH, pp. 139–169.

[14] L. Mordfin and M. J. Kerper, "Strength Testing of Ceramics-A Survey," pp. 243-261 in *Mechanical and Thermal Properties of Ceramics*, ed. J. Wachtman, Jr., National Bureau of Standards Special Publication 303, May 1969.

[15] H. Guckel, D. Burns, C. Rutigliano, E. Lovell, and B. Choi, "Diagnostic Microstructures for the Measurement of Intrinsic Strain in Thin Films," *J. Micromech. Microeng*, **2**, 86 - 95 (1992).

[16] H. J. McSkimin and P. Andreatch Jr., "Measurement of Third Order Moduli of Silicon and Germanium, *J. Appl. Phys.*, **35** [11] 3312-19 (1964).

[17] E. R. Fuller, D. Xiang, G. D. Quinn, "Finite Element Analysis of Hexagonal and Circular Theta Specimens," presented at the Engineering Ceramics Division meeting of the American Ceramic Society, Cocoa Beach, 25 January 2005.

[18] C. P. Chen and M. H. Leipold, "Fracture Toughness of Si," *Am. Ceram. Soc. Bull.*, **59** [4] 469-72 (1980).

[19] G. D. Quinn, W. N. Sharpe, Jr., G. M. Beheim, N. N. Nemeth, and O. Jadaan, "Fracture Origins in Miniature Silicon Carbide Structures," to be publ. in *Fractography of Advanced Ceramics, II*, ed. J. Dusza, TransTech Publ., Zurich, 2005.

[20] W. N. Sharpe, Jr., O. Jadaan, N. N. Nemeth, G. M. Beheim, and G. D. Quinn, "Fracture Strength of Silicon Carbide Microspecimens," accept by *J. Microelectromechanical Systems*, 2004.

NONDESTRUCTIVE EVALUATION OF MACHINING AND BENCH-TEST DAMAGE IN SILICON-NITRIDE CERAMIC VALVES

J. G. Sun, J. M. Zhang
Argonne National Laboratory
9700 South Cass Avenue
Argonne, IL, 60439

J. S. Tretheway, J. A. Grassi, M. P. Fletcher, and M. J. Andrews
Caterpillar Inc.
Peoria, IL 61656

ABSTRACT

An automated laser-scattering system was developed for nondestructive evaluation (NDE) of ceramic engine valves. The system was used to evaluate ten NT551 (Norton-Saint Gobain) silicon nitride valves that were bench tested for 1000 hours. The bench test simulated impact and wear between the valve and the seat insert at ambient temperature. The NDE system scans the entire valve surface and generates a two-dimensional scattering image that is used to identify the location, size, and relative severity of subsurface damage. NDE data showed surface and subsurface damage from both machining and bench tests. These data were analyzed and compared with surface photomicroscopy results.

INTRODUCTION

Advanced ceramics are leading candidates for high-temperature engine applications that offer improved fuel efficiency and engine performance. Among them, silicon nitrides have been evaluated and used for valve train materials in automotive and diesel engines. Ceramic valves experience high impact stress and wear, in addition to thermal and pressure loadings in a corrosive environment. Therefore, surface and subsurface damage induced by machining and operation needs to be characterized to assure the valve reliability.

For silicon-nitride ceramics, surface and subsurface defects from abrasive machining processes are known to significantly degrade fracture strength and fatigue resistance (Ott, 1997). These defects are in the form of microstructural discontinuities such as spalls, cracks, and voids and are typically within 200 μm under the surface. Because ceramics are partially translucent to light, a laser-scattering method based on cross-polarization detection of optical scattering originated from the subsurface can be used for noncontact, nondestructive measurement of variations in subsurface microstructure. This measurement involves scanning the entire surface of a ceramic component and constructing a two-dimensional (2D) scatter image, from which subsurface defects can be readily identified as they exhibit excessive scattering over the background, and their type and severity can be analyzed (Sun et al., 1999).

In this study, an automated laser-scattering system was developed for nondestructive evaluation (NDE) of an entire valve for subsurface damage (Sun et al., 2004). This system employs two rotation and two translation stages to align and focus the laser beam on the valve surface during the entire scan, and the resulting 2D scattering image is used to identify the location, size, and relative severity of subsurface defects. Laser-scattering images were obtained for ten Norton-Saint Gobain NT551 silicon nitride valves that were subjected to cyclic-impact loads at ambient temperature for 100, 500, and 1000 hours. These data were analyzed and compared with surface photomicrographs.

AUTOMATED LASER-SCATTERING SYSTEM

An automated laser-scattering NDE system was developed for scanning the entire surface of a valve (Sun et al., 2004). Figure 1 is a schematic diagram of the experimental setup. A precision lathe unit is used to rotate and translate the valve, and a translation-rotation-stage unit controls the position and orientation of the optical detection train so that the laser beam is always focused normal on the valve surface. The data acquisition program controls the motions of all four stages to follow the valve-surface profile and synchronously collects laser-scattering data during a scan. The detector signal represents subsurface-backscatter intensity recorded at every pixel of the valve surface, and it is plotted as a 2D optical-scattering image (Sun et al., 1999, 2004). Because optical scattering is stronger at regions with subsurface defects and damage, such as cracks and pores, the 2D scattering image can be used to identify the location, size, and relative severity of subsurface defects. Surface features, such as stains or deposits, may reduce the incident optical intensity into the subsurface, so they are typically shown with lower scattering intensity in the image.

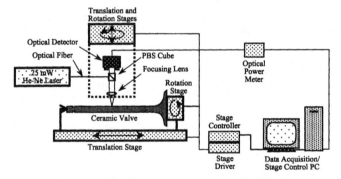

Fig. 1.   Schematic diagram of automated laser-scattering system for valve scan.

BENCH-TESTED CERAMIC VALVES

Figure 2 is a photograph of the ten NT551 silicon-nitride ceramic valves. Each valve has a head section of varying radius and a shaft section of constant radius. At the top of the head is a 4-mm-long conical surface of a constant slope (either 30 or 45 degree) along the axis. This surface contacts with the matching surface of a metal valve-seat insert. The lower portion of the valve head (the fillet) is a curved smooth surface of varying radius from the contact surface to the shaft surface. A keeper notch (groove) is located near the end of the shaft stem as seen in Fig. 2. During normal operation, the valve contact surface is in cyclic impact with the metal seat insert. The impact stress propagates through the fillet radius into the shaft, with the maximum principal tensile stress occurring at the fillet radius region. Therefore, material defects and processing (machining) damages in the entire valve surface and subsurface may affect the valve reliability.

128

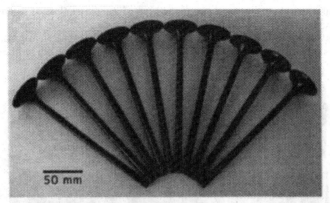

Fig. 2.   Photograph of ten NT551 silicon nitride valves.

The ten NT551 silicon nitride valves were tested for 1000 hr in a valve bench rig with cyclic-impact loads at ambient temperature. The bench tests simulated impact and wear between the valve and the seat insert. The temperature and pressure loadings on the valve end surface in real engine operations were not simulated in these bench tests. These valves were seated with metallic seat inserts having matched or mismatched angles, or with eccentric seat inserts. It was anticipated that the mismatch-angled and eccentric seat inserts would induce higher impact stress and wear to the ceramic valve. These valves were examined intermittently by the automated laser-scatter system at accumulated bench-test durations of 100, 500, and/or 1000 hr. For each valve, its NDE data at different test durations were compared and analyzed to determine any new damage or growth of existing defects in the contact and fillet radius surfaces.

EXPERIMENTAL RESULTS

All ten NT551 silicon-nitride valves were successfully tested for 1000 hr. Upon visual inspection of the valve surfaces, the only apparent damage was two wear scars around the circumference in the contact surface region. These scars were the result of repeated impact with the metal valve-seat insert, similar to those observed in 100-hr tested valves (Sun et al., 2004). Figures 3a and 3b show photomicrographs of typical wear scars in the contact area for NT551 valves CV19 after 500- and 1000-hr bench tests, respectively. Except for these scars, all bench-tested valves appeared to be undamaged. Therefore, NDE data are presented below for only one NT551 valve, identified as CV19.

Valve CV19 has a contact-surface angle of 45 degrees and was matched with a 45-deg seat insert. It was examined by the NDE system after 500- and 1000-hr accumulated test durations. Laser-scattering scans were performed for the entire valve-head surface, from the top of the contact area to the end of the fillet radius. Each scan took about 6 hours. Figure 4 shows the laser-scatter images of the head section of valve CV19 at 500- and 1000-hr test durations. The scanned axial length was 30 mm (the top 4-mm region is the contact area), which is aligned in the vertical direction of the images. The scan resolution was 10 μm in the axial direction and 11 μm in the circumferential direction. Typical to all NT551 valves, NDE data for CV19 indicated no significant accumulated damage from the bench tests, except surface wear within the contact surface. The reduction of scattering intensity within the wear scars (two darker horizontal stripes) is due to surface damage of increased roughness (loss of ceramic grains on surface) and contamination (embedded metal particles from the metal seat insert). A few darker lines are observed in the images; those are scratch stains of contaminated material attached on the surface. As indicated above, any subsurface damage would appear with increased scattering intensity.

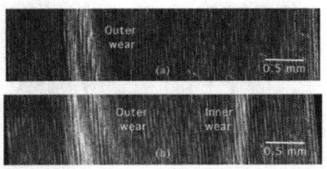

Fig. 3. Photomicrographs of wear scars on the contact surface of NT551 valve CV19 after (a) 500-hr and (b) 1000-hr bench tests.

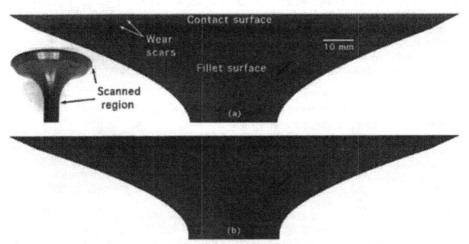

Fig. 4. Laser-scattering scan images for NT551 valve CV19 after (a) 500-hr and (b) 1000-hr bench tests.

Figure 5 shows the detailed image of the contact surface of CV19 after the 1000-hr test. Several segmented subsurface flaws of high optical-scattering intensity near the top edge are indicated in the figure. One of the detected edge flaws is shown in Fig. 6 and compared to surface photomicrograph. This flaw is a subsurface damage (~0.3 mm long) that was likely initiated during valve machining. Other flaws appearing as high optical-scattering spots are <80 µm, and they are likely high-porosity pores or inhomogeneities within the material's subsurface (Andrews, 1999; Sun et al., 2004). In addition, a subsurface damage near a scratch was detected after the 1000-hr test (Fig. 7), which was not there after the 500-hr test. However, none of the flaws grew during the second 500-hr test, suggesting that the maximum stresses generated from the bench rig are well within the strength limit of the NT551 silicon nitride.

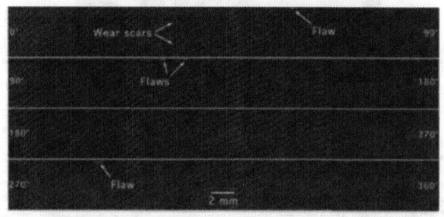

Fig. 5.   Laser-scattering image of the contact surface of NT551 valve CV19 after 1000-hr bench test.

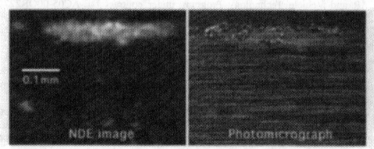

Fig. 6.   Detailed laser-scattering image and photomicrograph of a subsurface edge flaw from Fig. 5.

131

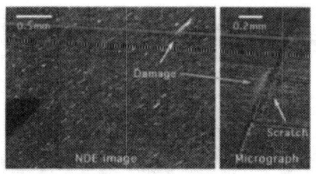

Fig. 7. Detailed laser-scattering image and photomicrograph of a subsurface scratch damage from Fig. 4b.

CONCLUSION

An automated laser-scatter system was developed to evaluate ten NT551 silicon-nitride valves at accumulated bench-test durations of 100, 500, and/or 1000 hr. For each valve, its NDE data at different test durations were analyzed to determine any new damage or growth of existing defects.   In general, no significant subsurface damage from the bench tests was detected. Existing flaws, such as machining damage and inherent material defects, did not grow under the various test conditions and durations. Damages from the bench tests are mainly in the valve contact surface showing as wear scars.

ACKNOWLEDGMENT

Research sponsored by the Heavy Vehicle Propulsion Materials Program, DOE Office of FreedomCAR and Vehicle Technology Program, under contract DE-AC05-00OR22725 with UT-Battelle, LLC.

REFERENCES

M. J. Andrews, 1999, "Life Prediction and Mechanical Reliability of NT551 Silicon Nitride," Ph.D. Thesis, New Mexico State University, Las Cruces, NM.

R. D. Ott, 1997, "Influence of Machining Parameters on the Subsurface Damage of High-Strength Silicon Nitride," Ph.D. Thesis, The University of Alabama at Birmingham, AL.

J. G. Sun, W. A. Ellingson, J. S. Steckenrider, and S. Ahuja, 1999, "Application of Optical Scattering Methods to Detect Damage in Ceramics," in *Machining of Ceramics and Composites*, Part IV, Chapter 19, Eds., S. Jahanmir, M. Ramulu, and P. Koshy, Marcel Dekker, New York, pp. 669-699.

J. G. Sun, J. M. Zhang, and M. J. Andrews, 2004, "Laser Scattering Characterization of Subsurface Defect/Damage in Silicon-Nitride Ceramic Valves," presented at the American Ceramic Society's 28th International Cocoa Beach Conference & Exposition on Advanced Ceramics & Composites, Cocoa Beach, FL, Jan. 25-30, 2004.

NDE FOR CHARACTERIZING OXIDATION DAMAGE IN REINFORCED CARBON-CARBON USED ON THE NASA SPACE SHUTTLE THERMAL PROTECTION SYSTEM

Don J. Roth,[*] Nathan S. Jacobson, Joseph N. Gray, Laura M. Cosgriff, James R. Bodis, Russell A. Wincheski, Richard W. Rauser, Erin A. Burns, Myles. S. McQuater
National Aeronautics and Space Administration
Glenn Research Center
21000 Brookpark Road
Cleveland, Ohio 44135

ABSTRACT
In this study, a coated reinforced carbon-carbon (RCC) sample of the same structure and composition as that from the NASA space shuttle orbiter's thermal protection system was fabricated to have predictable oxidation damage. The sample was fabricated by drilling holes to the bottom of the coating and oxidizing the sample to create cavities in the carbon substrate underneath the coating as oxygen reacted with the carbon and resulted in its consumption. The cavities varied in diameter from approximately 1.4 to 3 mm (as measured optically). The sample was used as a standard to determine capabilities of state-of-the-art nondestructive evaluation (NDE) methods for detecting and sizing oxidation damage beneath a mostly intact coating in RCC. Both one and two-sided NDE methods are used. One-sided methods are practical for on-wing inspection, while two-sided methods might be applicable in situations where RCC panel removal is possible. Comparisons of diameter and depth obtained from NDE measurements with those from optically-measured void sizes are shown.

INTRODUCTION
Reinforced carbon-carbon (RCC) with a silicon carbide coating for oxidation resistance is used on the NASA space shuttle orbiter's wing leading edge and nose cap for thermal protection during re-entry. Oxidation damage to RCC can occur if the silicon carbide coating is itself damaged but still intact such that hot gases have access to the carbon beneath the coating.[1] In such cases, it would be critical to evaluate the extent of the oxidation damage underneath the intact SiC coating. Even small breaches in the RCC coating system have recently been identified as potentially serious. In this study, an RCC sample having small amounts of well-characterized oxidation damage in the form of approximately hemi-spherical holes underneath the silicon carbide coating was inspected using various nondestructive evaluation methods.

The nondestructive evaluation (NDE) techniques employed included state-of-the-art x-ray, computed tomography, ultrasonics, eddy current, and thermographic methods. NASA Glenn Research Center led this investigation that had some of the top NDE specialists/facilities from NASA Glenn, NASA Langley, and Iowa State University inspect this sample with the various NDE methods. The sample was then metallographically sectioned to evaluate the true dimensions and morphology of the holes. The controlled oxidation damage provides standards for investigating the effectiveness of various nondestructive evaluation techniques for detecting and sizing oxidation damage in this material.

EXPERIMENTAL
RCC Material
The RCC material evaluated in this study was the same as that used for the wing-leading-edge of the shuttle. It is shown schematically in figure 1 and described in detail in reference 3. Briefly, this material

---

[*] Corresponding author: Phone: 216–433–6017, Fax: 216–977–7150, E-mail: donald.j.roth@nasa.gov

---

consists of layers of a graphitized rayon fabric, which are repeatedly impregnated with a liquid carbon precursor to fill the voids and then the precursor is converted to carbon by pyrolysis. A diffusion conversion coating of silicon carbide is then grown on this substrate at high temperatures by a siliconizing pack cementation process. Next a coating of tetraethyl orthosilicate (TEOS) is applied via a vacuum impregnation to form silicon dioxide on the surface to plug surface cracks and fissures. The final coating is a 'Type A' sealant which consists of sodium silicate and small particles and fibers of SiC. The difference in thermal expansion of the SiC coating and the carbon/carbon substrate leads to cracks in the SiC and the fluid glass from the TEOS and Type A coatings is intended to fill these cracks.

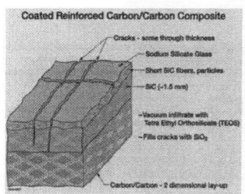

Figure 1. Schematic of SiC-protected carbon/carbon used in this study.

A sample in the form of an approximately 70-mm-diameter disk with approximately 1.05 mm total coating thickness was used in this study. To create the damage, seventeen 0.5 mm diameter holes, spaced 13 mm apart, were ultrasonically drilled through the coating system to the depth of the SiC coating, followed by oxidation treatments. The oxidation treatments were conducted in static (non-flowing) air at 1 atm. and 1200 °C in a box furnace. A series of drillings followed by oxidation treatments was performed; first Batch 1 was drilled, then the disk was oxidized for 10 minutes, then Batch 2 was drilled, then the disk was oxidized for 10 minutes (giving 20 minutes on Batch 1 and 10 minutes on Batch 2), and so forth. The sample underwent seven drillings, and oxidation treatments varying from 10 to 70 minutes. Figure 2 shows the disk, the various hole locations, and the oxidation treatment grouping.

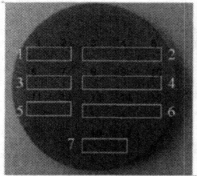

Figure 2. Schematic of the RCC disk showing holes and oxidation treatments. The white numbers signify the batch. The black numbers are the assigned hole number for the plots shown in figures 8 and 9. Batch

134

1 experienced 70 minutes total of oxidation, batch 2 experienced 60 minutes, batch 3 experienced 50 minutes, etc. while batch 7 experienced 10 minutes of oxidation. The three holes at the top of the disk were not drilled thru the coating and were used for orientation purposes.

In the oxidation model controlled by diffusion developed by Jacobson, et al.,[1] the damaged region is assumed hemispherical, and it was verified that the carbon/carbon substrate below the drilled hole oxidized to form an approximate hemispherical void (fig. 3). The model was used to obtain predicted void sizes in this study: the holes were predicted to vary in diameter from approximately 3.6 mm (10 minute oxidation time) to 7.2 mm (70 minute oxidation time). This controlled oxidation damage provides standards for investigating the effectiveness of various nondestructive evaluation techniques for detecting and sizing oxidation damage in this material.

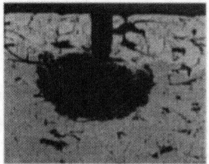
_____ 1 mm

Figure 3. Cross-section of cavity resulting from oxidation treatment.

Nondestructive Evaluation (NDE)
    Various state-of-the-art NDE methods were used to characterize the disk and oxidation damage. These methods included water-coupled, air-coupled, and guided wave ultrasonics; eddy current, flash thermography, digital and film xray, and xray computed tomography (CT). Several of the top NDE experts/facilities (at NASA Glenn Research Center, NASA Langley Research Center, and Iowa State University) in the country participated in this study to insure fair evaluation of each method. In some cases, the same method was performed at different facilities. References 4 – 6 provide basic principles for the various methods, and more details on the most promising experimental methodx are described in the appendix. Representative images from the different methods are shown.

    Apparent void diameter was estimated for each of the techniques except CT by making direct measurements off of original, contrast enhanced, or binarized and thresholded images. For CT, The diameters were obtained by looking at a small volume of interest around each hole. As the slices were swept by, the largest diameter cross section was extracted and pixels counted with 85 microns assigned per pixel. Many slices were taken. A fixed circle was used as a reference. After NDE was completed, the sample was sectioned, ground to the midplane in the oxidation hemisphere, polished, and examined with optical microscopy to characterize actual dimensions of the voids.

RESULTS AND DISCUSSION
    Figures 4 to 7 show selected images of the disk obtained from the most useful NDE methods. Major results are indicated in the figure captions. In some cases, batch 7 holes are demarcated to aid the reader. In general, none of the ultrasonic methods were capable of unambiguously resolving voids of these sizes

135

in RCC so no images from ultrasonic methods are shown here. Figure 8 shows apparent diameter results from the various NDE methods and comparison to optically-measured diameters. Figure 9 shows apparent depth of the holes beneath the coating from Xray CT and its comparison to depth measured from optical measurements. Table 1 shows the average error of the measurements for each of the NDE methods. Error is obtained by adding the standard deviation of the measurement of diameter and the pixel dimension for each measurement and obtaining an average. This error is somewhat optimistic when using binarization and thresholding to aid in size measurement because fine tuning the threshold at times resulted in greater than one pixel difference in size.

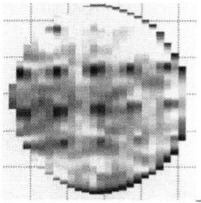

Figure 4. Eddy current lift-off image using Jentek MWM system (see appendix) at 6.3 MHz. This method detects the voids but it is difficult to measure void diameter due to the coarse scan resolution. Estimates of void depth are possible from (measured lift-off at hole) - (measured nominal lift-off) and are shown in figure 9. Nominal lift-off value for eddy current measurement estimated at approximately 1.3 mm.

Figure 5. Thermographic (derivative) image revealing most of the voids. a) Unprocessed image. B) After thresholding to demarcate voids. In the thresholding process, a decision has to be made as to where to set exact threshold; this results in some voids not being revealed, or the appearance of non-void areas that

mask the void areas. This effect makes voids in batches 5 – 7 difficult to detect/characterize and is also consistent with them experiencing the least oxidation compared to batches 1 - 4.

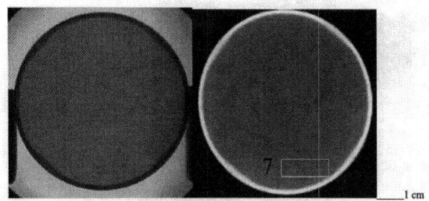

Figure 6. Xray images. A) Digital xray B) Film Xray.

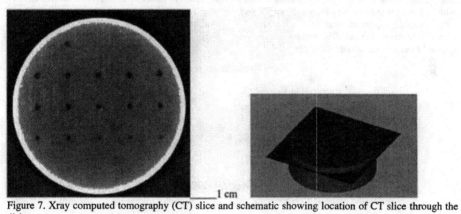

Figure 7. Xray computed tomography (CT) slice and schematic showing location of CT slice through the disk.

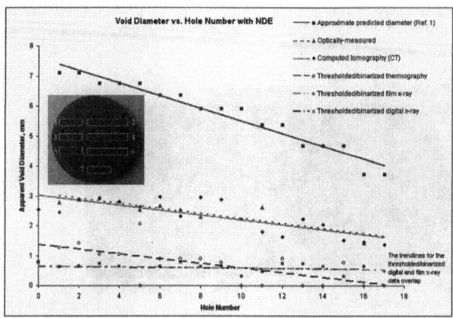

Figure 8. Apparent Void Diameter vs. Hole Number from the different NDE methods and from optical measurement after sectioning. Disk schematic shows hole numbers (black) and oxidation batch numbers (white). Linear trend lines are shown. Xray CT and thermography reveal the trend of void size increasing with increasing oxidation treatment (with Xray CT providing the best results). Digital and Film Xray were able to detect but not differentiate the void sizes. This is most likely due to the fact that the drilled hole and its extension into the cavity provided the dominant xray contrast.

138

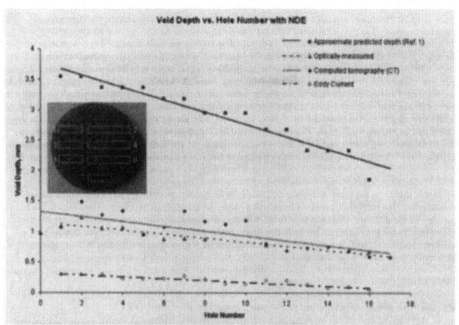

Figure 9. Apparent Void Depth vs. Hole Number from different NDE methods and from optical measurement after sectioning. Disk schematic shows hole numbers (black) and oxidation batch numbers (white). Linear trend lines are shown. Eddy current and CT measurements revealed the trend of void depth increasing with increasing oxidation treatment (with Xray CT providing the best results). Eddy current measurements will (and did) underestimate the actual void depth for small void diameters as unflawed material near the void will be averaged into the impedance measurement of the relatively large (0.1" x .05") coils.

Table 1. Average Measurement Error for NDE and Optical sizing methods in this study

NDE Method	Measurement Error ($\pm$ mm)
Xray CT	0.15 (diameter and depth)
Film Xray	0.36 (diameter)
Digital Xray	0.36 (diameter)
Thermography (Derivative)	0.30 (diameter)
Eddy Current	0.013 (depth)
Optically-Measured*	**0.32 (diameter and depth)**

*The error in the optically-measured value was conservatively estimated due to the possibility of over-polishing past maximum and also because of the uncertainty in exact choice of maximum diameter on the metallograph. X-ray CT allows the ability to sweep through various cross-sections and find the maximum diameter nondestructively and thus showed a potentially lower error.

SUMMARY

In this study, a reinforced carbon-carbon (RCC) disk sample having holes in its coating system ultrasonically drilled to the depth of the SiC coating underwent a series of oxidation treatments to create void-like damage in the carbon-carbon substrate underneath the coating. The oxidation treatments varied from 10 to 70 minutes to create cavities varying in diameter from approximately 1.4 to 3 mm (as measured optically) as oxygen reacted with the carbon underneath the coating and resulted in its consumption. State-of-the-art nondestructive evaluation (NDE) methods were used to detect and size the damage. Both one and two-sided NDE methods were used. One-sided methods are practical for on-wing inspection, while two-sided methods might be applicable in situations where RCC panel removal is possible. Comparisons to optically-measured void sizes were shown. Xray CT and thermography reveal the trend of void size increasing with increasing oxidation treatment time. Digital and Film Xray were not able to differentiate the void sizes. Xray CT and eddy current reveal the trend of void depth increasing with increasing oxidation treatment time. Xray CT is the most accurate NDE method with regards to void sizing and measuring void depth. Thermography and eddy current show promise for practical inspection due to the one-sided nature and some ability to differentiate the small differences in oxidation damage extent. Xray CT might be applicable without panel removal in a partial-angle configuration.

ADDITIONAL NOTES

In this study and the prior study,[1] the difficulties in ultrasonically drilling such a small hole lead to non-uniform hole radii and variations from hole to hole. Additionally, from observations of drill hole profiles for the different oxidation treatments, it is possible that increased erosion of the SiC coating might be occurring with increasing oxidation time. Both situations might affect oxygen flow to the substrate and thus oxidation rate. Void coalescence may be responsible for differences between predicted and optically-measured void diameters.

FUTURE STUDY

It is desired to perform a similar study using different types of damage in the coating system such as slits and other damage types deemed as potentially impacting RCC durability.

APPENDIX – MORE DETAIL ON PROMISING NDE METHODS

Eddy Current inspection of the RCC sample was accomplished using a Jentek Meandering Winding Magnetometer (MWM).[7] The system incorporates a flexible thin film array sensor and spring loaded encoding wheel to produce C-scan images over the area of interest. In the current work, a 37 channel sensor with 0.1" coil spacing was used to image an area of 3.7" width in a single pass. The system components consist of the probe head, flexible array coil, 39 channel impedance instrument, and laptop computer for data acquisition, processing, and archival. Previous experimental work has shown that the optimum excitation frequency for RCC inspection is around 2.5 MHz, with higher frequencies providing a better characterization of the interfacial region between coating and substrate and lower frequencies providing a better depth of penetration for subsurface flaw detection. For the current work a frequency of 6.3 MHz was used to provide optimum characterization of the small pin-hole oxidation voids beneath the coating. An air calibration was incorporated and impedance measurements converted to conductivity/lift-off values via the Gridstation Method (registered trademark of Jentek Sensors) incorporated in the MWM system. The differences in conductivity/lift-off values are displayed in various shades of gray in the c-scan image.

A pulsed thermographic NDE method utilizing flash lamps and a high speed camera was used to obtain images of the RCC sample.[8] The system consists of two high energy xenon flash lamps, each capable of producing a 1.8 kJ flash with a 5msec duration. The flash lamps were placed at locations that provide a relatively uniform distribution of heat across the surface of the specimen. The transient thermal response of the specimen after flashing was captured using high-speed infrared camera. The camera used in the study is a 640 by 512 InSb focal plane array type with a 14 bit dynamic range. The camera operates in the 3-5 micron wavelength range and is capable of capturing thermal data at rates of 30 Hz for the full array size. For this study, a 320 by 256 portion of the full array was utilized in order to increase the frame rate to approximately 120 Hz. Flash initiation, data collection, storage and processing were all performed using software on the acquisition computer. Experimental data was collected using the following procedure. The specimen was placed in front of the IR camera at a distance that allowed the sample to fill most of the active focal plane and then focused. Flash lamps were set at a distance of approximately 300 mm. from the sample at an angle of 45 degrees. Along with the images captured after the flash, 6 preflash images were collected. Instantaneous and derivative images (from relative temperature vs. time) were obtained, and the operator normally selected the best images for analysis using a subjective process of selecting frames of maximum contrast. The differences in surface temperature are displayed in various shades of gray in the image.

X-ray computed tomography (CT) was accomplished using a Kevex Model P13006 X-ray source and a Varian PaxScan 2520 Amorphous-Silicon Detector.[4] The sample was placed on a micropositioner between the source and detector which allowed positioning and rotation to obtain the slice images. The differences in x-ray density are displayed in various shades of gray image.

REFERENCES
[1]Jacobson, N.S., Leonhardt, T.A., Curry, D.M. and Rapp, R.A., "Oxidative attack of carbon/carbon substrates through coating pinholes," Carbon 37 (1999) 411 - 419.
[2]Jacobson, N.S and Rapp, R.A., "Thermochemical degradation mechanisms for the reinforced carbon/carbon panels on the space shuttle," NASA TM-106793, 1995.
[3]M. P. Gordon, "Leading Edge Structural Subsystem and Reinforced Carbon-Carbon Reference Manual," KL0-98-008, Contract NAS9-20000, Boeing, Oct. 1998.
[4]**Metals Handbook** (Nondestructive Evaluation and Quality Control), Volume 17, Metals Park, OH: American Society of Metals, 1989.
[5]Birks, A. and Green, R. **Nondestructive Testing Handbook (Ultrasonic Testing)**. Second Edition, Volume 7. United States of America: American Society for Nondestructive Testing, Inc., 1991.
[6]Roth, D.J., Cosgriff, L.M., Verilli, M.J., and Bhatt, R.T., "Microstructural and Discontinuity Characterization in Ceramic Composites Using an Ultrasonic Guided Wave Scan System," **Materials Evaluation**, Vol. 62, No. 9, pp. 948-953.
[7]Goldfine, N.J., "Magnetometers for Improved Materials Characterization in Aerospace Applications," Neil J. Goldfine, **Materials Evaluation,** Vol. 51, No. 3, March 1993, pp. 396-405.
[8]Shepard, S.M., Lhota, J.R., Rubadeux, B.A. and Ahmed, T., "Onward and inward: Extending the limits of thermographic NDE," Proc. SPIE Thermosense XXII, Vol. 4020, 2000, p 194.

DETERMINATION OF ELASTIC MODULI AND POISSON COEFFICIENT OF THIN SILICON-BASED JOINT USING DIGITAL IMAGE CORRELATION

Matthieu Puyo-Pain and Jacques Lamon
Laboratoire des Composites Thermostructuraux (LCTS)
UMR 5801 CNRS / Snecma / CEA / Université Bordeaux 1
3 Allée de La Boëtie
F-33600 Pessac, France

François Hild
Laboratoire de Mécanique et Technologie (LMT-Cachan)
ENS Cachan / CNRS-UMR 8535 / Université Paris 6
61 avenue du Président Wilson
F-94235 Cachan Cedex, France

ABSTRACT

Optical extensometry based on a digital image correlation technique (DIC) is an interesting approach to the determination of displacements and strains in thin joints. The test specimens consisted of two monolithic SiC bars assembled by using a silicon-based braze (BraSiC). They were tested in four-point bending. Elastic properties of BraSiC are extracted from joint strain state measured by using DIC and computed using finite element analysis. They are compared to results of tensile, off-axis compression and nanoindentation tests carried out on similar samples.

INTRODUCTION

Joining of materials is a well-known and widely used technique to make structures from simple elements. In the typical case of ceramics and ceramic matrix composites that operate under severe conditions of temperature and environment, silicon-based adhesives such as BraSiC[1] are required. Data on the mechanical behavior and properties of joint materials are a prerequisite to the design of reliable parts[2,3].

The determination of joint properties is generally difficult because of the small size of joints (thickness around 100 μm in the present paper) and because test specimens of bulk material are not available. In previous papers, use of a DIC technique in the measurement of joint deformations was initiated[4,5]. Pertinent test conditions and parameters were established[6]. In the present paper, a method for determination of elastic constants of BraSiC is proposed. It is based on four-point bending tests, SiC-BraSiC-SiC test specimens, DIC technique, finite element computation of deformations and a deviation function. The estimates of elastic constants were then compared to those obtained using different test conditions and methods of data analysis.

EXPERIMENTAL PROCEDURES

Specimen preparation

The α-SiC bars were produced by Boostec Inc. The elastic properties of α-SiC are $E_{SiC} = 420$ GPa and $\nu_{SiC} = 0.14$. Test specimens were machined out of plate ($100 \times 150 \times 8$ mm^3). Dimensions of specimens were $100 \times 7.5 \times 6.1$ mm^3. Width of the joint was 85 ± 5 μm. One face of the specimens was carefully polished in order to meet optical requirements[6].

Testing procedure

SiC-BraSiC-SiC test specimens were tested in four-point bending (inner span: 30 mm and outer span: 80 mm) using an Instron electromechanical testing machine (model 4501) with a 5 kN load cell. Displacement rate was 2.5×10^{-2} mm/min. An optical microscope was mounted on the testing machine. It was connected to a 12-bit digital camera and to a computer. Images of the specimen surfaces were taken at various loading steps, at various locations in the SiC substrate and in the joint, and at various magnifications. A DIC software, Correli[LMT 7,8], was then used to extract the local displacements and the strain fields.

Figure I. Schematic diagram showing the testing procedure using a DIC technique

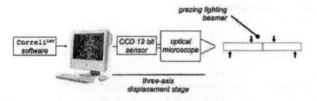

Digital Image Correlation principle

Digital image correlation determines the displacement field from digital images of the surface of test specimens taken at different steps during a mechanical test. It is based on the construction of a numerical mesh referred to as Region Of Interest (ROI) or correlation grid. ROI is a $m \times n$-pixel digital image. Construction of the mesh depends on the distribution of gray levels. In the present case, this distribution is given by the natural texture of the material microstructure under impinging light[6], which must display a random pattern so that groups of pixels can be matched at different load levels. Displacements are evaluated from the locations of $l \times l$ sets of pixels referred to as Zones of Interest (ZOI), separated by δ pixels. In Fig. II, c_{ij} represents the gray level associated to each pixel within a ZOI and C is the center of the ZOI. Successive locations of C during loading give the displacements $\underline{U}$. Average strains can then be calculated from the displacement field gradient on an $L_x \times L_y$-pixel gauge. For each load level, the displacement field, given at each center C in the $L_x \times L_y$ gauge is interpolated as follows,

$$\begin{cases} U_x(x,y) = K_{0,x} + K_{xx}x + K_{xy}y \\ U_y(x,y) = K_{0,y} + K_{yx}x + K_{yy}y \end{cases} \tag{1}$$

where $\underline{K}_x$ and $\underline{K}_y$ are parameters adjusted using a least square fitting method. The corresponding strains are given by,

$$\bar{\varepsilon}_{xx,exp} = K_{xx}, \ \bar{\varepsilon}_{yy,exp} = K_{yy} \text{ and } \bar{\varepsilon}_{xy,exp} = \frac{1}{2}(K_{xy} + K_{yx}) \tag{2}$$

Appropriate parameters were defined from the results of a campaign of experiments aimed at evaluating the resolution and the accuracy of the method and the system[6]. The parameters that were selected are summarized in Table I.

Figure II. Schematic diagram showing the principle of digital image correlation technique

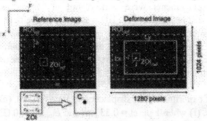

Table I. Four-point bending experimental conditions

Test	ROI (pixels)	l (pixels)	δ (pixels)	L_x (pixels)	L_y (pixels)	Number of gauge
Bending	960 × 1248	32	32	960	192	1

EXTRACTION OF ELASTIC PROPERTIES
Approach

Description of strain and stress states in the joint is not straight forward. Finite element computations were carried out for two dimensional plane stress conditions using the MSC.Marc 2000 finite element code and the MSC.Mentat 2000 pre-processor. The finite element mesh coincided exactly with the correlation grid. A refined 3-node element mesh was used in the joint area. The elastic constants of SiC substrates were taken to be $E_{SiC} = 420$ GPa and $\nu_{SiC} = 0.14$. The zone referred to as ROI* was identical to the experimental ROI (Fig. III). Joint elastic constants correspond to the minimum of J_{4B} that represents the deviation between experimental and computed displacement fields.

$$J_{4B}\left(E_{joint}, \nu_{joint}\right) = \sum_{F}\sum_{\{x,y\}}\left[\tilde{U}_{exp}(x,y,F) - \tilde{U}_{FEM}\left(x,y,F,\frac{E_{joint}}{E_{SiC}},\nu_{joint},\nu_{SiC}\right)\right]^2 \quad (3)$$

with $\tilde{U}_i = U_i - U_{0,i}$, where $U_{0,i}$ is the rigid body motion of node i. x and y are coordinates (Fig. III), F is the applied force. E_{joint} and ν_{joint} are respectively elastic modulus and Poisson ratio of joint.

Figure III. Schematic diagram showing the finite element mesh (ROI*) and the ROI selected on the four-point bending specimens

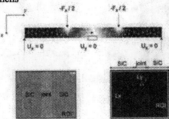

Results

Computations of displacements were carried out at various applied forces. Only U_y components (longitudinal displacements, Fig. III) were used in the determination of J_{4B}, since U_x displacements (lateral displacements, Fig. III) cannot be measured accurately[6]. Even for the maximum load level, U_x displacements do not exceed 0.03 pixel, which is of the order of the system uncertainty, whereas U_y displacements are much greater and reach 0.1 pixel (Fig. IV). The correlation parameters were l = 32 pixels and δ = 32 pixels (Table I).

Figure IV. Comparison of experimental U_y displacement field (a), (b), (c) and computed U_y displacement field (d), (e), (f) when 1 pixel = 0.33 μm and E_{joint} = 150 GPa and ν_{joint} = 0.4

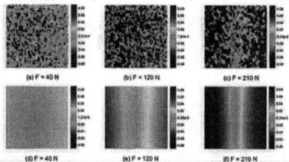

Minimum of J_{4By} was obtained with E_{joint} = 166 ± 35 GPa and ν_{joint} = 0.43 ± 0.05. Fig. V(a) shows that it is easy to identify a single couple of elastic constants (E_{joint}, ν_{joint}). Dependence of U_y on E_{joint} and ν_{joint} is significant in bending. Besides, E_{joint} and ν_{joint} were identified for different ROI sizes such that 480 × 288 pixels < ROI size < 960 × 1248 pixels, while the image size was 1024 × 1280 pixels and the BraSiC joint size is about 1024 × 384 pixels. The estimated values E_{joint} and ν_{joint} depend on the ROI size. For small ROI sizes, E_{joint} = 130 GPa and ν_{joint} = 0.40, and E_{joint} = 160 GPa, ν_{joint} = 0.42 for large ROI sizes. This difference can be attributed to increasing contribution of SiC as ROI becomes larger.

Figure V. Dependence of J_{4By} on E_{joint} and ν_{joint} (a), and ROI size (b)

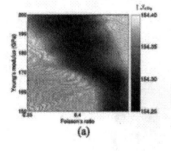
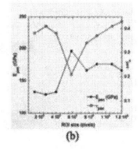

(a)   (b)

DISCUSSION

Table II. compares the elastic constants determined using various test conditions (tension, off-axis compression and nanoindentation[5,6]) and various methods of data analysis (closed form equations, finite element computations, deviation function[6]). Young's modulus estimates can be classified into two groups:

- the lower values (around 116 GPa), obtained using closed form equations of deformations and stresses,
- the larger ones (around 160 GPa), extracted using finite element computations of displacements and a deviation function. Note that comparable Young's modulus estimates were obtained using nanoindentation[5].

Furthermore, it is worth pointing out that Poisson's ratio was estimated only in the following cases: (a) tension and off-axis compression when influence of lateral deformations induced by Poisson effect in SiC substrate were taken into account. As a consequence, longitudinal deformation was expressed in terms of ν_{joint}. (b) Tension when finite element analysis and a deviation function were employed. (c) Four-point bending. Scatter in ν data estimated in four-point bending (J_{4By}) and in $\underline{J}_{3D}$ analysis is significantly small. In both cases, longitudinal deformations depend on ν_{joint}. Since they are larger than the resolution threshold, satisfactory E_{joint} and ν_{joint} estimates can be extracted using a deviation function. When finite element is used, the amount of deformation data available for estimation of elastic constants is very large: 14040 in tension ($T01 - J_{Tx}$), and 15210 in four point bending. On the contrary, it was much smaller when closed form equations of stresses were used (Table II.). This is because the average of strains in the gauge at each force is considered. It is compared to the corresponding strain given by theory. Therefore, the approach based on four-point bending tests seems to be the most satisfactory. Longitudinal deformations depend on Poisson's ratio. Furthermore, testing is easier when compared to tension or off-axis compression for which much care must be taken to eliminate misalignment of specimen with respect to loading direction. However, contribution of SiC substrate needs to be taken into account. In summary, the following elastic constants may be considered to be appropriate for the BraSiC material, namely, E_{joint} = 166 ± 35 GPa, ν_{joint} = 0.43 ± 0.05 and a shear modulus G_{joint} = 58 GPa.

Table II. BraSiC elastic parameters determined by different local techniques

Test	Reference	Strain state description	Number of data	E (GPa)	ν
Tension	T0i-xxx-J	Analytical	13	118-216	-
Off-axis compression	C45-128-J	Analytical	13	116 ± 88	-
Tension and off-axis compression	$\underline{J}_{3D}$	Analytical	39	113 ± 18	0.42 ± 0.05
Tension	$T01 - J_{Tx}$	FEM	14040	150-210	0.08-0.35
Four-point bending	J_{4By}	FEM	15210	166 ± 35	0.43 ± 0.05
Nanoindentation	-	-	-	179 ± 9	

SUMMARY AND CONCLUSIONS

Digital Image Correlation technique allowed deformations in a thin silicon-based joint (joint thickness < 100 μm) to be determined. Estimation of elastic constants was found to be dependent of joint deformations with respect to the uncertainty of the DIC technique[6]. Lateral deformations were generally too small. Therefore, a sound method of estimation E_{joint} and ν_{joint} could rely only on longitudinal deformations. Four-point bending tests coupled with finite element analysis of deformations and with a deviation function appeared to be an appropriate approach. Estimates of joint elastic modulus E_{joint} were in good agreement with those obtained using nanoindentation. Estimation of E_{joint} and ν_{joint} is also possible with other tests such as tension or off-axis compression. Longitudinal deformations depend on ν, as a result of the influence of Poisson effect in the SiC substrate.

ACKNOWLEDGMENTS

The authors gratefully acknowledge the support of Snecma, CNRS and CEA. The authors also wish to thank Bruno Humez for valuable contributions to mechanical tests.

REFERENCES

[1] F. Moret, P. Sire and A.Gasse, "Brazing of SiC and SiC Based Materials using the BraSiC® Process Chemical and Thermal applications", *International Conference on Joining of Advanced Materials,* Rosemont-USA, Oct. 12-15, 1998.

[2] J. Martinez Fernandez, A. Muñoz, F.M. Valeria Feria and M. Singh, "Interfacial and thermomechanical characterization of reaction formed joints in silicon carbide-based materials", *Journal of the European Ceramic Society,* **20** 2641-2648 (2000).

[3] H. Serisawa, C.A. Lewinsohn and H. Murakawa, "FEM analysis of experimental measurement technique for mechnical strength of ceramic joints", *Ceramic Engineering and Science Proceedings,* **22** [4] 635-642 (2001).

[4] M. Puyo-Pain and J. Lamon, "Determination of elastic properties of a ceramic-based joint using a digital image correlation method", *Ceramic Engineering and Science Proceedings,* **25** [4] 247-253 (2004).

[5] M. Puyo-Pain and J. Lamon, "Determination of elastic properties of a Si-based joint using local techniques", Proceedings of the 5th conference on *High Temperature Ceramic Matrix Composites,* Edited by M. Singh, R.J. Kerans, E. Lara-Curzio and R. Naslain, American Ceramic Society, Weterville, Ohio, 551-556 (2004).

[6] M. Puyo-Pain, "Comportement mécanique d'assemblages de composites 2D SiC/SiC brasés par un joint à base-silicium : mesures de champ par corrélation d'images numériques", *PhD thesis,* Université Bordeaux 1, N° 2905, (2004).

[7] F. Hild, "Correli^{LMT}: a software for displacement field measurements by digital image correlation", *Internal Report* N° 254, LMT-Cachan, 2002.

[8] F. Hild, B. Raka, M. Baudequin, S. Roux and F. Cantelaube, "Multiscale displacement field measurements of compressed mineral-wool samples by digital image correlation", *Applied Optics,* **41** [2] 6815-6827 (2002).

USE OF RUPTURE STRENGTH TESTING IN EXAMINING THE THERMAL CYCLE BEHAVIOR OF VARIOUS TYPES OF PLANAR SOLID OXIDE FUEL CELL SEALS

K. S. Weil, J. S. Hardy, G-G. Xia, and C. A. Coyle
Pacific Northwest National Laboratory
P.O. Box 999
Richland, WA 99352

ABSTRACT

A significant challenge in developing and manufacturing planar solid oxide fuel cells is in hermetically sealing the electrochemically active ceramic PEN to the metallic body of the device. One means of doing this is by chemically bonding the electrolyte surface of the PEN to the metal frame, thus forming a fixed or static seal. Typically a glass or glass-ceramic material is used in making this type of seal. However because of their brittle nature, there are number of manufacturing and operational problems that are encountered when using these materials. In an effort to efficiently screen potential solutions to these problems, a seal rupture test was developed. Details of the test procedure, the test device, and its use in thermal cycle testing various planar solid oxide fuel cell sealing concepts will be discussed.

INTRODUCTION

The development of a hermetic seal for planar solid oxide fuel cell (pSOFC) stacks is one of the key issues in the commercial viability of these devices. Hermeticity ensures that efficiency and fuel utilization are maximized and that "hot spots" due to localized fuel combustion across a leak do not arise in the device. Among the various concepts under consideration for pSOFC sealing, the development of a rigid joint, either by glass sealing or brazing, offers several advantages over compressive sliding seals, i.e. of the type made using a gasket material such as mica.[1,2] Unfortunately, rigid sealing generally requires that each component in the joint matches with respect to thermal expansion, a condition which is not easy to design for given the other requirements of the high-temperature cell and interconnect materials. The primary concern is that if the respective coefficients of thermal expansion (CTEs) do not match within a narrow range (~5-10%), high thermal stresses will develop in the sealing region during stack heat-up and/or cool-down and cause fracture of the cell or seal. The use of thermomechanical modeling can provide insight into how a specific joint design will respond under a given set of conditions; e.g. rapid thermal cycling, which is encountered in transportation applications such as automotive and truck auxiliary power units (APUs). However, computational models require accurate material data on the various components, not only as a function of temperature but also as a function of time at temperature (because the materials are expected to evolve in microstructure during operation). Although some of this information is available on the cell materials and on a handful of the candidate interconnect alloys, only scant data has been published on the various sealing materials.

We have developed an alternative seal evaluation approach: a semi-quantitative mechanical test (the rupture test) that allows us "bridge" the gap in our knowledge of what may constitute an effective pSOFC seal. Essentially a modified version of an adhesive blister testing technique,[3] the test was developed specifically for the purpose of screening the vast number of material, processing, and operational parameters that may impact the efficacy of a given pSOFC sealing technique.[4] As shown in Figure 1, it is conducted by placing a sealed disk specimen in a test fixture and slowly pressurizing the backside of the sample until seal rupture occurs, which gives a measure of the maximum pressure that the specimen can withstand. That is, the specimen is subjected to an accelerated stress test, using air pressure to generate high levels of stress within the seal. In this way, the test makes it possible to identify the weakest constituent in the sealing system, i.e. the ceramic substrate, the braze material, the metal substrate and associated oxide scale, or any of the interfaces in between, so that the seal can potentially be improved in the next round of development. Here we discuss the use of this characterization technique in investigating the strength of glass sealed and air brazed joints as a function of the number thermal cycles to which each was exposed.

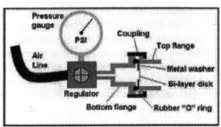

FIGURE 1 Schematic illustration of the rupture test apparatus.

EXPERIMENTAL
Materials

As shown in Figure 2, a miniaturized version of the cell-to-frame component is the key test specimen employed in this study. Each sample was prepared by brazing a 25mm diameter ceramic bilayer directly to a ferritic stainless steel washer (of the type under consideration for use in pSOFC stacks) that measures 44mm in outside diameter with a 15mm diameter concentric hole. The bi-layer coupons are composed of a thick NiO-5YSZ anode layer co-sintered to a thin YSZ electrolyte, each of which was fabricated by tape casting and co-sintering techniques that are identical to those used in fabricating full-scale, anode-supported pSOFC cells.[5] The average electrolyte and anode thicknesses respectively measured ~8µm and 600µm in these samples. In several series of tests, thicker bilayers were employed measuring ~1200µm in thickness. The metal washers were fabricated from Crofer-22 APU (23% Cr, 0.45%Mn, 0.08% Ti, 0.06% La, bal. Fe; Krupp-Thyssen) and FeCrAlY (22% Cr, 7% Al, 0.1%La+Ce, bal. Fe; Engineered Materials Solutions). Prior to joining, each washer was de-burred and cleaned in acetone.

The sealing material compositions employed in this study were both in-house PNNL-developed formulations. The sealing glass, designated as G-18, is barium calcium aluminosilicate based glass formed from the following mixture of oxides (by weight percent): 56.4% BaO, 22.1% SiO_2, 5.4% Al_2O_3, 8.8% CaO, and 7.3% B_2O_3.[6] Once melted, homogenized, and cooled, the material was attrition milled to an average particle size of ~20µm and mixed with a proprietary binder system to form a paste that could be dispensed onto the substrate surfaces at a uniform rate of 0.075g/linear cm using an automated syringe dispenser. In this manner, the glass paste was dispensed onto the YSZ side of the bilayer discs. Each disc was then concentrically positioned on a washer specimen, loaded with a 50g weight, and heated in air under the following sealing schedule: heat from room temperature to 850°C at 10°C/min, hold at

850°C for one hour, cool to 750°C at 5°C/min, hold at 750°C for four hours, and cool to room temperature at 5°C/min. The filler metal used in air brazing was formulated by ball-milling in methanol the appropriate amounts of copper (99%, Alfa Aesar) and silver powder (99.9%, Alfa Aesar) required to achieve an as-oxidized target composition of 4mol% CuO in silver. Previous results from X-ray diffraction indicate that the copper powder will fully oxidize *in-situ* during a typical air brazing heating schedule to form CuO, the wetting agent employed in this method of air brazing.[7] After milling and drying, the powder was mixed with a proprietary binder to form a paste that was dispensed at a rate of 0.1g/linear cm on the YSZ side of the bilayer discs. The discs were again positioned onto metal washers, loaded with a 100g weight, and heated in air under the following schedule: heat from room temperature to 1000°C at 10°C/min, hold at 1000°C for 15min, and cool to room temperature at 5°C/min.

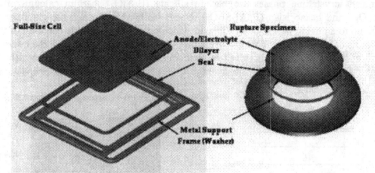

FIGURE 2   Comparison of the full-size SOFC window frame component to the rupture test specimens (not shown to size).

Testing and Characterization

Rupture strength specimens fabricated for thermal cycle testing were first tested for hermeticity by pressurizing to a modest pressure (10psi in the case of the glass sealed samples and 20psi in the brazed specimens), closing the isolation valve on the test apparatus, and examining the rate of pressure decay over a period of 10min. Samples that displayed no decay were used in subsequent testing. Thermal cycling was conducted in a pancake-shaped infared furnace containing a horizontal 3" thick x 15" diameter chamber that was lined on top and bottom with quartz windows through which a series of high-intensity quartz lamps could radiate and thereby induce rapid rates of heating with great reproducibility. The specimens were heated to 750°C at 75°C/min and held at temperature for 10min followed by cooling at 75°C/min to 400°C, after which the samples were cooled to 70°C in approximately 25 min. Testing was generally conducted in 20cc/min dry air out to a maximum of 50 rapid cycles, after which the samples were tested for hermeticity and rupture strength. A maximum pressure of 130psi and a minimum of six specimens for each joining condition were employed in testing. Subsequent microstructural analysis was conducted on polished cross-sectioned samples using a JEOL JSM-5900LV scanning electron microscope (SEM) equipped with an Oxford energy dispersive X-ray analysis (EDX) system.

RESULTS AND DISCUSSION
Glass Seals

Shown in Figure 3(a) are rupture strength results for seals formed against Crofer-22 APU and FeCrAlY as a function of the number of rapid thermal cycles from ≤70°C to 750°C. Note that in both cases, significant degradation in seal strength is observed. A typical failed glass-sealed specimen is shown in Figure 3(b). Fracture appears to initiate along the glass/metal scale interface, eventually propagating through the glass seal itself. There are at least two reasons for the observed degradation in joint strength, both related to microstructural changes in the seal during heating. Presented in Figure 4(a) is a plot of crystalline phases that form within the devitrifying glass as a function of time held at 750°C. This graph was constructed from a series of quantitative XRD measurements conducted on the aged glass-ceramic samples. Note for example that just after sealing at 850°C the glass has devitrified by ~50% on a volumetric basis. The dominant crystalline phases are barium silicate ($BaSiO_3$), barium calcium orthosilicate ($Ba_3CaSi_2O_8$), and hexacelsian ($BaAl_2Si_2O_8$). Under isothermal heating, the glass continues to crystallize up to an aging period of ~120hrs, at which point the crystalline make-up of the glass-ceramic levels off at ~70vol%. That is, during high-temperature exposure both the microstructure and CTE of the resulting bulk glass-ceramic change. As shown in Figure 4(b), over a temperature range of $25 - 750°C$ the average CTE of the evolving material drops with time at

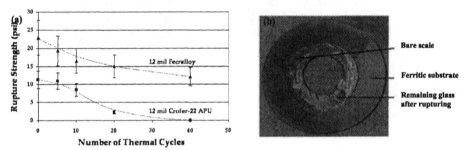

FIGURE 3 (a) Rupture strength of glass-sealed specimens as a function of the number of thermal cycles between room temperature and 750°C. (b) Planar view of a typical ruptured glass-sealed specimen.

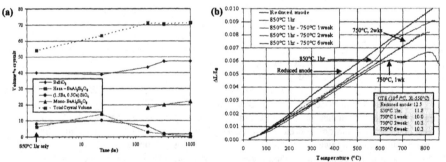

FIGURE 4 (a) Quantity of crystalline phases formed in G-18 as a function of time held at 750°C. (b) Thermal expansion of G-18 as a function of temperature after thermal aging in 750°C air.

152

temperature. The consequent reduction in CTE eventually leads to the development of residual stresses at the glass/metal interface, which can weaken the strength of the seal upon thermal cycling as observed in our thermal cycle testing results.

In addition, the glass/metal interface also evolves during high temperature exposure. Cross-sectional SEM examination of the glass/metal interfaces in the specimens demonstrates that regardless of the composition of the stainless steel substrate, a reaction zone develops between the metal's oxide scale and the bulk glass-ceramic as shown in Figure 5. This layer thickens with time at temperature. For example in Crofer-22 APU specimens, the barium chromate interfacial product observed in the joint grows from 10μm thick in the as-joined condition to 44μm when held for 100hrs at 750°C and to 76μm when held for a total of 200hrs at temperature. As observed in other studies of layered material structures, increasing the thickness of an intermediary layer between two thermally mismatched co-joined layers will increase the overall residual stress in each.[8] Again the resulting effect is a weakening of the glass seal during repeated thermal cycling.

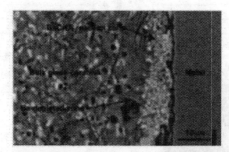

FIGURE 5   Glass/metal interface against a representative chromia former.

Braze Seals

Plotted in Figures 6(a) and (b) are the respective rupture strengths of the thermally cycled FeCrAlY- and Crofer-based air brazed sealing specimens as a function of the number of cycles to which the specimens were exposed. The results indicate that rapid thermal cycling causes no measurable degradation in rupture strength of this sealing material. Unlike the glass-sealed joints, all of the braze specimens fail through the ceramic bilayer membrane, as shown in Figure 6(c), indicating that this remains the weakest component in the brazed seal upon thermal cycling. The average strength of the 600μm thick bilayers is 187±39MPa, as measured at room temperature using ball-on-ring biaxial flexure testing (ASTM F 394-78). The corresponding range of rupture strength values, indicated by the shaded regions in Figures 6(a) and (b), are the effective pressures at which failure is expected to occur in the cell; i.e. the type of failure observed in Figure 6(c). To verify this, additional testing was conducted using the 1200μm thick bilayer discs. Samples of this type that were exposed to 0, 10, and 20 thermal cycles in air could all be rupture tested to the maximum pressure of 130psi without failure.

Shown in Figure 7 are a series of SEM micrographs that display the YSZ-braze interface in specimens that have undergone 5, 10, and 50 rapid thermal cycles. Examination reveals that after 5 and 10 thermal cycles, Figures 7(a) and (b) respectively, no significant change in interfacial microstructure occurs. However after 50 cycles, Figure 7(c), signs of delamination along the

153

YSZ-braze interface appear. Small patches, ~5 – 20μm in length, are observed where the braze has pulled away from the YSZ to form a thin void. The practical effect of this phenomenon on rupture strength and mode of specimen failure under pressurization is essentially nil, i.e. the strength of the ceramic membrane remains the dominant factor in defining failure in the partially delaminated seals. However with added cycling, the resulting strength of this interface may become the limiting factor in air braze joint strength. In an effort to identify the operational limit for thermal cycling in this type of seal, testing is being continued to determine the point at which rupture strength degradation is eventually observed.

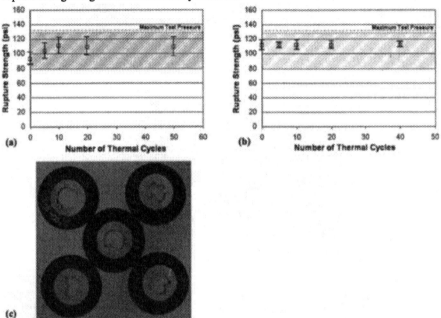

FIGURE 6   Rupture strength as a function of the number of thermal cycles between room temperature and 750°C for (a) the bilayer/Ag4CuO/FeCrAlY specimens and (b) the bilayer/Ag4CuO/Crofer-22 specimens. (c) Planar view of typical ruptured brazed specimens.

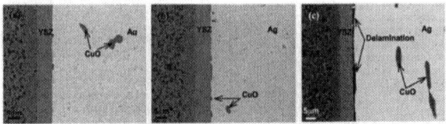

FIGURE 7   Cross-sectional SEM micrographs of the YSZ-braze interfaces within three rupture specimen that were thermally cycled in air at 75°C/min between room temperature and 750°C: (a) for 5 cycles, (b) for 10 cycles, and (c) for 50 cycles.

CONCLUSIONS

This series of initial experiments to examine the effects of thermal cycling on glass and air brazed seals revealed the following:

- Through the use of the rupture test, we have determined that alumina-forming ferritic steel substrates offer greater bond strength with the barium aluminosilicate based-glass employed here in both the as-joined and thermally cycled conditions. The dominant factor in joint strength appears to be the composition and thickness of the reaction zone that forms in between the metal's oxide scale and the bulk glass. Regardless of the composition of the metal substrate, fracture appears to initiate within the interfacial reaction zone between the metal scale and bulk glass-ceramic.

- Air brazed specimens that were thermally cycled at 75°C/min showed little change in microstructure out to 10 cycles. However after 50 cycles, signs of delamination along the YSZ-braze interface were present.

- Subsequent rupture testing of the cycled brazing specimens indicates that no accompanying degradation in rupture strength takes place. Unlike what was observed with the glass-sealed specimens, the ceramic membrane is the weakest constituent in the brazed joint.

ACKNOWLEDGMENTS

The authors would like to thank Nat Saenz, Shelly Carlson, and Jim Coleman for their assistance. This work was supported by the U. S. Department of Energy as part of the SECA Program. The Pacific Northwest National Laboratory is operated by Battelle Memorial Institute for the United States Department of Energy (U.S. DOE) under Contract DE-AC06-76RLO 1830.

REFERENCES

[1] S-B. Sohn, S-Y. Choi, G-H. Kim, H-S. Song, and G-D. Kim, G-D, *J. Non-Cryst. Solids,* **297,** 103 (2002).

[2] S.P. Simner and J.W. Stevenson, *J. Power Sources,* **102,** 310 (2001).

[3] G. P. Anderson, S. J. Bennett, and K. L. De Vries, *Analysis and Testing of Adhesive Bonds,* Academic Press, 1977.

[4] K. S. Weil, J. E. Deibler, J. S. Hardy, D. S. Kim, G-G Xia, L. A. Chick, and C. A. Coyle, *J. Mater. Eng. Perf.,* **13,** 316 (2004).

[5] S. Mukerjee, S. Shaffer, J. Zizelman, L. Chick, S. Baskaran, C. Coyle, Y-S Chou, J. Deibler, G. Maupin, K. Meinhardt, D. Paxton, T. Peters, V. Sprenkle, and S. Weil, *8th Int. Symp. Solid Oxide Fuel Cells,* Volume 2003, The Electrochemical Society, 2003.

[6] K. D. Meinhardt, J. D. Vienna, T. R. Armstrong, and L. R. Pederson: U. S. Pat. 6430966, 2002.

[7] K. S. Weil, J. Y. Kim, and J. S. Hardy, *Electrochem. and Sol. St. Lett.,* in press.

[8] V. Teixeira, M. Andritschky, W. Fischer, H. P. Buchkremer, and D. Stover, *Surf. and Coat. Tech.,* **120-121,** 103 (1999).

CHARACTERISTICS OF SEALED PARTS UNDER INTERNAL PRESSURE IN SUPER HIGH PRESSURE MERCURY DISCHARGE LAMPS

Masahiko Kase
ORC Manufacturing CO.,LTD.
Tamagawa 4896, Chino-Shi
Nagano, Japan. Zip-Code: 391-0011

Toshiyuki Sawa
Department of Mechanical Engineering
Hiroshima University
Kagamiyama 1-4-1, Higashi Hiroshima-Shi
Hiroshima, Japan. Zip-Code: 739-8527

Yuichiro Iwama
ORC Manufacturing Co.,LTD.
Tamagawa 4896, Chino-Shi
Nagano, Japan. Zip-Code: 391-0011

ABSTRACT
        Super high pressure mercury discharge lamps have been used as a UV light source in photolithography exposure processes such as patterning LCDs, PCBs and semiconductors. It is well known that burst of the lamps during operation is a major problem. Rupture initiation has been found empirically in the sealed parts of lamps and an internal pressure is assumed as a dominant factor for the rupture in the sealed parts. Thus, the characteristics of rupture and improvement methods for the sealed parts under internal pressure have been investigated. Static water pressure tests were carried out in order to investigate the rupture pressure of the sealed parts. In addition, photoelasticity was employed to observe the stress distribution of the sealed parts. Finite-element method (FEM) calculations were done to obtain the maximum principal stress σ_1 distribution of the sealed parts under internal pressure. From the results, it was found that the rupture pressure of sealed parts was increased as debonded length between the foils and quartz was decreased. In addition, it was observed that the rupture of the sealed parts occurred at about 4 MPa internal pressure and a stress concentration exists near the end of molybdenum foil and quartz glass. The trends of the numerical results were in good agreement with that of the experimental ones.

INTRODUCTION
        Super high-pressure mercury discharge lamps have been used as UV and visible light sources in optical system for the exposure process of photolithography such as patterning PCB, LCD, and semiconductors. It is estimated that their internal pressure in operation is more than 1.0 MPa and the outer temperature of bulb is over 500℃. Thus, the lamp during operation, fragments of the bulb can cause great damage in optical parts such as mirrors and lens.
        Thouret[1] investigated optimal shape of the bulb in the lamps under internal pressure and

thermal loading with numerical calculations. Erdogan et al[2]., calculated stress distributions in sealed parts using theory of elasticity. However, it is difficult to directly apply these results to actual design of lamps.

It is known empirically that rupture initiations often found in sealed parts where metal foils and glass parts are joined together[3]. Generally, specific factors have influenced the burst of lamps in operation. Among the factors, internal pressure loading is assumed as one of the most dominant. Thus, the objectives of this study were to examine the rupture characteristics of sealed parts under internal pressure and to propose methods for improving pressure resistance of the sealed parts.

Structure of lamps

Figure 1 shows a schematic illustration of the lamp and its sealed parts. The lamps generally can be divided three components as follows.

(1)Bulbs: These are made of quartz glass and their thickness is about 3.5 mm. A pair of electrodes are arranged in the bulb. In addition, inert gas (such as argon and xenon) and mercury are enclosed in the bulb to emit UV and visible light. (2)Sealed Parts: These are linked to both ends of the bulb. They are made of quartz glass, metal parts (molybdenum, tantalum, and tungsten are generally used) and six molybdenum foils are used. It has been well known that the foils and quartz are in contacted and bonded adhesively[3] during the sealing process. The functions of sealed parts are to keep the bulb airtight and pass electric current between the inside and outside of the lamps. (3)Base: These are generally made of metal materials. The bases are glued at the end of sealed parts by a heat-resistant adhesive. The function of the base is to fix the lamps in the optical system.

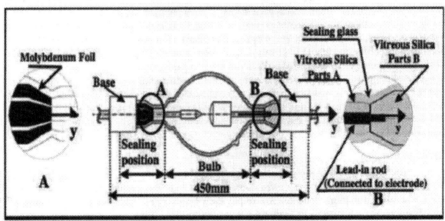

Fig. 1 Schematics of lamp and their sealing position

METHODS

Experimental Method

Figure 2 shows the schematic in the static water pressure tests[4]. Figure 3 shows detail of a test specimen. The test specimens imitated the actual sealed parts in the lamps. They were manufactured with the same parts of the actual lamps. In the experimental system, they are linked to a pump by quartz tubes and pressurized. Pressurizing conditions were kept constant at 6.0MPa/5min in the experiments. Rupture pressures for each test specimen were compared. The test specimens showed that the stress distributions changed as the pressure increased Cracks were observed by photoelastic systems shown in Fig. 2.

In fabricating the test specimens, burner gas-fluent (H_2 and O_2 gas) conditions in the sealing process was changed by 20% to investigate the effect of the sealing process condition on the pressure resistance of the sealed parts. That is, the first one is the normal process, the second one is increased by 20%, the third one is decreased by 20%.

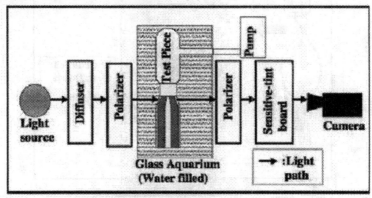

Fig. 2 Schematic of static water pressure test

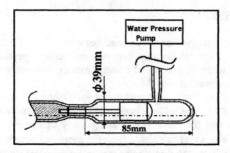

Fig. 3 Schematic of a specimen in static water pressure test.

159

FEM calculations

In the FEM calculations, a part of the sealed parts was calculated and the maximum principal stress σ_1 distributions under 3 MPa internal pressure were obtained. It is well known that ruptures of brittle materials are generally explained by the maximum principal stress theory. Quartz glass is one of the most typical brittle materials[5].

Figure 4 shows details of FEM model. A one-sixth symmetry FEM model was employed because six-sheets of molybdenum foils were arranged symmetrically about the y-axis in the sealed parts as shown in Fig 4. In the FEM calculations, the number of the elements was between 30000 and 31000. The FEM code name employed was "Marc".

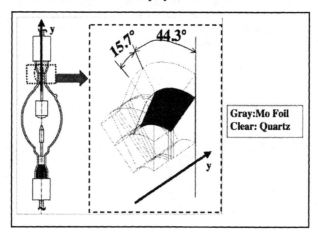

Fig. 4 Detail of model for FEM calculations.

RESULTS AND DISCUSSION
Experimental results

It was found that test specimens ruptured between 3.5 and 5.0 (MPa) internal water pressure. Figure 5 shows the relationship between the rupture pressure of the twelve test specimens and their frequency. It was found that the rupture pressure was increased as the gas fluent in the sealing process was increased.

Figure 6 shows an example of ruptured test specimens and the measured dimensions. The lengths A , B and C are representing debonded length between molybdenum foil and quartz, the total length of crack and the crack length along the y-axis respectively. Moreover, θ is representing crack angle and it was calculated from the measured lengths A, B and C. The angle of crack θ is assumed to indicate the direction of maximum principal stress σ_1 because the crack growth direction in brittle materials is generally perpendicular to the direction of maximum principal stress σ_1[5].

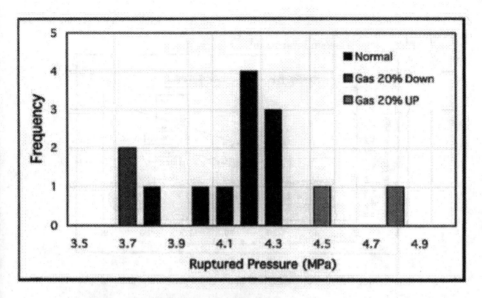

Fig. 5 Rupture pressure of the test specimens.

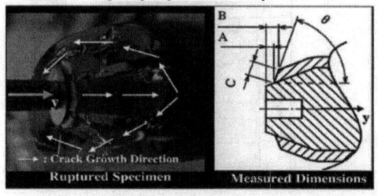

Fig. 6 An example of ruptured test specimen and measured dimensions.

Figure 7 shows the relationship between the angle of crack θ and the rupture pressure of the test specimen. It was found that the conditions of sealing process influenced the rupture pressure and the pressure increased linearly with increasing crack angle θ.

Figure 8 shows relationship between the length A shown in Fig.6 and the rupture pressure. It was found that the rupture pressure of the test specimen increased as the debond length A decreased. Thus, it can be assumed that the debond length A is dependent on the sealing process condition.

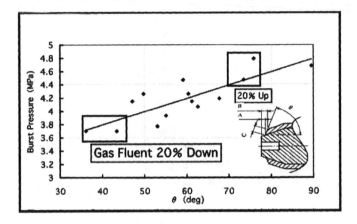

Fig. 7 Relationship between crack angle θ and rupture pressure of the test specimens.

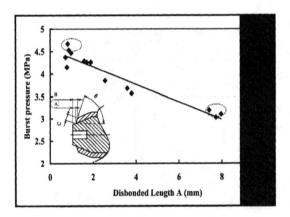

Fig. 8 Relationship between debond length A and rupture pressure.

Thus, it can be assumed that the conditions of the sealing process are some of the most dominant factors for the pressure resistance of the sealed parts. In addition, it also can be assumed that the pressure resistance of the sealed parts increases as the debond length A decreases.

Numerical Results

Figure 9 shows an example of calculated maximum principal stress σ_1 distribution as contours. In this figure, the stress in tension is represented as positive value. It was found that the stress concentration occurs in the quartz where the molybdenum foils and quartz were bonded. It can be estimated from the results that the initiation of rupture occurs near the point where the end of the foils and quartz are bonded.

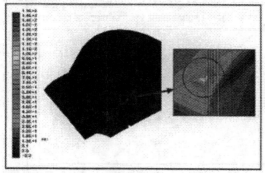

Fig. 9 An example of calculated maximum principal stress distribution as contours.

Figure 10 shows the relationship between the debond length A and the maximum value of maximum principal stress σ_1. It was found that the maximum value of maximum principal stress σ_1 increased as the debond length A in the sealed parts increased. Thus, it can be assumed that the pressure resistance of the sealed parts increases as the debond length A in sealed parts decreases.

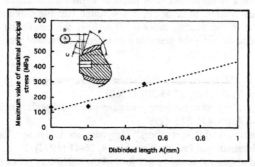

Fig. 10 Relationship between debond length A and calculated maximum value of σ_1.

Figure 11 shows the relationship between the debond length A and the estimated crack angle θ from the maximum principal stress directions[5]. It was found that the estimated crack angle θ increased as the debond length A decreased. Thus, it can be assumed that the crack angle θ increased as the debond length A decreased. In other words, it can be assumed that the crack angle θ increased as the pressure resistance of the sealed parts increased. The trend of the experimental results were in good agreements with numerical results.

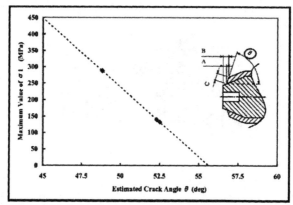

Fig. 11 Relationship between estimated crack angle θ and maximum value of σ_1.

CONCLUSIONS
This paper has dealt with the rupture characteristics of sealed parts under internal pressure. The following conclusions can be drawn: the pressure resistance of the sealed parts increased as the debond length A decreased. The trend of experimental results were in good agreement with the numerical results. In addition, it was found that the debond length A was influenced by the conditions of the sealing process. Therefore, the conditions of the sealing process should be considered to improve the pressure resistance of the sealed parts.

REFERENCE
[1] Wolfgang E. Thouret , "Tensile and thermal stress in the envelope of high brightness high pressure discharge lamps.", Illu, Eng ., 295-302 (1960)
[2] F.Erdogan, P.F.Joseph , "Fracture mechanics of quartz-to-metal seals. Engineering Fracture Mechanics.", Vol.44, 4, pp. 491-513 (1993) .
[3] G.Leichtfried, G.Thurner, and R.Weirather, " Molybdenum Alloys for Glass-to-Metal Seals." Proceedings of 14th International Pransee Seminar., 4, 26-41 (1997)
[4] JIS S-2302 "Method of internal pressure test for carbonated beverage bottles."(In Japanese)., JIS Handbook, 33., JAS(2001)
[5] R.C.Bradt and R.E.Tressler, "Fractography of Glass.", 111-184 (1994) Plenum Press New York.

Processing and
Properties of Ceramics

SYNTHESIS OF TITANIUM ALUMINATE -ALUMINA COMPOSITIONS FOR LOW TEC APPLICATIONS

James A. Geodakyan, Aram K. Kostanyan, Karen J. Geodakyan
Scientific-Research and Production Enterprise of Materials Science
17 Charents Str., Yerevan 375025, Republic of Armenia

W. Roger Cannon
Rutgers State University of New Jersey
607 Taylor Rd., Piscataway, NJ 08854-8065, USA

ABSTRACT
By choosing appropriate stabilizing additions and optimizing the thermal treatment, aluminum titanate with a low thermal expansion coefficient (TEC) (minus $3.38 \cdot 10^{-6} K^{-1}$ in the interval of 20 – 200 °C) and hysteresis (0.07 % at 400 °C) was obtained. The material was used as an additive for decreasing the TEC of alumina. Destructive thermal stresses were overcome by optimizing the milling and homogenization technique. The sintering rates, phase content and microstructure are described.

INTRODUCTION

Thermal shock resistance depends on multiple parameters. It depends not only on size and shape of samples, but also on thermal, -physical and mechanical properties of the material [1]. In order to increase thermal shock resistance of materials they can be reinforced by strong filament monocrystals (whiskers) or be strengthened by other methods. [2-3]. According to the research, carried out at Oak Ridge National Laboratory [4], alumina ceramics, reinforced with SiC whiskers or powder, can resist up to 200 thermal cycles between 1200 °C and cold airflow, without losing appreciable strength. For comparison the thermal shock resistance of ordinary alumina ceramics cannot withstand an 800°C quench into a cold airflow.

Despite of these positive results, research aimed at increasing thermal shock resistance of ceramics by reinforcements suffers an obvious disadvantage, it does not seek to decrease the level of stresses from thermal shock. According to [5], the level of thermal shock stresses that appear in a solid body is directly proportional to the thermal expansion coefficient (TEC) and Young's module and increases with the Poisson's ratio. Decreasing the last two values is neither easily accomplished nor desirable. Decreasing the TEC, on the other hand, is desirable in every respect, as long as it does not affect other properties, and can be achieved by introducing additives that possess low or negative values of TEC into the initial ceramic. Unfortunately, thermal stresses, appearing along phase interfaces between components with high differences in TEC, significantly exceed those caused by thermal shock. This fact creates serious difficulties increasing thermal shock resistance by this method.

Our previous research shows that the problem of incompatibility of components with high difference in TEC can be solved by optimizing the dispersion of components to achieve highly homogeneous compositions [6, 7]. The reason is likely due to decreases in the phase dimension and, therefore, the likelihood of fracture. For example, an alumina composition, containing 30 mass % of β-eucryptite glass-ceramic TEC of $5.61 \cdot 10^{-6} K^{-1}$ has been developed by the first three authors. The flexure strength is 120 MPa and critical thermal shock parameter, ΔT, is higher than 1200 °C in running water. However β-eucryptite melts at 1390±10 °C [6], which limits working temperature of the composition.

In order to increase refractoriness of the composite, this research examined the possibility of decreasing the TEC of Al_2O_3 by introducing Al_2TiO_5 as a second phase The melting temperature of Al_2TiO_5 is 1860 ±10°C.[8].

Aluminum titanate crystallizes in the rhombic system. Its crystal structure is the pseudo-brookite [8, 9]. It forms well-crystallized grains of Al_2TiO_5, which are usually prismatic shaped. The low cleavage strength and high anisotropic TEC of monocrystals lead to poor sinterability [10, 11]. During reaction sintering, the volume of material increases 14.8 % which retards sintering [12]. Though less durable spodumene may be a better choice [13, 14].

In the interval of 20-1000 °C the TEC of Al_2TiO_5 monocrystals on the "*a*", "*b*" and "*c*" axis are 11.8, 19.4 and $-2.6 \cdot 10^{-6} K^{-1}$ respectively. Low and inconstant values of TEC, as well as hysteresis of the thermal expansion curves for polycrystalline Al_2TiO_5, have led most authors to conclude that this anisotropy leads to microcracking [8, 9, 11, 15]. Furthermore, [8, 10] the TEC of the Al_2TiO_5 increases with successive measurements, while the area under the hysteresis loop decreases. Various researchers describe the TEC of polycrystalline Al_2TiO_5 in the interval of 20-700 °C as varying from -0.44 up to $+ 5.3 \cdot 10^{-6} K^{-1}$ [8, 10, 16-19]. That this is a result of Al_2TiO_5 microcracking is supported by thermodynamic calculations [20] and radiographic tests [10, 16, 17].

According to [10], Al_2TiO_5 is stable at temperatures above 1300 °C. In the interval of 750-1300 °C it decomposes to rutile and corundum. Below 750 °C decomposition is not observed due slow kinetics. Additives such as MgO, BeO, CaO, SiO_2, Fe_2O_3, ZrO_2, kaolin, etc are used to stabilize Al_2TiO_5. [10, 12, 19, 21-27, 33].

The influence of several additives on the basic properties of Al_2TiO_5 was also investigated in the laboratory of one of the authors of the present paper in the 70-s [23, 24]. According to this research, the change in the TEC of Al_2TiO_5 in presence of additives, containing Mg^{+2}, Ca^{+2}, Fe^{+3} ($MgTi_2O_5$, CaO, Fe_2TiO_5) ions, suggests the formation of substitutional solid solutions, where larger ions, replacing Al^{+3} increase the Me-O interionic distance and decrease the bonding strength of the structural units, therefore increasing the TEC. In contrast the comparatively small Si^{+4} ions (r = 0.37 Å) easily fits into the empty space of the Al_2TiO_5 lattice. Up to 23 mol % [Si^{+4}] additions form a continuous range of solid solutions and due to the fact Si^{+4} can form only tetrahedrons [SiO_4] with small interionic distance (about 1.6 Å). In comparison to the octahedral environment of Me-O (1.7-2.25 Å), its introduction to an Al_2TiO_5 crystal lattice leads to additional compression of octahedrons [TiO_6] and [AlO_6] and, consequently, to decrease of TEC. Corresponding lattice parameters also change.

In presence of Li_2O the TEC of Al_2TiO_5, containing SiO_2, decreases even more. Drawing the analogy to the β-eucryptite structure [28] it is possible to assume, that Li^{+1}ions, fits in the double spiral of [TiO_6], [AlO_6], promoting the maximum compression of spirals and correspondingly maximum decrease in the TEC. Thus, Al_2TiO_5 has been made with the TEC equal $-0.82x10^{-6}K^{-1}$ in the interval 20-700°C in a molar ratio $SiO_2 : Li_2O$ = 10:1.

The current work resulted in synthesis of Al_2TiO_5 with a low value of TEC and acceptable stability in the interval of 750-1300 °C. The current research also examines the influence of Al_2TiO_5 on sintering, TEC, mechanical and dielectric properties, structure and thermal shock resistance of Al_2O_3. Compositions in the system Al_2O_3–Al_2TiO_5 are also reported in this study [13, 14, 29-32].

MATERIALS AND TESTING METHODS

The α-Al_2O_3 (A), used in this study was obtained by dehydrating aluminum hydroxide, precipitated from an $Al(NO_3)_3$ solution by addition of ammonia. The impurity level in the alumina powder was no more than 0.05 %, and 82 % of the particles were less than 2 microns. Dependence of properties of Al_2TiO_5 on purity was studied by using as a source of TiO_2 chemicals of (1)"High-purity" (HP), (2) "Analytical purity," and(3) "Technical" grade . The total impurity content of (1) is $\leq 10^{-3}$, (2) is ≤ 0.1, (3) is ≤ 1.5 mass %. As stabilizing additives for Al_2TiO_5 we used SiO_2 (amorphous HP silica or refined quartz sand, containing no more than 0.01 % Fe_2O_3) and Li_2O (Li_2CO_3 of H-p type).

Initial components, except Al_2O_3, were preliminary ground in a planetary mill (water+ethyl alcohol for 4 hours) to obtain an average particles size equal to 2-3 microns.

Mixing and additional grinding of components was carried out in the same mill under the same conditions. After drying the initial charge was formed into billets 15x10x65 mm by cold pressing (80 MPa).

In order to synthesize stabilized Al_2TiO_5 (AT), billets were fired in an air atmosphere (muffle electric furnace of Nabertherm mark, HT 16/17, Germany) at temperature of 1500 and 1550 °C in several stages.

Between stages the samples were crushed, ground, dried and formed into billets again. This was done in order to fully react the components by creation of fresh contact surfaces between components, and to investigate the TEC and other properties of the material, as a function of depth of transformation. Samples were heated from room temperature to the maximum firing temperature at 7-15 °C /min., soaked for 3-5 hours, furnace cooled at a speed of 5-20 °C/min.

The influence of the Al_2TiO_5 additions on the basic properties of Al_2O_3 was investigated, employing compositions that were prepared by the above technique. Some specimens were synthesized according to the methodology, described above except that before the final firing the compositions were mixed with 10 % polyvinyl alcohol, used as a plasticizer. Then compositions were fired at 1500 °C for 3 hours and 1650 °C for 4 hours.

Other firing schedules were 1550 °C for 2 hours; 1600 °C for 1 hour; 1650 °C for 1 hour; 1450 °C for 2 hour + 1650 °C for 1 hour. We investigated compositions, containing from 3 up to 80 mass % of Al_2TiO_5. Purity of initial materials was measured by chemical and emission-spectroscopic analysis and phase content by x-ray and crystal optical analyses. Optical microscopy together with SEM were used to evaluate particle sizes in powders.

For property and compositional analysis we used: spectrograph DFS-8, LOMO (Russia); polarizing microscope Polam, LOMO (Russia); x-ray diffractometer DRON 4, LOMO (Russia); scanning electronic microscope VEGA TS 5130 MM, TESLA (Czech Republic); dilatometers DKV-4, GIS (Russia) and L76/1550B Linseis (Germany); derivatograph Q-1500, MOM (Hungary); tearing machine ZD-10/90 (Germany).

RESULTS AND DISCUSSION

a) Synthesis and properties of aluminum titanate

We studied the influence of composition, synthesis conditions and processing conditions on the TEC and TEC hysteresis of Al_2TiO_5 (Table I). The lowest values of TEC and least values area under the hysteresis curve were sought.

Table I. TEC and hysteresis of stabilized Al_2TiO_5 under different conditions

No	Initial components *)		Firing Temp., °C, soak time, hours each stage	TEC, $10^{-6}K^{-1}$		Maximum Difference on hysteresis curves		
	TiO_2	SiO_2		20-400 °C	20-700 °C	Temp, °C	Δl heating - Δl cooling	
							micron	% of L
1.	High Purity	High purity amorph.	I, 1550, 3	-1.18	-0.19	500	19	0.039
			II, 1550-, -3-	-1.49	-0.69	400	23	0.047
			III, 1550, 3	-2.16	-1.52	400	30	0.062
2.	High Purity	quartz sand	I, 1550, 3	-0.99	-0.15	500	23.5	0.048
			II, 1550-, 3	-1.19	-0.23	500	13	0.026
3.	Analytical purity	High purity amorph.	I, 1550, 3	-0.71	+0.52	-	-	-
			II1550,3	-0.84	+0.25	300	12	0.024
			III, 1550,3	-1.33	-0.61	400	12	0.025
4.	Technical	High purity amorph.	I, 1550, 3	-0.16	+0.86	-	-	-
			II, 1550, 3	-0.38	+0.26	-	-	-
			III,1550, 3	-0.93	-0.06	300	10	0.023
5.	High purity	High purity amorph.	I, 1500, 5	-0.83	-0.19	-	-	-
			II,1500, 5	-1.12	-0.51	-	-	-
			III, 1500, 5	-1.65	-1.06	400	22	0.045
			IV,1500,5	-1.65	-0.98	-	-	-
			V,1500, 5	-1.66	-1.08	-	-	-

*) Initial alumina (Σ of impurity did not exceed 0.05 %, particles with sizes less than 2 microns 82 %) and lithium carbonate (high purity) are the same for all the compositions.

On the basis of the most negative (Table I) the best compositions contained high purity TiO_2 and amorphous silica. Further studies showed an improved firing condition was 1500 °C for 5 hours in three stages with cooling after first two stages in water, after the third furnace cooling. Under these processing conditions the Al_2TiO_5 possessed the following properties:
- Linear shrinkage after firing: 12.6-14.2 %,
- Water absorption: 0.145 – 0.167 %,
- Density of sintered samples: 2.877 – 3.016 g/cm^3,
- Picnometer density of powder: 3.339 g/cm^3,
- Porosity of sintered samples after 3rd firing: 9.67 % (open : 0.44 %)
- TEC at 20-300 °C: - $3.29 \cdot 10^{-6}$; 20-700 °C : $-1.79 \cdot 10^{-6}$, minimum 20-200 °C : $-3.38 \cdot 10^{-6}$
- Maximum hysteresis when rate of heating and cooling was 3–4 °C/min between 20°C and 400°C, 3.5microns or 0.07 % of the length of a sample (Fig. 1, curve 1)
- Phase structure:
  - according to x-ray analysis: 80 mass % of Al_2TiO_5, 20 β-spodumene
  - crystal optical analysis: Al_2TiO_3 + β-spodumene - 97 %, glass - 3 %.

We also investigated the stability of Al_2TiO_5 with Al_2O_3 (A-AT), and Al_2TiO_5 with Al_2O_3 and spodumene (A-AT-S) during a long-term (100 hour) soak at 1100 °C. The latter two compositions were obtained by three stage firing consisting of twice at 1500 °C for 3 hours + once at 1650 °C for 4 hours and twice at 1300 °C for 2 hours + once at1320 °C for 1 hour, respectively. Samples were furnace cooled. In addition the A-AT-S samples was annealed 2 hours at 1250 °C for crystallization of spodumene. Heat treatment of samples at 1100 °C was carried out in air in electric muffle furnaces. At periodic intervals samples were furnace cooled and analyzed by x-ray diffraction. Monolithic samples were subjected to dilatometric analysis, as well. Results of these tests are found in Table II.

Table II. X-ray and TEC results for soaking stabilized aluminum titanate of various compositions at 1100 °C

Composition of composites, mass %	Ageing at 1100 °C, Hour	*)Most intense x-ray reflection												TEC, $10^{-6}K^{-1}$ at 20-700 °C
		Al₂TiO₅						TiO₂ (rutile)				**)3		
		1		2		3		1		2				
		d, Å	I/I₀, %	d, Å	I/I₀, %	d, Å	I/I₀, %	d, Å	I/I₀, %	d, Å	I/I₀, %	d, Å	I/I₀, %	
100 AT	0	4.71	60	3.351	93	2.653	100	-	-	-	-	1.69	10	-1.79
	10	4.72	26	3.351	100	2.652	80	3.23	4	2.484	2	1.687	15	-0.64
	20	4.695	40	3.351	100	2.659	75	3.243	25	2.488	10	1.684	30	+0.82
	30	4.72	15	3.351	48	2.659	35	3.243	100	2.488	35	1.687	35	2.36
	40	4.72	20	3.351	54	2.652	36	3.232	100	2.485	45	1.687	51	+4.25
	60	-	-	-	-	-	-	3.243	100	2.488	50	1.687	58	+6.41
60 Al₂O₃ + 40 AT (A-AT)	0	4.695	17	3.351	39	2.653	40	-	-	-	-	-	-	2.63
	10	-	-	-	-	-	-	3.232	60	2.485	27	1.684	30	6.31
	60	-	-	-	-	-	-	3.255	50	2.488	25	1.687	25	6.48
50.3 Al₂O₃ + 26.3 AT + 23.4 β-spodumene (A-AT-S)	0	4.71	8	3.350	24	2.660	20	-	-	-	-	-	-	4.21
	10	4.71	7	3.351	30	2.659	20	3.232	30	2.481	15	1.684	20	4.36
	40	4.72	5	3.351	25	2.660	10	3.243	30	2.481	20	1.687	35	4.21
	60	4.71	3	3.350	20	2.654	10	3.243	50	2.488	30	1.687	40	4.30
	100	4.71	2	3.351	15	2.637	9	2.255	23	2.495	13	1.689	15	4.29

*) According to the ASTM X-ray file the most intense reflections for Al₂TiO₅ are 4.71Å–42 %, 3.36Å–100 %, 2.65Å–64 % (card 26-40); TiO₂: 3.25Å–100 %, 2.49Å–50 % and 1.69Å–60 % (card 21-1276).
**) Matched with reflection of Al₂TiO₅ – 1.688Å–10 %.

It follows from Table II, Al₂TiO₅ decomposes during the early stages of annealing and after 60-hour soak time has almost completely transformed to TiO₂ and Al₂O₃. In compositions containing Al₂O₃, Al₂TiO₅ reacts rather rapidly with Al₂O₃, though Al₂TiO₅ is still present in small amounts even after 60-hour of soak time. In A-AT-S composition stability of Al₂TiO₅ is much lower.

b) Influence of aluminum titanate on sintering and basic properties of Al₂O₃.

Table III. Sintering and TEC of compositions Al₂O₃+ stabilized Al₂TiO₅

No	Contents of AT, Mass %	Linear shrinkage, Δl,%	Density d_v, gr/cm³	Water absorption, %	Porosity, %		Bulk density, d_d, gr/cm³	TEC, $10^{-6}K^{-1}$, at 100-700 °C
					open	general		
*)1.	0	24.2	3.320	1.5	4.98	17.6	4.030	8.16
2.	3	9.4	2.670	2.52	6.73	-	-	7.81
3.	10	9.38	2.584	2.39	6.18	34.8	3.963	6.89
4.	20	9.23	2.536	2.90	7.35	35.0	3.899	5.31
5.	30	11.39	2.862	0.700	2.00	25.5	3.842	4.01
6.	40	12.31	3.045	0.481	1.46	17.1	3.674	2.85

*) Soaking conditions Al₂O₃ – 1650 °C for 2 hours.

The influence of soak time and Al₂TiO₅ content on concentrations ranging from 3 to 40 mass % of Al₂TiO₅, is shown in Table III.

The Al_2TiO_5 concentration dependence on the TEC in the A-AT system (Fig. 2) is linear and may be described empirically by the following equation:

$$\alpha_{100-700} = (8.16 - 0.1328 \text{ x C}) \cdot 10^{-6}\text{K}^{-1} \dots\dots (1)$$

where $\alpha_{100-700}$ is the TEC in the interval of 100-700 °C, $8.16 \times 10^{-6}\text{K}^{-1}$ is the TEC of Al_2O_3 in interval of 100-700 °C, and C is the concentration of AT in mass %.

The TEC results shown in Fig. 2 decrease faster than the law of mixtures would dictate. If equation (1) is extrapolated to 100% AT the TEC would be -5.12×10^{-6} °K^{-1} compared with the TEC of AT in the temperature range quoted above ($\alpha_{AT}=1.97 \times 10^{-6}$ °K^{-1}.) Though the TEC of pure AT is almost half an order magnitude less negative, than TEC of β-eucryptite glass ceramics, its influence on TEC of the A-AT composite is greater [6]. This can be explained as follows:

- There is an abrupt decrease in the absolute value of the TEC of β-eucryptite glass ceramics as a result of fine grinding and re-crystallization after melting,
- There is a decrease in concentration of β-eucryptite as a result of its interaction with SiO_2 (milling),
- There is an additional decrease of TEC in composition of Al_2O_3 with Al_2TiO_5 as a result of interaction of Al_2O_3 with SiO_2 (milling) with the formation of mullite.

In addition to corundum (α-Al_2O_3), theolite (Al_2TiO_5) and β-spodumene ($Li_2O \cdot Al_2O_3 \cdot 4SiO_2$), x-ray diffraction showed compositions of Al_2O_3+AT also contain mullite ($3Al_2O_3 \cdot 2SiO_2$).

DTA curves of stabilized Al_2TiO_5 and of the initial mix of Al_2O_3 + 40 mass % of AT composition (Fig.3 curve1,2) clearly shows an exothermal peak (~ 970 °C) and two endothermic peaks (1300 and 1380 °C). Absence of corresponding effects on TGA and DTG curves suggests that they are caused either by polymorphic transformations, or by crystallization and fusion. The presence of spodumene in the structure of AT, as well as presence of a second endothermic peak on the DTA near its fusion temperature (1380 ± 10 °C) suggests crystallization of glass with the spodumene structure. The first endothermic peak coincides with temperature of fusion of β-eucryptite (1300 ± 10 °C) and suggests that there is a certain amount of β-eucryptite in the AT. Absence of x-ray peaks for β-eucryptite suggests the quantity of these two is insignificant.

Table IV. Flexure strength dielectric properties and estimation of thermal cycling resistance of sintered samples of Al_2O_3, containing stabilized Al_2TiO_5

No	Contents of TA, Mass %	σ_{flex}, MPa	σ_{comp}, MPa	Log ρ, Ohm·cm, at 150 °C	ε at 10^6 Hz and 25 °C	σ_{flex}, (MPa) after 5-fold thermo-cycling to 1000, 1100, 1200 °C then into running water
1.	0	221	-	15.8	10.0	failed above 1000 °C
2.	10	42.1	327	13.6	7.8	15-20
3.	20	24.0	181	13.8	8.7	≤ 15
4.	30	29.8	351	13.5	9.8	≥ 25
5.	40	29.8	436	13.4	9.8	≥ 20

DTA, TG and DTG curves of a sintered sample with "Al_2O_3 + 40 mass % of AT" show the absence of any of these peaks up to 1500 °C. Such a result favors good thermal shock resistance. The process of mullite formation during firing proceeds in a solid phase, and is characterized by low rates; consequently it is not observed on DTA curve of the initial mix. The microstructure of an annealed sample A + 40 AT (Fig. 4) is quite homogeneous. With 100 x magnification (Fig. 4a) one can observe numerous pores with sizes up to 200 microns, distributed in the volume of a sample almost in regular intervals. At high magnification (Fig. 4, b and c) we see the homogeneous mass without internal structure, which suggests undetectable crystallite (sizes less than 0.5 microns).

Mechanical and dielectric properties, as well as preliminary results on thermal resistance of sintered samples of Al_2O_3, containing Al_2TiO_5, are shown in Table IV.

It follows from Table IV that the flexure strength of compositions, containing AT, are rather poor. This is, of course, due to their high porosity and probably large number of microcracks, caused by hysteresis. The thermal cycling resistance of these materials is quite good likely due both to the low thermal expansion coefficient and their very porous nature. This must also mean that there is reasonably good thermal compatibility between components with high difference of TEC (about $10 \cdot 10^{-6}K^{-1}$) when then are homogeneously mixed. Further development of this material to obtain high density ceramics is the next step in this project and will determine whether highly dense composites will also have high thermal cycling resistance.

CONCLUSIONS AND ADDITIONAL OBSERVATIONS:

1. By using high purity oxides, choosing the proper conditions of synthesis, and also by combining reactive components of Al_2TiO_5 SiO_2 and Li_2O in appropriate amounts, it is possible to achieve material with quite low TEC $- -.79 \cdot 10^{-6}K^{-1}$ at 20-700 °C.

2. Li_2O and SiO_2 added to Al_2TiO_5 form spodumene. In the process of firing of alumina compositions the spodumene melts, covering parts of Al_2O_3 with a film of Al_2TiO_5.

3. The TEC of compositions of $Al_2O_3 + Al_2TiO_5$ decreases linearly with a rising concentration of Al_2TiO_5.

4. Mechanical strength, dielectrical and other properties of composites worsen in comparison with Al_2O_5. Sintering and thermal shock resistance of the compositions sharply improve.

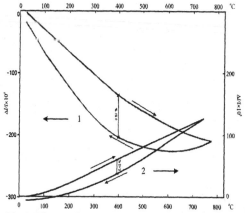

Fig. 1. Dilatometric curves of (1) stabilized Al_2TiO_5 and of (2) Al_2O_3+40 mass % Al_2TiO_5 during heating and cooling.

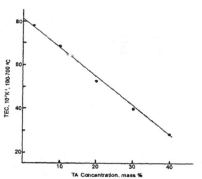

Fig. 2. TEC of composition in Al_2O_3 - stabilized Al_2TiO_5 system.

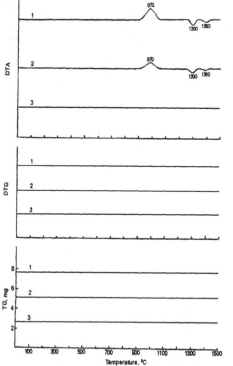

Fig. 3. The DTA, TGA, DTG of powdery samples:
1 - Stabilized Al_2TiO_5 (AT),
2 - Initial mixture of 60 mass % Al_2O_3 + 40 mass % AT,
3 - Sintered composition 60 mass % Al_2O_3+40 mass % AT.

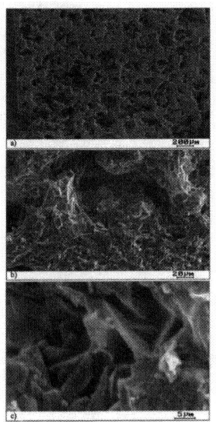

Fig. 4. SEM pictures of sintered compositions containing 60 mass % Al_2O_3+40 mass % AT.

174

REFERENCES

[1]W. D. Kingery, H. K. Bowen, D. R. Uhlmann, "Introduction to Ceramics", New York: John Wiley and Sons, 1976.

[2]T. N. Tiegs and P. F. Becher, "Thermal Shock Behaviour of Alumina-SiC Whisker Composite", *J. Am. Ceram. Soc.* 70(5) C 109 - C 111 (1987).

[3]G. Schneider and G. Petzov (edc), "Thermal shock and Thermal Fatigue of Advanced Ceramics", Kluwer Academic Publishers, Pordrecht, The Netherlands, p.p 37-48, 1993.

[4]Joachim H. Shneibel, Stephen M. Sabol et al., "Cyclic Thermal Shock Resistance of Several Advanced Ceramics and Ceramic Composites", *J. Am. Ceram. Soc.*, 81(7), 1888-1892 (1998).

[5]Л. Д. Ландау, Е. М. Лифшиц, "Теория упругости ", т. 7, М., "Наука", с. 246, 1987.

[6]J. Geodakyan, A. Kostanyan, K Geodakyan, S. Sagatelyan, "The Influence of β-Eucryptite Glass Ceramic on the Structure and Main Properties of Alumina Ceramic", The 28th International Conference & Exposition on Advanced Ceramics & Composites, Abstract Book, p. 103, 2004.

[7]Д. А. Геодакян, А. К. Костанян, К. Д. Геодакян, "Преодоление термической несовместимости компонентов с высокой разностью ТКЛР в керамических композициях", Тезисы докладов XVII научно-технической конференции, стр. 45, 12-14 октября 2004г.

[8]S. M. Lang, C. L. Fillmore, L. H. Maxwell, JOM. Kes. Nat. Bur. St. 48[4], 1952, 298.

[9]A. E. Austin, S. M. Schwaitz, Acta Cryst. 6(10), 1953.

[10]А. С. Бережной, Н. В. Гулько, Сб. научн. работ по химии и технологии силикатов, М., Госстройиздат, с. 217-234, 1956.

[11]А. С. Бережной и Н. В. Гулько, Укр. хим. журн. XXI [2], 158, 1955.

[12]В. А. Брон, "Огнеупоры" (7), 312, 1951.

[13]D. Asmi, I. M. Low, Use of Spodumene for Liquid-Phase-Sintering of Alumina / Calcium-hexaluminate Composites to be Presented at Powder Metallurgy World Congress and Exhibition 2000, Nov. 12-16, 2000, Kyoto, Japan.

[14]D. Asmi, P. Manurung, R. D. Skala, I. M. Lowo, B. H. O'Connor, S. J. Kennedy, Depth - Profiling of Phasa Compositions and Developments in Graded Alumina / Calcium-Hexaluminate and Alumina / Aluminium-Titanate Composites to be presented at Powder Metallurgy World Congress and Exhibition, 2000, Nov. 12-16, 2000, Kyoto, Japan.

[15]I. J. Kim, "Thermal shock resistance and thermal expansion behaviour of Al_2TiO_5 ceramics prepared from electrofused powders", Journal of Ceramic Processing Research, Vol. 1, No 1, p.p. 57-63 (2000).

[16]Д. С. Белянкин, В. В. Лапин, ДАН СССР, LXXX [3], 421, 1951.

[17]А. А. Русаков, Г. С. Жданов, ДАН СССР, LXXVII[3], 411, 1951.

[18]Г. С. Жданов, А. А. Русаков, ДАН СССР, LXXXII[6], 901, 1952.

[19]W. R. Bussem, N. R. Thielke, R. V. Sarakauskas-Cer. Age. 60[5], Nov. 38, 1953.

[20]Н. Е. Филоненко, В. И. Кудрявцев, И. В. Лавров, ДАН СССР, LXXXVI[3], 561, 1952.

[21]A. M. Lejus, D. Goldberg and A. Revcolevschi, C. R. Seances, Acad. Sci. Ser. C, 263[20], 1223-1226 (1966).

[22]M. Hamelin – Compt. Kend, 238, 1896, 1954.

[23]Бабаян С. А., Костанян К. А., Геодакян Дж. А., "Дилатометрическое изучение твердых растворов на основе титаната алюминия", Армянский Химический Журнал, XXVI №7, стр. 549, 1973.

[24]Бабаян С. А., Костанян К. А., Геодакян Дж. А., "Получение материалов с отрицательным коэффициентом термического расширения", Тезисы докладов Всесоюзной Конференции по синтезу и исследованию термостойких неорганических соединениов на основе окислов металлов, часть I, стр. 78, 1972.

[25]DWP 29.794, (16/I-1962, 25/IX-1964), H. Beng.

[26]DAS 1.238.376 (29/I-1964, 6/IV-1967), A. Rosental, Sh. Erfinder, P. Bock.

[27]W. Hannes, Silakattechnik, 9, 304, 1970.

[28]A. I. Lichtenstein, R. O. Jones, H. Xu and P. J. Hoaney, "Anisotropic Thermal Expansion in the β-eucryptite: A Neutron Diffraction and Density Functional Study", PHYS. REV. B 58, 10, 6219 (1998).

[29]T. Kosmac, A. Daskobler, Josef Stefan Institute, Slovenia Meeting Guide, p. 102, 2004.

[30]O. Kubashevski and C. B. Alcock, Int. Ser. On Materials Science and Technology, vol. 24, 5th ed, 449 p., Pergamon Press / Elsevier Science Ltd, Oxford, United Kingdom, 1979.

[31]D. Goldberg, Rev. Int. Hautes Temp. Refract., 5[3], 181-194, (1968).

[32]Rafael Uribe and Carmen Baudin "Influence of a Dispersion of Aluminum Titanate Particles of Controlled Size on the Thermal Shock Resistance of Alumina", *J. Am. Cer. Soc*, v. 8, N 5, (2003).

[33]Masahide Takahashi, Masahiro. Fukuda, Masaaki. Fukuda, Hisato Fukuda, and Toshinobu Yoko, "Preparation, Structure, and Properties of Thermally and Mechanically Improved Aluminum Titantate Ceramics with Alkali Feldspar," *J. Am. Ceram. Soc.* 85 [12] 3025-30 (2002).

STRUCTURING CERAMICS USING LITHOGRAPHY

Michael Schulz, Jürgen Haußelt, Richard Heldele
Institute of Materials Research III
Research Center Karlsruhe
P.O. Box 3640
Karlsruhe, 76021
Germany

Albert-Ludwigs-Universität Freiburg
Institute of Microsystem Technology
Georges-Köhler-Allee 102
Freiburg, 79110
Germany

ABSTRACT

Due to their outstanding chemical stability the application of ceramic micro components is becoming more and more important. Different replication techniques like low or high pressure injection molding of ceramic feedstocks have been established in the industrial process. The success of microelectronics and microsystem technologies is based on the development of various lithographic methods. These can be used to produce very precise micro patterned components from plastic, ceramic, or metal with minimal line width.

Silicon organic compounds containing the elements silicon, nitrogen, and carbon in the polymer backbone can serve as precursors for Si_3N_4, SiC or Si-C-N ceramics depending on the pyrolysis atmosphere. These so-called preceramic polymers react as negative type resists in lithography. By applying lithography, ceramic micro components can be produced using established photo resins filled with ceramic powder as well as preceramic polymers. To guarantee dimensional reproducibility, substrates made from the same material were manufactured. By adding ceramic powder to the polymeric matrix and adapting the dispersion method the properties of the resulting ceramic microparts were optimized. Different ceramic micro test structures were produced from unfilled and ceramic powder filled preceramic polymers. High aspect ratios with smallest structural details of several micrometers were achieved depending on the lithographic method (UV- or X-ray) and the masks used. An overview over the process and the current results will be given.

INTRODUCTION

There are only a few techniques, besides electrophoretic deposition, low- and high pressure micro injection molding, that offer the potential for manufacturing ceramic microparts. Various lithographic methods are known from microsystem technologies that allow for the fabrication of high aspect ratio components with minimal line width. This technique is the first step in the well-known LIGA process that is used for the manufacturing of highly accurate mold inserts with best surface properties needed e.g. for injection molding. To transfer a pattern, radiation must strike a photosensitive material changing the properties of that material in such a way that a replica of the mask is left on the surface of the substrate after the lithography process is complete. The photosensitive compound used is called photoresist or simply resist.

Former work demonstrated that ceramic microstructures can be achieved using both, established SU-8 photoresist filled with ceramic powder [1,2] as well as preceramic polymers [3,4], in lithographic processes. It was shown that the investigated ceramic precursors react as a negative resist under X-ray lithographic conditions [3]. The knowledge of the cross linking mechanism for optimization of lithographic parameters is essential. The photochemical cross linking was investigated by various spectroscopic methods [4].

Ceramic precursor materials often show massive shrinkage during pyrolysis. This causes the destruction of the microstructures and coatings on substrates (e.g. Al_2O_3, Si) by crack formation [5,6]. Hence, to guarantee dimensional reproducibility, new substrates made from the same material were manufactured using casting processes. Adding silicon nitride powder to the preceramic polymeric matrix leads to lower shrinkage in the pyrolysis step. On the other hand, the depth of penetration for UV-light decreases extremely due to scattering and absorption by the filler particles, i.e. this type of radiation is not suitable. The X-rays produced by a synchrotron radiation source [7] are able to penetrate the powder particles since the mass absorption coefficient of Si_3N_4 is relatively low [8]. It was shown that the way of conditioning the dispersions strongly influences the properties of the concluding microparts. Depending on the lithographic masks used in the process different micro test structures were produced.

Established Lithographic Process

The established UV- and deep X-ray lithography process allows for the microstructuring of commercially available photo resists (e.g. SU-8) with smallest structural details in the micrometer range.

The SU-8 resin is a so-called high contrast negative photoresist designed for applications where tall, high aspect ratio microstructures are desired. This photoresist is sensitive in the near UV (350-400 nm) and E-beam regions and is also being used in X-ray and deep X-ray lithography. Layer thickness up to 2 mm can be processed and aspect ratios up to 100 were demonstrated with X-ray lithography [9]. The resist consists of three basic components, an epoxy matrix, a photoinitiator, and a solvent that is used to adjust the viscosity to achieve the required resist thickness through the coating process. Composites based on this material system were used to produce microstructures made from Al_2O_3 with aspect ratio up to 16 recently [1,2].

The steps in a photolithographic process are listed in Fig. 1. The substrate is coated with photoresist by various casting or coating techniques after it was pretreated to remove contamination. After coating the solvent is evaporated from the resist in a softbake step. The material is then exposed and subsequently developed in an appropriate solvent. In case of SU-8 resist an additional so-called post exposure bake (PEB) is performed to initiate the cross linking after exposure. After drying the resist can be stabilized by a further hardbake step. Negative type photoresists like SU-8 and preceramic polymers react by cross linking in the exposed areas during exposure to the electromagnetic radiation used. Further information on resist processing in given in the literature [10].

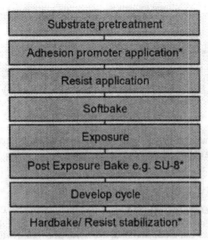

Fig. 1: Typical process flow in a photolithography, *marks optional steps.

Preceramic Polymers as Resist Material

Several silicon containing polymers that serve as precursors for Si-C-N ceramic were developed in the past [5,11]. The photo induced cross linking of a selection of these preceramic polymers was investigated. The selection of suitable precursors was based on the criteria listed in Tab. 1. From Tab. 2 it can be seen that none of the examined polymers fits all requirements. Several other polymers were examined but are not mentioned here because they were not applicable at all (amongst them Si-B-C-N precursors).

Tab. 1: Required physical and chemical properties with respect to lithography.

a) solid aggregate state under ambient conditions; in case of UV-lithography an adjustable highly viscous state was acceptable
b) soluble in standard aprotic organic solvents
c) formation of stable films on substrate surfaces
d) processability at ambient conditions for several hours, i.e. improved chemical stability against oxygen and moisture
e) ready and selective solubility in developing solvents
f) high sensitivity against photo chemical induced cross linking throughout the presence of chromophoric side groups
g) availability and small batch variation

179

Tab. 2: Investigated ceramic precursors and granted properties with reference to Tab. 1; ++ = very good suited, - not suited.

	Vendor	a)	b)	c)	d)	e)	f)	g)
NCP-200 [3,12]	Nichimen Corp.	+	+	+	+	+	+	-
SLM465012VP [12]	Wacker	+	+	+	+	+	+	-
Ceraset [3,13]	KiON Corp.	-	+	-	+	+	+	+
VL20 [3,13]	KiON Corp.	-	+	-	+	+	+	+
ABSE [3,4,13]	IMA Bayreuth, Germany	+	+	+	++	-	-	0

It is possible and reasonable to raise the photo chemical sensitivity of the preceramic polymers by adding a photo active compound as the photo initiator mentioned for SU-8.

EXPERIMENTAL

Substrates from Preceramic Polymer and Ceramic Filler

The massive shrinkage of the ceramic precursor during pyrolysis often leads to the destruction of microstructures adherent to the substrate. By mixing the polymer with $30 - 70$ wt.% of α-Si_3N_4-powder (Fa. UBE SN-E10, $d_{50} = 0.5$ µm) related to the polymer content the ceramic yield was increased. Two different preceramic polymers were used in the experiments, the liquid Ceraset[©] and VL20 respectively and the solid ABSE precursor in a 50 wt.% solution in n-octane. Powder particles were deagglomerated by using both, a dissolving stirrer ensuring high shear rates and an ultrasound disintegrator (Fa. Branson Sonifier II W-450). Particle size measurements were performed on ethanolic suspensions with a Coulter LS230 laser diffractometric instrument. No preceramic polymer was added to prevent the equipment from damage.

Delamination of the structures occurred during the concluding pyrolysis step because the alumina-substrates used were not able to follow the shrinkage. To handle this effect, substrates were produced from composites containing Si_3N_4-powder in the preceramic polymeric matrix using a laboratory tape casting device.

The construction of the tape casting device is comparable to the doctor blade principle with a moveable storage slide and a static carrier foil. For the experiments the gap for the emerging suspension was set to 800 µm. The thickness of the resulting tape depends among others on the hydrostatic pressure in the storage tank. Therefore, the level of the suspension was kept constant at ca. 6 mm by steady addition. The take-off speed of the slide was set constantly to 2 mm·s^{-1}. A PTFE foil that was carefully cleaned with acetone served as carrier tape and was supported by a flat metal plate. After tape casting the assembly consisting of the metal plate support, the PTFE foil and the green tape was put in a chamber oven at 50 °C to accelerate the drying. To prevent the substrates from swelling or dissolving during the developing step of the lithographic process, the tapes were thermally cross linked at 250 °C. The influence of the ultrasound disintegration on density and porosity was investigated using the hydrometer method. Vickers hardness was determined using a micro hardness test equipment (Fa. Paar Physika MHT-10). Surface roughness was measured with the FRT Microglider System using a CHR-150 sensor based on a chromatic coded principle.

The substrates were cut to the recommended measure in the green state and polished after the thermal cross linking using grinding machines (Fa. Buehler Phoenix Beta and Pheonix 4000). By this, plane parallel platelets with a thickness between 0.5 and 1 mm were obtained. The platelets were coated with either a solution of $45 - 85$ wt.% ABSE in n-octane or the above mentioned suspensions of silicon nitride in a solution ABSE respectively. Spincoating or casting was, therefore, used allowing for resist thickness between 20 and 500 µm.

After exposure of the specimen using a lithographic test mask and an appropriate synchrotron beam line (for further details see [4]) the microstructures were developed by rinsing with n-octane or acetone. Subsequently the microstructures could be pyrolysed at temperatures up to 1200 °C.

RESULTS

Influence of Suspension Composition and Treatment

The treatment of the suspensions is of great influence on the porosity. Fig. 2 compares the open porosity of pyrolysed specimens when the starting suspensions were deagglomerated with and without ultrasound, respectively. The ultrasonic treatment reduced the residual open porosity significantly.

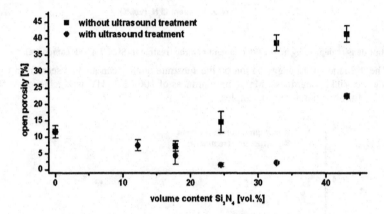

Fig. 2:   Influence of the treatment on the porosity of pyrolysed samples (Si_3N_4 / ABSE composite).

The raw density depending on the filler content and the treatment is shown in Fig. 3. For comparison the variation of the theoretical density of the Si-C-N/Si_3N_4-ceramic is plotted. For calculation a theoretical density of the amorphous Si-C-N-ceramic $\rho = 2.0$ g·cm^{-1} and of the crystalline α-Si_3N_4 $\rho = 3.2$ g·cm^{-1} was assumed. If no ultrasound was applied to the starting suspension the raw density decreased far below the theoretical density with increasing filler content.

181

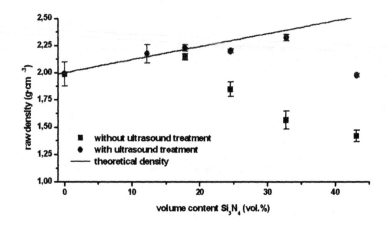

Fig. 3: Raw density depending on the filler content and the treatment (Si₃N₄ / ABSE composite).

The influence of the aggregation on the substrate quality can also be seen from Fig. 4 by comparing the Vickers hardness. Maximum hardness of 1068 ± 36 HV was achieved at 50 wt.% (24.5 vol.%) for ultrasound treated samples.

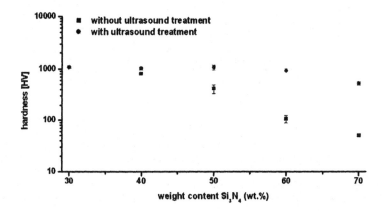

Fig. 4: Dependency of the micro hardness on filler content and treatment (Si₃N₄ / ABSE composite).

DISCUSSION

By dispersion of the highly agglomerated Si₃N₄-powder in ethanol with an ultrasound disintegrator the average particle size could be lowered by a factor of 10. This caused a reduction of the sedimentation velocity by a factor of 100 (Eq. 1).

182

$$v_s = \frac{2 \cdot r^2 \cdot (\rho_p - \rho_l) \cdot g}{9 \cdot \eta_l}$$

Eq. 1: Sedimentation velocity v_s of a spheric particle (radius r, density ρ_p) in a liquid (viscosity η_l, density ρ_l) assuming laminar flow (g = acceleration due to gravity) [14].

The reduction of the particle size is expected to be smaller in a solution of preceramic polymer in n-octane. The ABSE polymer chains act as a binder enclosing the particles and, therefore, tend to stabilize the dispersion sterically (Zipper-Bag-Theory [15]). Ultrasonic treatment of the suspensions allowed for the minimization of the deflection of the platelets after the pyrolysis that was caused by sedimentation of the filler particles during the drying period of the casted tapes.

The average surface roughness of the substrates in the green state (R_q = 0.2 µm) was promising and did not increase significantly after pyrolysis. The extended deagglomeration caused a significant decrease in the porosity (compare Fig. 5). For a filler content of 32.5 vol.% (60 wt.%) the porosity was lowered by a factor of 16. The raw density almost corresponded to the theoretical values for filler contents below 43 vol.% (70 wt.%). Considering the residual open porosity the maximum Vickers hardness achieved is acceptable compared to dense sintered silicon nitride (SSN HV_{10} = 1400).

 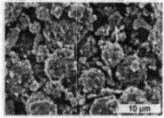

Fig. 5: SEM top view of the pyrolysed substrate surface. Starting tape casting suspensions treated with (right) and without ultrasound (left) (50 wt.% Si_3N_4-powder in ABSE matrix).

A sample microstructure of unfilled resist material on a substrate made from ABSE + 50 wt.% Si_3N_4 by tape casting is shown in Fig. 6. It can be seen that the deflection of the star structure on top of the substrate is caused by the mismatch of shrinkage that is still present during pyrolysis up to 1200 °C. The good adhesion between the substrate surface and the preceramic resist hindered the structure from delamination. Thus the cracks in the substrate surface especially appear at the tips where the occurring tensions are large. The low surface roughness and the excellent edge precision[**] are obvious, too.

[**] edge precision - the sidewall of the structures are perpendicular to the surface of the substrate and the pattern respectively.

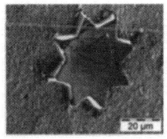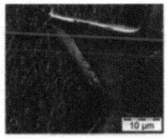

Fig. 6: Crack formation of the unfilled structured resist material (pure ABSE) on a powder filled substrate after pyrolysis (SEM picture, left: overview, right: detail view).

CONCLUSION

Suspension treatment aiming at the optimized deagglomeration is indispensable for lithographic microstructuring of resists based on composites of a photosensitive polymer and ceramic filler. To guarantee an accurate reproduction of details the particle size must be as small as possible. Substrates that allow for the shrinkage of the resist without crack formation, delamination or deflection during the concluding pyrolysis step are necessary. Composites based on preceramic polymers and Si_3N_4-powder were produced allowing for both, the use as substrates (Ceraset© or ABSE matrix) and as a resist material (ABSE matrix). Resist layers with a thickness up to 500 µm were produced and lithographically microstructured. It was shown that aspect ratios up to 60 and structural details around 5 µm could be realized.

ACKNOWLEDGEMENT

The authors would like to thank M. Boerner from the Institute of Microstructure Technology of the Research Center Karlsruhe for support with the X-ray lithography, and G. Motz from IMA Bayreuth for providing the ABSE material.

REFERENCES

[1] C. Mueller, T. Hanemann, G. Wiche, C. Kumar, J. Goettert, „Fabrication of Ceramic Microcomponents Using Deep X-ray Lithography", 5th Intern. Workshop on High-Aspect-Ratio Micro-Structure Technology, 15.-17.06, 2003, Monterey, CA, USA.

[2] C. Mueller, T. Hanemann, G. Wiche, C. Kumar, J. Goettert, „Fabrication of Ceramic Microcomponents Using Deep X-ray Lithography", Microsystem Technologies, **2005**, in press.

[3] M. Schulz, M. Boerner, J. Goettert, T. Hanemann, J. Hausselt, G. Motz, "Cross Linking Behaviour of Preceramic Polymers Effected by UV- and Synchrotron Radiation", *Advanced Engineering Materials*, **6**(8), 676-80 (2004).

[4] M. Schulz, M. Boerner, J. Hausselt, R. Heldele, „Polymer derived ceramic microparts from X-ray lithography – cross-linking behavior and process optimization", *Journal of the European Ceramic Society*, **25**(2–3), 199-204 (2004).

[5] Kroke, Li, Konetschny, Lecomte, Fasel, Riedel, "Silazane derived ceramics and related materials", *Materials Science & Engineering*, **R26**(4-6), (2000).

[6] Y. Li, E. Kroke, R. Riedel, C. Fasel, C. Gervais, F. Babonneau, "Thermal cross-linking and pyrolytic conversion of poly(ureamethylvinyl)silazanes to silicon-based ceramics", *Applied Organometallic Chemistry*, **15**(10), 820-32 (2001).

[7] H. Winick (Editor), "Synchroton Radiation Sources"; *World Scientific Publishing Company Pte. Ltd.*, Singapur (1994)

[8] P. Meyer, J. Schulz, L. Hahn, *Rev. Sci. Instrum.*, **74**(2), 1113-19 (2003)

[9] A. Bogdanov, S. Peredkov, "Use of SU-8 photoresist for very high aspect ratio X-ray lithography", *Microelectronic Engineering*, **53**, 493-96 (2000).

[10] S.A. Campbell, "The Science and Engineering of Microelectronic Fabrication", *Oxford University Press*, (1996).

[11] G. Motz, J. Hacker, G. Ziegler, "Special Modified Silazanes for Coatings, Fibers, and CMCs", *Ceramic Engineering and Science Proceedings*, **4**(21), (2000).

[12] T. Hanemann, J. H. Hausselt, "UV- & Deep X-ray Lithography on Preceramic Polymers", *Polymer Preprints*, **39**(2), 659 (1998).

[13] T. Hanemann, M. Ade, M. Börner, G. Motz, M. Schulz, "Microstructuring of Preceramic Polymers", *Advanced Engineering Materials*, **4**(11), 869 (2002).

[14] H.H. Hahn, „Wassertechnologie - Fällung, Flockung, Separation", *Springer-Verla,* Berlin, Heidelberg, New York, (1987)

[15] R.E. Mistler, E.R. Twiname, "Tape Casting – Theory and Practice", *The American Ceramic Society* (2000)

THE NOTION OF DENSIFICATION FRONT IN CVI PROCESSING WITH TEMPERATURE GRADIENTS

Gerard L. Vignoles,
Lab. des Composites ThermoStructuraux (LCTS),
Université Bordeaux 1 − 3, Allée La Boëtie,
F 33600 Pessac, France

Nathalie Nadeau, Claude-Michel Brauner,
Mathématiques Appliquées de Bordeaux (MAB),
Université Bordeaux 1 − 351, Cours de la Libération,
F 33405 Talence Cedex, France

Jean-François Lines, Jean-Rodolphe Puiggali,
TRansferts Ecoulements FLuides Energetique (TREFLE),
ENSAM − Esplanade des Arts et Métiers,
F 33405 Talence, France

ABSTRACT

Many variants of the Chemical Vapor Infiltration (CVI) process make use of a thermal gradient, which allows to localize the densification reaction first in the parts of the preform which are most difficult to reach by precursor transport. When the process is well controlled, a densification front starts from these regions and moves towards the outside of the preform ; however, the control of this front is difficult.

We propose a simple 1D model of this front, based on the local energy and mass balances. The analytical and numerical study yield existence conditions and relate the front characteristics (*i.e.* width, velocity, residual porosity) to control parameters. The model results are validated with respect to experimental data, and process control guidelines are given.

INTRODUCTION

Chemical Vapor Infiltration is one of the main routes for the processing of carbon and ceramic matrices in thermostructural composites[1,2] . Its principle rests on the heterogeneous deposition reaction of a gaseous precursor inside the preform, a porous medium constituted by the fibers which are intended to reinforce the future material. The first implementations of CVI − still in use today − were carried out in isothermal conditions. It has been shown experimentally are theoretically that isothermal CVI (I-CVI) involves the competition between gas diffusion and reaction. A straightforward Thiele modulus analysis implies then that moderate temperature and low pressure are adequate conditions to avoid premature pore plugging and thus to preserve the material quality, in terms of matrix deposition homogeneity. The drawback of such conditions is that the chemical kinetics are dramatically slow, turning I-CVI a very lengthy and expensive process.

A variety of CVI modifications have then been designed in order to overcome this limitation. Most of them rely on the use of a thermal gradient, because this can help monitoring the chemical reaction and make it happen at the right place first − that is, at the

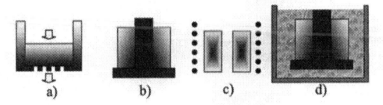

Figure 1: Thermal-gradient variants of CVI. a) F-CVI ; b) "Rapid process" ; c) RF-CVI, d) "Film-boiling" process.

center of the preform, or at its extremity which lies farthest from the impinging gas flux[3,4] . They are summarized at fig. 1. The forced CVI (F-CVI) process was initially based on the intent of accelerating chemical deposition by forcing the gases through the pores with a pressure drop[2,5-7] , but it has been rapidly found out that the superposition of a thermal gradient would enhance the quality and processing time. Golecki *et al.* have implemented a "rapid densification" process[8] based on isobaric conditions and the creation of a "hot side" and "cold side" on the preforms. Impeding the gas entrance by the hot side helps in densifying the pore bottom at first and the pore mouth at last, as wanted. In practice, one uses a central inductor or resistor around which the preforms are settled. Thermal-gradient isobaric CVI (TG-CVI) has also been considered with in-situ heat production, like microwave heating (MW-CVI)[9-11] or radio-frequency induction heating (RF-CVI)[12-14] , with the idea of creating a hot deposition zone directly at the preform center. Houdayer *et al.*[15] have implemented a "film-boiling" process, also called Kalamazoo, which consists in having the preform, heated as in the preceding processes, merged into boiling precursor. This helps to maintain its surface temperature at a constant, well-known value, and ensure a very strong thermal gradient[16] . All these CVI variants have the advantage of allowing to work with higher pressures, higher temperatures, and consequently higher rates and lower processing times. Moreover, the engineers experience is that such processes were difficult to control, and that undesired behaviors would show up.

PROCESS BEHAVIOR

The urgent need for CMC parts with constant density has motivated numerous modelling works and experimental studies. In many cases, it has been possible to model successfully the processing cycle and propose optimization guidelines. However, some cases in CVI with thermal gradients do not behave optimally. Trying to understand these ill-behaved cases may be a very useful tool for a better overall comprehension and optimization of TG-CVI in general.

Indeed, reports on the "rapid process"[17] evidenced that, after a correct initial phase where densification would start from the hot side as expected, the density would eventually stop growing there, while keeping on increasing closer to the cold side. The resulting material has not an optimized density. Some runs of computed and measured infiltrations by the Georgia Tech team [18] also gave nonuniform density gradients, with a density maximum lying somewhere between the hot and the cold zone – if not sometimes at the cold side extremity. In the case of CVI with volume heating, a similar behavior occurred. Morell *et al.*[10]

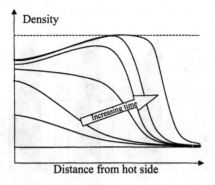

Figure 2: Density evolution with time in many thermal-gradient CVI experiments

report a correct start of densification in the middle of the preform -as can be expected from the temperature field which results from the balance between volumic heating and surface radiative losses- followed by the apparition of a density maxima, located in a circular region, surrounding a less dense center, and finally of a densification front propagating towards the preform periphery. This behavior has been found again by Leutard *et al.*[14] in another RF-CVI setup, both numerically and experimentally. An MW-CVI model[9] also showed the same feature. Figure 2 schematizes this rather frequent scenario. Again, the resulting material is not adequate. Reports on the forced-CVI [5] and on the "film-boiling" process have mostly shown well-densified materials [16] . Indeed, the densification front seems to exist since the beginning and until the end of the process. However, this process has other drawbacks, chiefly the strong energy consumption that underlies the presence of extreme gradients, and the necessity to handle flammable liquids together with a strong energy source. A detailed modelling study[20] was carried out on this process, and validated with respect to experimental results in resistive-heating configuration[21] . Fig. 3 is an illustration of the comparison between numerical prediction of the density evolution, and experimental determination using X-ray radiographs. Then, the model has been utilized in order to optimize the process with respect to the thermal gradient : if it is too large, then the process consumes unnecessary amounts of energy, but if it is too small, the density of the piece is not optimal. Fig. 4a) is plot of the overall mass uptake after 5h. processing as a function of the cold side temperature, the hot side temperature being held constant ; fig.4b) displays density profiles after 5h. for three selected cold side temperatures. There are indeed three possible behaviors : *i*) a full densification front when the thermal gradient is strong enough , *iii*) the I-CVI situation when the thermal gradient is not strong enough to counteract against diffusional limitations, and *ii*) an intermediate behavior that is indeed the same as depicted at fig. 2. An analysis of the heat flux during the process shows that the thermal gradient always increases, since the temperature at the cold side of the front is always the same, the temperature at the cold side of the preform is always the same, and the distance between these two points is decreasing. Evidently there is a moment at which the front begins to settle, when a critical heat flux (or thermal gradient) is achieved. The idea that arises from this study is that a total

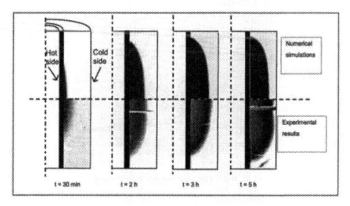

Figure 3: Validation of the numerical simulation of the "film-boiling" process

infiltration without time limitations is feasible, at the expense of controlling the densification front – especially in taking care that this front settles immediately at the beginning of the experimental run. Accordingly, a model which pays special attention to this front should be a useful tool for predicting experimental conditions that ensure a maximal density.

A 1D FRONT MODEL : EXISTENCE CRITERIA

The simplified model[22] is restricted to a 1D description of the front itself, as defined by the region where the reaction rate :

$$R = \sigma_v(\varepsilon).k(T).C^\alpha \tag{1}$$

is not zero, or at least, is non-negligible. This leads to approximate the Arrhenius equation for the rate constant $k(T)$ by a modified law in which an ignition temperature explicitly appears. In practice, this has the advantage to give a finite front zone width, while all other results are not perturbed. Also, eq. (1) clearly indicates that the hot side of the front has to be defined either by the depletion of reactant concentration ($C^\alpha \to 0$) (note that α is the reaction order) or by the lack of reacting surface area σ_v. The latter quantity is a function of open porosity ε. Two fronts are thus possible, as sketched at fig. 5. As suggested by eq. 1, the simplest model of a densification front (without forced mass transfer, thus excluding F-CVI) has to solve balance equations for three unknowns, for instance, the temperature T, the reactant concentration C, and the open porosity ε. They have the following form :

$$\rho C_p(\varepsilon, T)\frac{\partial T}{\partial t} + \nabla \cdot (-\lambda(\varepsilon, T)\nabla T) = S_{th}(\varepsilon, T) \tag{2}$$

$$\varepsilon\frac{\partial C}{\partial t} + \nabla \cdot (-D(\varepsilon, T)\nabla C) = -R(x, t) \tag{3}$$

$$-\frac{\partial \varepsilon}{\partial t} = -\Omega_s R(x, t) \tag{4}$$

where ρC_p, λ, S_{th}, D, and Ω_s are respectively the heat capacity and heat conductivity of the material, the volume heat source, the effective gas diffusivity, and the solid molar volume.

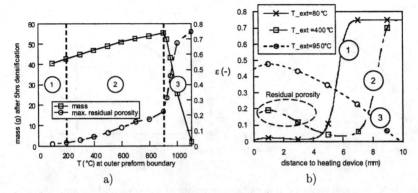

Figure 4: Different regimes of densification. a) Deposited mass after 5 h processing and maximal residual porosity as a function of the external imposed temperature. b) Typical porosity profiles in the three identified regimes.

This set of equations may be greatly simplified under some assumptions : i) the reaction order α is unity; ii) the transient times for eqs. 2 and 3 are short with respect to the densification time scale appearing in eq. 4, the corresponding terms may be neglected; iii) the volume heating term in the (narrow) front zone may be neglected with respect to the heat flux q brought from the hot side ; as a consequence, eq.2 may be immediately integrated once; iv) convection is negligible with respect to heat conduction and to mass diffusion inside the front zone ; v) the thermal dependances of heat conductivity and mass diffusivity are small with respect to the thermal dependance of the reaction rate, so they may be neglected, vi) the heat conductivity is mainly linked to the solid phase, and a law of mixtures is assumed, vii) the effective gas diffusivity is assumed to be a power law of the open porosity, as in ref. [23] , where Archie's law has been generalized :

$$D(\varepsilon) = D_0 \varepsilon^{1+m}$$

and $viii$) the surface area is a function of ε with a strong decrease [10] :

$$\sigma_v \varepsilon = A \varepsilon^{1/n}(1-\varepsilon)$$

Summarizing all hypotheses and simplifications, and defining a coordinate frame that moves with the front velocity v, a set of three ODEs is obtained :

$$-\lambda_0(1-\varepsilon)\frac{dT}{dx} \;=\; q \tag{5}$$

$$\frac{d}{dx}\left(-\varepsilon^{m+1}D_0\frac{dC}{dx}\right) \;=\; -A\,k(T_{ref})\varepsilon^{1/n}(1-\varepsilon)f(T)C \tag{6}$$

$$v\frac{d\varepsilon}{dx} \;=\; \Omega_s A\,k(T_{ref})\varepsilon^{1/n}f(T)C \tag{7}$$

Summing up eqs. 6 and 7 gives a total mass balance which can also be integrated once. The integration constant $\mathcal{K}$ is a function of the velocity and of the reactant mole flux J that

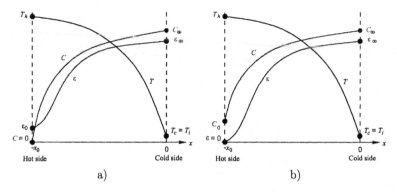

Figure 5: Two possibilities for a densification front. a) incomplete, b) complete.

feeds the front. One faces now a set of three ODEs that can be easily solved. However, the particularity of this system is that it contains more boundary conditions than necessary, and three internal parameter which are until now unspecified : the front velocity v, the front width x_0, and the residual porosity ε_0 if the densification is incomplete, or the residual precursor concentration C_0 in the converse case. This constitutes a "shooting problem" or a non-linear eigenvalue problem, which has to be tackled with a specific algorithm. The numerical computations have been validated with respect to the more complete simulation case and to experimental data.

A preliminary analysis yields a first result concerning the structure of the porous medium : if the exponents m and n are such that $m + 1/n - 1 < 0$, then both kinds of front are possible. In the converse case, only incomplete densification fronts are possible. Ranges for m and n are $0 \leq m \leq 2$ and $1 \leq n \leq 3$. Typical values for fibrous media are $m \approx 1$ and $n \approx 2$. Accordingly, a complete densification front is not theoretically possible. The attention has then been focused on the incomplete densification case.

A dimensional analysis has shown that there are only two independent parameter sets that control the front, and that in the two-dimensional parameter space, there is a front existence domain sketched at fig. 6. The front existence is indeed linked to the following criterion : the mass flux must be at least equal to what the chemical reaction needs (lower branch of the frontier); however it is in itself limited by diffusion (upper branch of the frontier). Since the amount of reaction and diffusion fluxes is computed with respect to a length scale that is natural to the problem :

$$L_{ref} = \frac{\lambda_0 T_h}{q} \frac{\mathcal{R} T_h}{E_a} \tag{8}$$

, then there exists a minimal heat flux which allows to fulfil the mass flux criterion :

$$q \gtrsim \frac{\lambda_0 \mathcal{R} T_h^2}{E_a} \sqrt{\frac{A k(T_h)}{D_0}} \tag{9}$$

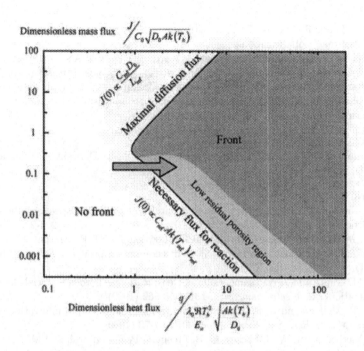

Figure 6: Domain of front existence in scaled parameter space, as obtained numerically with $m = n = 2$.

This means that the critical flux decreases when : i) the activation energy increases, ii) the gas diffusion coefficient increases (with a pressure decrease unless a Knudsen regime is attained), iii) the hot side temperature decreases, iv) the solid phase conductivity decreases, and v) the fiber diameter increases (since $A \approx 4/d_f$) [10]. The diagram of fig. 6 helps to understand how the undesired infiltrations of fig. 2 appear : indeed, when processing starts, the heat gradient is not strong enough and the critical condition 9 is not fulfilled ; later on, as the process proceeds, the heat flux increases and eventually traverses the existence boundary, as shown by the arrow on the diagram. Also, it has been shown that the residual porosity tends to zero when the system lies close to the lower frontier of the existence diagram : in practice, it will be difficult to distinguish between such a behavior and a total infiltration. In this case, the velocity is approximately equal to $\Omega_s C_0 A k(T_h) L_{ref}$, and in any case the front width is roughly equal to L_{ref}.

An important consequence of this result are guidelines for Thermal-Gradient CVI processing : i) it is important to bring enough heat flux at the very beginning of the process, and ii) it is unnecessary to increase the heat flux after the front has started to settle.

CONCLUSION

Many Thermal-Gradient CVI modalities have a large potential to overcome the traditional isothermal, isobaric process ; its largest advantage is that a densification front may settle and ensure full attainable density with low processing times. A survey of many experimental and modelling results indicates that the critical issue is indeed the set-up of this front. A detailed modelling helped understand the various behaviors that are possible in unforced TG-CVI. Then, a simplified 1D model allowed to produce an existence diagram in parameter space. In particular, a minimum heat flux is given.

ACKNOWLEDGEMENTS

The authors wish to thank Région Aquitaine for a Ph. D. grant to N. N. and CEA for a Ph. D. grant to J.-F. L.

REFERENCES

[1] R. Naslain and F. Langlais. "Fundamental and practical aspects of the chemical vapor infiltration of porous substrates", *High Temperature Science*, **27**, 221-235 (1990).

[2] T. M. Besmann , B. W. Sheldon, R. A. Lowden and D. P. Stinton. "Vapor-phase fabrication and properties of continuous-filament ceramic composites", *Science*, **253**, 1104-1109 (1991).

[3] S. M. Gupte and J. A. Tsamopoulos. "Densification of porous materials by chemical vapor infiltration", *J. Electrochem. Soc.*, **136**, 555-562 (1989).

[4] R. R. Melkote and K. F. Jensen. "A model for chemical vapor infiltration of fibrous substrates", *Mater. Res. Soc. Symp. Proc.*, **168**, 67-72 (1990).

[5] S. Vaidyaraman, W. J. Lackey, G. B. Freeman, P. K. Agrawal, and M. D. Langman. "Fabrication of carbon-carbon composites by forced flow-thermal gradient chemical vapor infiltration." *J. Mater. Res.*, **10**, 1469-1477 (1995).

[6] T. L. Starr, A. W. Smith and G. F. Vinyard, "Model-assisted control of chemical vapor infiltration for cermaic composite fabrication", *Ceram. Eng. Sci. Proc.*, **12**, 2017-2028 (1991).

[7] T. M. Besmann and J. C. McLaughlin. "Scale up and modeling of forced CVI", *Proceedings of the 18th Annual conference on Composites and Advanced Materials - B*, **15**, 897-907 (1994).

[8] I. Golecki, C. Morris, and D. Narasimhan. US Patent no. 5 348 774 (1994).

[9] D. Gupta and J. W. Evans. "A mathematical model for CVI with microwave heating and external cooling", *J. Mat. Res.*, **6**, 810-818 (1991).

[10] J. I. Morell, D. J. Economou, and N. R. Amundson. "Pulsed-power volume-heating chemical vapor infiltration", *J. Mater. Res.*, **7**, 2447-2457 (1992).

[11] D. J. Devlin, R. P. Currier, R. S. Barbero, and B. F. Espinoza. "Microwave assisted chemical vapor infiltration", *Mater. Res. Soc. Symp. Proc.* **250**, 245-250 (1992).

[12] D. J. Devlin, R. S. Barbero, and K. N. Siebein, "Radio frequency assisted chemical vapor infiltration", *The Electrochemical Society Proceedings Series*, **PV 96-5**, 571-577 (1996).

[13] V. Midha and D. J. Economou. "A two-dimensional model of chemical vapor infiltration with radio frequency heating", *J. Electrochem. Soc.*, **144**, 4062-4071 (1997).

[14] D. Leutard, G. L. Vignoles, F. Lamouroux, and B. Bernard. "Monitoring density and temperature in C/C composites elaborated by CVI with radio-frequency heating", *J.*

Mater. Synth. and Proc., **9**, 259-273 (2002).

[15]M. Houdayer, J. Spitz, and D. Tran Van. US Patent no. 4 472 454 (1984).

[16]E. Bruneton, B. Narcy, and A. Oberlin. "Carbon-carbon composites prepared by a rapid densification process", *Carbon*, **35**, 1593-1611 (1997).

[17]I. Golecki, R. C. Morris, D. Narasimhan, and N. Clements. "Carbon-carbon composites inductively-heated and rapidly densified by thermal-gradient chemical vapor infiltration : Density distributions and densification mechanism", *Ceram. Trans.* **79**, 135-142 (1996)

[18]S. Vaidyaraman, W. J. Lackey, P. K. Agrawal, and T. L. Starr. "1-D model for forced-flow -thermal gradient chemical vapor infiltration process for carbon/carbon composites", *Carbon*, **34**, 1123-1133 (1996).

[19]D. Rovillain, M. Trinquecoste, E. Bruneton, A. Derré, P. David, and P. Delhaès. "Film-boiling chemical vapor infiltration. An experimental study on carbon/carbon composites", *Carbon*, **39**, 1355-1365 (2001).

[20]J.-F. Lines. *"Modélisation et optimisation du procédé de densification de composites carbone/carbone par caléfaction"*, PhD thesis, Université Bordeaux 1 (2004).

[21]P. Delhaès, M. Trinquecoste, J.-F.Lines, A. Cosculluela, and M. Couzi. "Chemical vapor infiltration of C/C composites : fast densification process and matrix characterization", in *Proc. Intl. Conf. On Carbon 2003*.

[22]N. Nadeau. *"Modélisation mathématique et numérique d'un front de densification lors de l'élaboration d'un matériau composite carbone/carbone"*, PhD thesis, Université Bordeaux 1 (2004).

[23]M. M. Tomadakis and S. V. Sotirchos. "Ordinary, transition and Knudsen regime diffusion in random capillary structures", *Chem. Eng. Sci.*, **48**, 3323-3333 (1993).

FABRICATION AND CHARACTERIZATION OF MGC COMPONENTS FOR ULTRA HIGH EFFICIENCY GAS TURBINE

Yoshiharu Waku
HPGT Research Association
573-3 Okiube
Ube City, Yamaguchi, 755-0001 Japan

Narihito Nakagawa
Ube Research Laboratory, Ube Industries, LTD
Kogushi,
Ube City, Yamaguchi, 755-0001 Japan

Kenji Kobayashi
Advanced Technology Department, Ishikawajima-Harima Heavy Industries, LTD
Nishitokyo City, Tokyo, 188-8555 Japan

Yasuhiko Kinoshita
Gas Turbine Research & Development Center, Kawasaki Heavy Industries, LTD.
Kawasaki-Cho,
Akashi City, 673-8666 Japan

Shinya Yokoi
HPGT Research Association
Nishitokyo City, Tokyo, 202-0023 Japan

ABSTRACT

Much attention has been paid to unidirectionally solidified ceramic composites as a candidate for a high-temperature structural material. We have recently developed eutectic composites, which are named as Melt Growth Composites (MGCs). The binary MGCs (Al_2O_3/YAG and Al_2O_3/GAP binary systems) have a novel microstructure, in which continuous networks of single-crystal Al_2O_3 phases and single-crystal oxide compounds (YAG or GAP) interpenetrate without grain boundaries. Therefore, the MGCs have excellent high-temperature strength characteristics, high creep resistance, superior oxidation resistance and thermal stability in the air atmosphere at very high temperatures.

Manufacturing processes for the MGCs are being examined under a Japanese national project, scheduled from 2001 - 2005. To achieve higher thermal efficiency for gas turbine systems, bowed stacking nozzle vanes have been fabricated on an experimental basis. The steady state temperature and thermal stress distribution on the surface of the bowed stacking nozzle have been analyzed. In addition, some properties of MGC components such as the thermal stability in an air atmosphere and the stability in water vapor at elevated temperatures are reported.

INTRODUCTION

In the advanced gas generator field, studies all over the world are seeking to develop ultra-high-temperature structural materials that will improve thermal efficiency in aircraft engines and other high-efficiency gas turbines. Research is being vigorously centered on the development of very high temperature structural materials that remain stable under use for prolonged periods in an oxidizing atmosphere at very high temperatures.

It has been reported that a unidirectionally solidified Al_2O_3/YAG eutectic composite has superior flexural strength, thermal stability and creep resistance at high temperature [1-3], and is a candidate for high-temperature structural materials. As a different eutectic ceramic oxide system, rods with small diameters of the Al_2O_3/$ZrO_2(Y_2O_3)$ system have been reported by Sayir et al [4] and Pastor et al [5] as having superior mechanical properties at high temperatures.

The recently developed MGCs are a new class of ceramic matrix composites made by melting and unidirectional solidification of raw material oxides using a eutectic reaction to precisely control the crystal growth. The MGCs have excellent high temperature characteristics due to their unique microstructure without eutectic colony boundaries [6-11]. In this paper, high temperature characteristics of the MGCs such as temperature dependence of flexural strength and creep resistance are briefly introduced. Furthermore, thermal stability of microstructure and strength of MGCs in an air atmosphere at very high temperature has been investigated and the stability of MGCs in combustion gas flow environments at elevated temperatures is being evaluated.

HIGH TEMPERATURE CHARACTERISTICS OF MGCs
Temperature Dependence of Flexural Strength
The change in flexural strength of Al_2O_3/YAG and Al_2O_3/GAP binary and Al_2O_3/YAG/ZrO_2 ternary MGCs as a function of temperatures is shown in Figure 1 with a Si_3N_4 advanced ceramic which was recently developed for high temperature structural materials [12]. For comparison, the change in tensile strength of superalloys (CSMX-4) [13] is also shown in Figure 1. Superalloys are excellent high temperature structural materials above about 1300 K, but their strength decreases rapidly above about 1300 K [13]. The Si_3N_4 advanced ceramics, on the other hand, has a higher flexural strength than those of the other ceramic materials at room temperature, but its strength decreases gradually with an increase of temperatures above approximately 1000 K [12] contributing to microgram superplasticity due to grain-boundary sliding or grain rotation at grain boundaries at high temperatures.

In contrast to the above mentioned, binary MGCs maintain their room temperature strength up to very high temperatures below their melting point, with a flexural strength in the range of 300~400 MPa for the Al_2O_3/YAG binary MGC [10] and 500-600 MPa for the Al_2O_3/GAP binary MGC. In addition, the flexural strength of the Al_2O_3/YAG/ZrO_2 ternary MGC increases gradually with a rise in temperatures and its average flexural strength at 1873 K is approximately 800 MPa, more than twice that of the Al_2O_3/YAG binary MGC [14]. This difference in the flexural strength's temperature dependence between binary and ternary MGCs is presumed to depend mainly on the dimensions of microstructure. Therefore, the higher

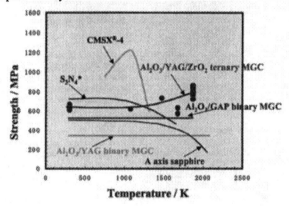

Fig. 1 Temperature dependence of flexural strength of the Al_2O_3/YAG, Al_2O_3/GAP binary and the Al_2O_3/YAG/ZrO_2 ternary MGCs in comparison with superalloys (CMSX-4) and a Si_3N_4 sintered advanced ceramic.

high-temperature strength increases, the finer dimensions of microstructure become. MGCs fracture in an intergranular mode, which is significantly different from the transgranular mode in which the sintered composite with the same chemical composition fractures [8,10].

MGCs have the unique microstructure of a three-dimensionally continuous network of single crystal phases (Al_2O_3 and YAG or Al_2O_3 and GAP or Al_2O_3 and YAG and ZrO_2) without eutectic colony boundaries. Therefore, binary and ternary MGCs have excellent high temperature strength up to very high temperatures below their melting points.

Creep Characteristics

Figure 2 shows the relationship between tensile creep rupture strength and Larson-Miller parameter, T(22+log t) [15, 16], for Al_2O_3/YAG binary MGC compared with that of polymer-derived stoichiometric SiC fibers: Hi-Nicalon Type S [16], Tyranno SA (1, 2) [17], and Sylramic (1) [17], those of silicon nitrides [18]. Here T is the absolute temperature; t is the rupture time in hours. For comparison, the Larson-Miller curve for a representative superalloy, CMSX®-10 [19], is shown in Figure 2 as well.

The relationship between tensile creep rupture strength and Larson-Miller parameter shows three broad regions. The Larson-Miller parameter for CMSX® -10 is 32000 or less in region I. This material is already being used for turbine blades in advanced gas turbine systems at above 80% of its melting point, and its maximum operating temperature is approximately 1273-1373 K. It is not envisioned that the heat resistance of this superalloy will be significantly improved in the future. On the other hand, advanced ceramics such as silicon nitrides and SiC fibers are found in region II where the Larson-Miller parameter is between 33000 and 42000. These materials are promising candidates for high temperature structural materials. They have better high temperature resistance than the superalloys. The creep strength of SiC fibers is approximately coincident with that of silicon nitrides.

In contrast, the Al_2O_3/YAG binary MGC is found in region III where the Larson-Miller parameter is between 44000 and 48000. The high temperature resistance of this MGC is superior to those of the silicon nitrides and the SiC fibers. The creep deformation mechanisms for the MGC can be assumed to follow the dislocation creep models, which are believed to be essentially different from the grain boundary sliding or grain rotation of the sintered ceramics.

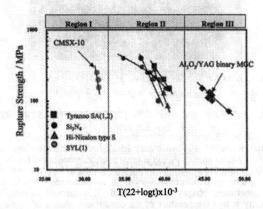

Fig. 2 Larson-Miller creep rupture strength of MGC compared to other heat-resistant materials.

199

MGC GAS TURBINE COMPONENT

The thermal efficiency of the gas turbine with uncooled turbine nozzle depends strongly on the turbine inlet temperature and the overall pressure ratio. The current efficiency of a 5MW-class gas turbine is around 29% at a TIT of 1373 K and a pressure ratio of 15. In contrast, the efficiency of the MGC gas turbine with the uncooled turbine nozzle is higher than that of the conventional gas turbine. For example, the 29% efficiency of the conventional 5 MW-class gas turbine increases to 38% at a TIT of 1973 K and a pressure ratio of 30 [20].

An Al_2O_3/GAP binary MGC with high temperature strength superior to that of an Al_2O_3/YAG binary is being examined as a candidate material for the bowed stacking nozzle. Figure 3 shows the bowed stacking nozzle manufactured from an Al_2O_3/GAP binary MGC ingot with 53 mm in diameter and 700 mm in length by machining using a diamond wheel.

For the bowed stacking nozzle, the steady state temperature and thermal stress distribution at a TIT of 1973 K have been analyzed. The maximum temperature is observed along the central vane section from leading edge to trailing edge at the surface of the bowed stacking nozzle. The maximum steady state thermal stress, generated at the trailing edge of the nozzle, is estimated at 117 MPa [20]. On the other hand, the maximum transient tensile stress in the bowed stacking nozzle during shut-down in one second from 1973 K to 973 K, generated at the leading edge near the mid-span location at 1373 K-1473 K, is estimated at 482 MPa [20]. In order to ensure the structural integrity of the MGC bowed stacking nozzle under the steady-state and thermal cycle conditions, a test rig at an inlet gas temperature level of 1973 K is being planned. Existing rig equipment is being improved for the 1973 K test to enable measurement of a continuous temperature distribution on the nozzle surface by using an infrared camera.

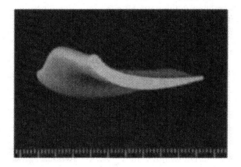

5 mm

Fig. 3 A bowed stacking nozzle manufactured from an Al_2O_3/GAP binary MGC ingot by machining using a diamond wheel.

THERMAL STABILITY OF MICROSTRUCTURE AND STRENGTH OF MGC COMPONENTS

Figure 4 shows the SEM images of microstructure of cross-section perpendicular to the solidification direction of the Al_2O_3/GAP binary MGC before and after exposure tests for 1000 hours at 1973 K in an air atmosphere. A slight growth of the phase width was observed after the exposure test for 1000 hours. However, the MGC was shown to be comparatively stable without void formation during lengthy exposure at very high temperatures of 1973 K even though the morphology of the constituent phases became round. This thermal stability resulted from the thermodynamic stability at that temperature of the constituent phases of single-crystal Al_2O_3 and single-crystal GAP, and the interface. The Al_2O_3/GAP binary MGC has about 500 MPa of the flexural strength after the exposure test for 1000 hours at 1973 K in an air atmosphere. The flexural strength of the MGC slightly decreases in a short time of 200 hours. After that period the flexural strength becomes constant. The Al$_2$O$_3$/GAP binary MGC exhibited good thermal stability in the microstructure and the strength at very high temperatures of 1973 K in an air atmosphere.

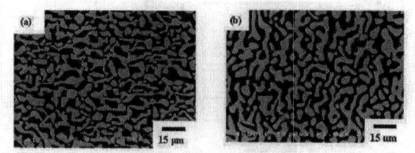

Fig. 4 SEM images showing microstructural changes of cross-section perpendicular to the solidification direction of the Al$_2$O$_3$/GAP binary MGC (a) before and (b) after the exposure test at 1973 K for 1000 hours in air atmosphere.

EXPOSURE TEST IN COMBUSTION GAS FLOW ENVIRONMENT

The exposure test in combustion gas flow environments was performed using the combustion gas flow testing apparatus (produced by Maruwa Electric Incorporated, Chiba, Japan) shown in Figure 5. Kerosene was used as fuel for the combustor. The dimensions of the exposure test specimen were 3 x 4 x 50 mm^3 and the surfaces were polished to a mirror condition. For comparison, single crystal Al$_2$O$_3$ (99.99%, Earth Chemical, Tokyo, Japan) and Si$_3$N$_4$ (Ube-Industries. LTD.) were used. Three specimens were placed in the specimen holder made of high purity sintered Al$_2$O$_3$ and exposed to the combustion gas flow conditions (T = 1500 °C, P = 0.1 - 0.3 MPa, V = 150 - 250 m/s, P$_{H2O}$ = 15 - 45 kPa) for 10 hours. The recession of specimens was evaluated by the change in surface roughness, Ra (μm), and strength.

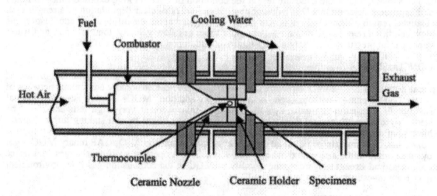

Fig. 5 Schematic drawing of the burner rig test equipment

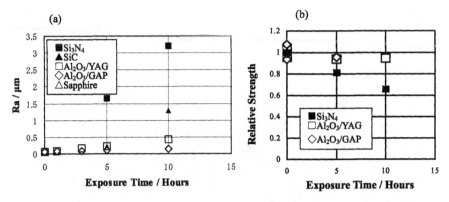

Fig. 6 The stability of the MGC binary systems in the combustion gas flow environments. (a) Changes in surface roughness, Ra, of Si_3N_4, SiC, binary MGCs and sapphire. (b) Relationship between relative strength and exposed time for Si_3N_4 and binary MGCs.

Fig. 6 (a) shows the changes in surface roughness of Si_3N_4, SiC, Al_2O_3/YAG and Al_2O_3/GAP binary MGCs, and sapphire. The Si_3N_4 shows distinct change of surface morphology revealed by an increase of Ra. This is due to the formation of substantial volume of SiO_2 scale on the surface of the Si_3N_4. In contrast to this, Al_2O_3/GAP binary MGC and sapphire show a little change in Ra. A little recession was observed only at the Al_2O_3 phase. No recession was observed at the GAP phases.

The residual strength after exposure tests to the combustion gas flow environment was evaluated by the change of four-point flexural strength. Figure 6 (b) shows the relationship between relative strength and exposed time. The relative strength is the ratio of the flexural strength after the exposure test to as-received strength of specimens. The relative strength of MGC decreases gradually as the exposed time increases. However the relative strength of MGC shows above 0.9 after the exposure test at the condition of H_2O=45 kPa, exposed time=10 hours and exposed temperature=1773 K. In contrast, Si_3N_4 indicates clear degradation of strength with relative strength of about 0.65. It is now being investigated about the influence of combustion gas flow condition (e.g. gas temperatures, exposure time, gas flow velocity, total pressure and water vapor partial pressure) on the high-temperature oxidation and volatilization of MGCs.

CONCLUSIONS

MGCs have many advantages over other ultra-high temperature structural materials because of their unique microstructure consisting in a three-dimensionally continuous network of single crystal phases without grain boundaries. In addition, MGCs display superior thermal stability of microstructure and strength in an air atmosphere at very high temperatures. The steady state temperature and thermal stress on the surface of the bowed staking turbine nozzle have been analyzed. To apply MGC to gas turbine components, MGC bowed stacking turbine nozzle has been manufactured on an experimental basis from the Al_2O_3/GAP binary MGC ingot by machining using a diamond wheel. Some practical properties such as changes in surface roughness and strength regarding the stability of MGC in the combustion gas flow environment have been investigated.

ACKNOWLEDGEMENTS

The authors would like to express their thanks to the New Energy and Industrial Technology Development Organization (NEDO) and the Ministry of Economy, Trade and

Industry (METI) for the opportunity to conduct our project on the Research and Development of Ultra-high Temperature Heat-resistant Materials MGC.

REFERENCES
[1]T. Mah and T.A. Parthasarathy, "Processing and Mechanical Properties of $Al_2O_3/Y_3Al_5O_{12}$(YAG) Eutectic Composite," *Ceram. Eng. Sci. Proc.,* 11, 1617-27 (1990).
[2]T.A. Parthasarathy, T. Mah and L.E. Matson, "Creep Behavior of an Al_2O_3-$Y_3Al_5O_{12}$ Eutectic Composite" *Ceram. Eng. Soc. Proc.,* 11, 1628-38 (1990).
[3]T.A. Parthasarathy, Mar, Tai-II and L.E. Matson, "Deformation Behavior of an Al_2O_3-$Y_3Al_5O_{12}$ Eutectic Composite in Comparison with Sapphire and YAG," *J.Am.Ceram. Sci.,* 76, 29-32 (1993).
[4]A. Sayir, A. and S. C. Farmer, "The effect of the microstructure on mechanical properties of directionally solidified $Al_2O_3/ZrO_2(Y_2O_3)$ eutectic," *Acta Mater.,* 48, 4691-97 (2000).
[5]J. Y. Pastor, P. Poza, J. LLorca, J.I. Pena, R. I. Merino and V. M. Orera, "Mechanical properties of directionally solidified Al_2O_3-$ZrO_2(Y_2O_3)$ eutectics," *Mater. Sci. Eng.,* A308, 241-49 (2001).
[6]Y. Waku, H. Ohtsubo, N. Nakagawa and Y. Kohtoku, "Sapphire matrix composites reinforced with single crystal YAG phases," *J. Mater. Sci.* 31, 4663-70 (1996).
[7]Y. Waku, N. Nakagawa, T. Wakamoto, H. Ohtsubo, K. Shimizu and Y. Kohtoku, "A ductile ceramic eutectic composite with high strength at 1873 K," *Nature,* 389, 49-52 (1997).
[8]Y. Waku, N. Nakagawa, T. Wakamoto, H. Ohtsubo, K. Shimizu and Y. Kohtoku, "High-Temperature Strength and Thermal Stability of a Unidirectionally Solidified Al_2O_3/YAG Eutectic Composite," *J. Mater. Sci.,* 33, 1217-25 (1998).
[9]Y. Waku, N. Nakagawa, T. Wakamoto, H. Ohtsubo, K. Shimizu and Y. Kohtoku, "The Creep and Thermal Stability Characteristics of a Unidirectionally Solidified Al_2O_3/YAG Eutectic Composite," *J. Mater. Sci.,* 33, 4943-51 (1998).
[10]Y. Waku, N. Nakagawa, H. Ohtsubo, A. Mitani and K. Shimizu, "Fracture and deformation behaviour of Melt Growth Composites at very high temperatures," *J. Mater. Sci.,* 36, 1585-94 (2001).
[11]Y. Waku, S. Sakata, A. Mitani, K. Shimizu, "A novel oxide composite reinforced with a ductile phase for very high temperature structural materials," *Materials Research Innovations,* 5, 94-100 (2001).
[12] M. Yoshida, K. Tanaka, T. Kubo, H. Terazone and S. Tsuruzone, "Development of Ceramic Components for Ceramic Gas Turbine Engine (CGT302)" in *Proceedings of the international Gas Turbine & Aeroengine Congress & Exibition,* Stockholm, Sweden, June 2-5, 1998, pp1-8 (The American Society of Mechanical Engineeers, 1998).
[13]M. J. Goulette, "The future costs less-high temperature materials from an aeroengine perspective," in *Proceeding of the eighth international symposium on superalloys 1996* (edited by R. D. Kissinger, D. J. Deye, D.L. Anton, A.D. Cetei, M.V. Natal, T.M. Pollock, & D.A. Woodford) 3-6 (TMS,Pennsylvania, 1996).
[14]Y. Waku, S. Sakata, A. Mitani, K. Shimizu, A. Ohtsuka and M. Hasebe, "Temperature dependence of flexural strength of $Al_2O_3/Y_3Al_5O_{12}/ZrO_2$ ternary melt growth composites," *J. of Mater. Sci.,* 37, 2975-82 (2002).
[15] FR Larson and J Miller, "A time-temperature relationship for rupture and strength," Transaction of ASME 74 765-71 (1952).
[16] Hee Mann Yun and James A. DiCarlo, "Time / temperature dependent tensile strength of SiC and Al_2O_3-based fibers," *Ceramic Transaction,* 74, 17-25 (1996).
[17]Y. H. Yun and J. A. DiCarlo, "Comparison of the tensile creep, and rupture strength properties of stoichiometric SiC fibers," *Ceram. Eng. Sci. Proc.,* 20, 259-72 (1999).
[18]R. F. Jr. Krause, W. E. Luecke, J. D. French, B.J. Hockey and S. M. Wiederhorn, "Tensile Creep and Rupture of Silicon Nitride," *J. Am. Ceram. Soc,* 82, 1233-42 (1999).
[19] G. L. Erickson, "The development and application of CMX®-10," in *Proceeding of the*

eighth international symposium on superalloys 1996 (edited by R. D. Kissinger, D. J. Deye, D.L. Anton, A.D. Cetei, M.V. Natal, T.M. Pollock, and D.A. Woodford) 35-43 (TMS, Pennsylvania, 1996).

[20]Y. Waku, N. Nakagawa, K. Kobayashi, Y. Kinoshita and S. Yokoi, "Innovate manufacturing process of 1MGC components for ultra-high efficiency gas turbine systems," ASME TURBO EXPO 2004 – Power for Land, Sea & Air, 14-17 June 2004, Vienna, Austria.

COMPARISON OF SODIUM CAPRYLATE AND SODIUM STEARATE AND THE EFFECTS OF THEIR HYDROCARBON CHAIN LENGTHS ON ADSORPTION BEHAVIOR AND ALUMINA PASTE RHEOLOGY

C.R. August
Rutgers University
Department of Ceramic & Materials Engineering
607 Taylor Rd
Piscataway, NJ 08854

R.A. Haber
Rutgers University
Department of Ceramic & Materials Engineering
607 Taylor Rd
Piscataway, NJ 08854

L.E. Reynolds
Rutgers University
Department of Ceramic & Materials Engineering
607 Taylor Rd
Piscataway, NJ 08854

ABSTRACT

Fatty acids with carbon chain lengths 8 and 18 were synthesized and used as lubricants in an alumina extrusion batch. These were also combined with alumina, washed and the supernatant tested for UV absorbance. Fatty acid adsorption to A16 alumina as a function of hydrocarbon dosage was examined using UV/visible spectrophotometry. Chain length was shown to effect total adsorption. Extrusion behavior was examined using capillary rheometry and was shown to vary with hydrocarbon chain length.

INTRODUCTION

Extrusion is a widely used formative technique that often involves multiple inorganic components and a variety of binders and lubricants. As the geometry of extrudates becomes more demanding, a greater understanding of the particle interactions in an extrusion body must be reached. By better understanding the effects of fatty acid chain length on the rheology of a paste, optimization of additives in an extrudate may be approached. In this exploration, the adsorption behaviors of sodium stearate and sodium carbonate were examined and their effect on alumina paste rheology was explored.

The importance of adsorption behavior of additives with regards to dispersion in oxide suspensions has been explored. It was found that, in some cases surfactants could create repulsive forces between particles thereby dispersing an oxide slurry, but adsorption of these additives could lower the zeta potential, causing aggregation of the particles. Polymeric additives have also been analyzed and found to enhance stabilization by means of steric as well as electrostatic mechanisms.[1]

Fatty acids are composed of a hydrophilic carboxyl group and a hydrophobic carbon chain. The saponification process exchanges the hydrogen for a sodium ion. This changes the nature of the bonding (from polar OH to ionic O-Na) and switches the solubility properties of the acid from being insoluble in water, soluble in methanol, before the saponification to the reverse afterwards.

The conversion of a fatty acid into a sodium salt can be achieved by the alkali refining of the acid using sodium hydroxide in a methanol solution.[2] This method has been used to create soap stock from a replicated soybean oil to be further converted into a fatty acid methyl ester.

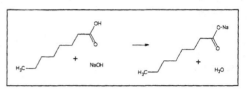

Hydrocarbon chain length has been found to have an effect on the amount of polymeric adsorption to inorganic particles.[3] In a study analyzing the adsorption of decyltrimethylammoniums of different chain lengths onto clinoptilolite, it was found that greater adsorption occurred for hydrocarbons with longer chain lengths. This study used adsorption isotherms and zeta potential to explore ion exchange and hydrophobic interactions at the clinoptilolite/water interface.

Torque rheometry is useful in examining the rheological properties of a paste. This technique focuses on the transition from dry ingredients and water, to a single paste system. On a graph resulting from torque rheometry, this transition is examined by using the peak torque and the time to peak. The peak torque is the maximum torque required to plasticize the paste. The time to peak torque indicates how long it takes for this transition to occur. After peaking, the torque relaxes to a stable value. Torque values are related to viscosity of the paste.

There are multiple techniques with which capillary rheometry can be used to analyze the rheology of a paste. The Bagley method involves using dies of cross slit and round geometries in order to evaluate the effects of die geometry on extrusion pressure.[4] The Benbow analysis utilizes dies of varying die length over diameter at a series of extrusion speeds to calculate certain constants. These constants can be related to properties of the initial paste. This method can then be used to predict extrusion behavior of a paste in more complicated systems.[5]

Benbow uses capillary rheometry results to calculate parameters which relate to the paste rheology. These parameters include the bulk stress and the yield stress of the material flowing through the die land as well as two parameters which relate to velocity of the paste in relation to the barrel wall and die land. These parameters are calculated using the equation:

$$P = 2(\sigma_0 + \alpha V^m)\ln(D/D_0) + 4(\tau_0 + \beta V^n)(L/D)$$ (1)

In this equation P is the extrusion pressure, V^m is the velocity of the paste in the die land, D is the diameter of the extrusion die, D_0 is the diameter of the capillary barrel, V^n is the velocity of the paste near the wall, and L is the length of the extrusion die. The parameters calculated from this equation are, σ_0, the bulk stress, τ_0, the yield stress, α, the velocity factor for the die land, and β the velocity factor for the barrel wall. The equations used to calculate these parameters contain variables taken by measuring capillary pressures for multiple extrusion speeds using multiple dies with varying length and the same diameter.[6]

It is important to understand the geometry of the capillary rheometer to fully comprehend parameters in the Benbow equation. Figure 1 is a diagram of the capillary rheometry barrel with the parameters needed for the Benbow equation labeled.

EXPERIMENTAL PROCEDURE

Saponification

Sodium based soap is synthesized by adding a fatty acid (ex. stearic acid) to sodium hydroxide in a solution of methanol. These fatty acids have a 1:1 chemical reaction with the sodium hydroxide. The general reaction is

$$NaOH + FA \rightarrow NaFA + H_2O \tag{2}$$

Sodium hydroxide is added to methanol and agitated at an increased temperature. Evaporated methanol is recycled via a reflux condenser. When the sodium hydroxide is completely dissolved, the fatty acid is added in proportion to the ratio of the molecular weight of the fatty acid over the molecular weight of the sodium hydroxide. Agitation of this mixture then continues until the reaction has completed and the fatty acid salt has precipitated out. The remaining methanol is then allowed to evaporate off, leaving the salt. Table I shows a range of fatty acids, their carbon chain lengths and molecular weights.

Adsorption Isotherms

An important characteristic of the interaction between inorganic and organic materials in a batch is the adsorption behavior. The acids chosen for this experiment were stearic and caprylic due to their variation in carbon chain length and large volume of precipitate formed during saponification. By using additives with the same basic chemical structure, competitive adsorption is minimized. Samples with varying concentrations of saponified fatty acids with set percents of A-16 alumina and water were created. A-16 alumina was chosen for the adsorption testing because it is a higher surface area alumina. These samples were centrifuged and the supernatant collected. This supernatant was then analyzed with UV spectrophotometry on a Lambda 9 dual beam spectrophotometer.

Calibration curves were created by analyzing the fatty acid salts in water at fixed concentrations. The supernatants were then analyzed and the calibration curves were used to determine the concentration of surfactant in the supernatant. The surface area of the alumina was then used to allow adsorption isotherms to be created.

Thermal Analysis

Thermal gravimetric analysis was performed on the saponified fatty acids, sodium caprylate and sodium stearate, using the Perkin-Elmer TGA7. These samples were examined as they were heated from room temperature (25°C) to about 800°C. A ramp setting of 5°C/minute was used during the sample analysis. Differential thermal analysis was also performed on the Perkin-Elmer DTA7. The samples were analyzed as the temperature was ramped from room temp to about 950°C at a rate of 10°C/minute. Both of these analyses were performed in an air atmosphere.

Torque Rheometry

Measurements were taken on a Haake Rheometry with a dual sigma blade configuration at 50 rpm for 15 minutes. Three batches were analyzed through torque rheometry, one containing sodium stearate lubricant, one with sodium caprylate, and one with a 50/50 blend of sodium stearate and sodium caprylate. Batches contained A-2 alumina, 3% weight methyl cellulose binder, 2% one or both of the saponified fatty acids, and 20% water. A-2 alumina was used in the rheological testing because it flows more easily than the A-16 and allows us to keep tested pressures in an observable range. These batches were pre mixed under low shear before being added to the rheometer.

Capillary Rheometry

The capillary rheometer was assembled on an Instron 4500 Series. This equipment consisted of a capillary barrel and support structure, pressure meter, transducer, plunger, ram, die head, and dies, Figure 1 shows the set-up. The barrel and support structure were fixed to the crosshead of the Instron so the equipment could be raised into the ram attached to the top of the Instron. The transducer was fixed into the extrusion barrel slightly above the die level and connected to the power meter.

Capillary rheometry was performed on batches of the same composition as those used in torque rheometry. Each batch was analyzed with two 50 mil dies of two different lengths and at two different speeds. The recorded pressure for these runs was the stable pressure and not the static yield pressure. Capillary rheometry results were then used to calculate the parameters from Benbow's equation.

RESULTS & DISCUSSION

The adsorption isotherms created for the sodium stearate, sodium caprylate and 50/50 blend of these saponified fatty acids can be seen in figure 2. It can be seen that the sodium stearate had a greater mass adsorb to the surface of the alumina than the sodium caprylate. The greatest adsorption behavior was seen in the blend of sodium caprylate and sodium stearate. This can be explained by assuming that competitive adsorption is minimal, which would allow both polymer chains to adsorb to the surface of the same particle. Because the chain lengths are so differently sized, a greater amount of adsorption can occur and a greater surface density of polymer achieved.

Thermal analysis was performed on the sodium caprylate and sodium stearate. The results of these analyses can be seen in figure 3. The thermal gravimetric analysis shows the sodium caprylate beginning to lose mass at about 125°C and ending mass loss abruptly at around 400°C. The sodium stearate exhibits a more vertical loss at around 300°C. The differential thermal analysis shows a small endotherm around 120°C (180°C for the caprylate), most likely caused be H_2O evaporation. The large exotherms, which begin at 230°C and 280°C for sodium stearate and sodium caprylate, respectively, are the result of an ignition phenomenon in which the carbon-carbon bonds break down and react with O_2 in the air. The result is CO_2 and CO gases with sodium left as a byproduct.

Torque rheometry curves can be seen in figure 4. The peak torque is largest for sodium caprylate at about 591 MG, followed by the sodium stearate at 507 MG, and the smallest for the blend of the two at 496 MG. The time to peak torque followed the reverse trend with the smallest time for the caprylate and the largest for the blend.

The extrusion pressures obtained through capillary rheometry can be seen in table II. Sodium stearate required the largest extrusion pressure, followed by the sodium caprylate and the blend for both dies used in the analysis at the slower extrusion speed. The Benbow parameters calculated from the extrusion pressures show the sodium caprylate as having the highest yield stress and the lowest bulk stress. The sodium stearate has a higher bulk stress than the stearate/caprylate blend, as well as a higher yield stress.

CONCLUSION

A comparison of the adsorption and rheological behavior of a sodium stearate with an 18 carbon chain length and a sodium caprylate with an 8 carbon chain length was performed. Adsorption isotherms with the additives in an aqueous alumina system show a greater amount of adsorption occurring when the two saponified fatty acids were combined than when either was tested individually. Rheological studies also show that a combination of the fatty acids used as a paste additive, generally lowers the viscosity of the paste as well as lowers the extrusion pressures. Parameters calculated to analyze the extrusion paste indicate that combining the two additives lowers the yield stress of the paste while maintaining a reasonably high bulk stress. This would be beneficial to extrusion as it would allow a green body to keep its form post-extrusion while not requiring higher extrusion pressures.

REFERENCES

[1]D.W. Fuerstenau, R. Herrera-Urbina, and J.S. Hanson, "Adsorption of Processing Additives and the Dispersion of Ceramic Powders," *Ceramic Powder Science*, 2, 331-351 (1988).

[2]M.J. Haas, S. Bloomer, and K. Scott, "Simple, High-Efficiency Synthesis of Fatty Acid Methyl Esters from Soapstock," *Journal of the American Ceramic Society*, 77, no 4, 373-9 (2000).

[3]B. Ersoy and M.S. Celik, "Effect of Hydrocarbon Chain Length on Adsorption of Cationic Surfactants onto Clinoptilolite," *Clays and Clay Minerals*, 51, no 2, 172-81 (2003).

[4]E.B. Bagley, "End Corrections in the Capillary Flow of Polyethylene," Journal of Applied Physics, 28, no 5, 624-7 (1957).

[5]J. Benbow, and J. Bridgwater, *Paste Flow and Extrusion*, Clarendon Press, Oxford, 29-33 (1993)

[6]C.R. August, R.A. Haber, "Benbow Analysis of Extruded Alumina Pastes," Ceramic Engineering Science and Proceedings,

FIGURES AND TABLES

Table I Fatty Acid Properties

Fatty Acid	# of Carbon Atoms	Molecular Weight	Ratio FA:NaOH
Caprylic	**8**	**144.22**	**3.61**
Capric	10	172.3	4.31
Lauric	12	200.32	5.01
Myristic	14	228.38	5.71
Palmitic	16	256.43	6.41
Stearic	**18**	**284.48**	**7.11**
Oleic	18	282.47	7.06
Linoleic	18	280.45	7.01
Arachidic	20	312.5	7.81

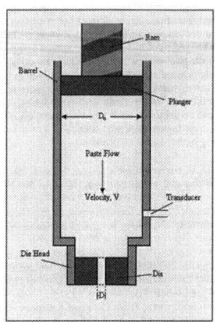

Figure 1 Diagram of Capillary Rheometer

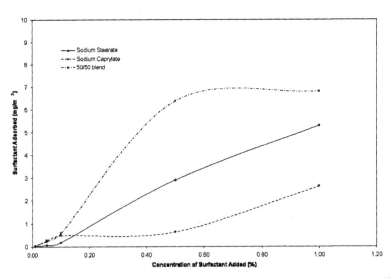

Figure 2 Adsorption Isotherms for Sodium Stearate, Sodium Caprylate, and a 50/50 Blend

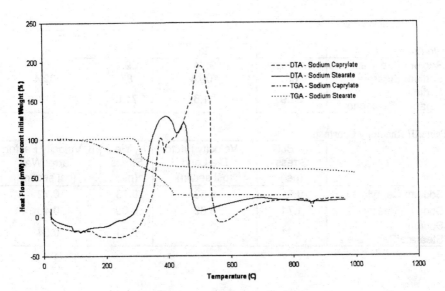

Figure 3 Thermal Analyses of Sodium Caprylate and Sodium Stearate

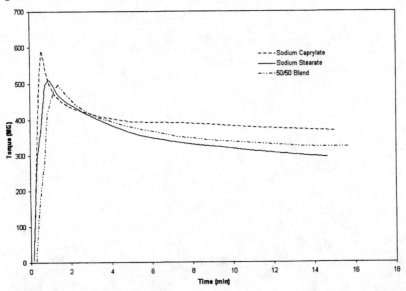

Figure 4 Torque Rheometer results for Sodium Caprylate, Sodium Stearate and a 50/50 Blend

Table II Extrusion Pressures

L/d (in/in)	0.25 in/min		0.5 in/min	
	10	20	10	20
Sodium Caprylate	57.1(psi)	97.4	69.4	110.9
Sodium Stearate	68	109.4	87	132.4
Sodium Stearate/Caprylate	55.9	91.6	71.9	113

Table III Rheology Constants

	σ_0, Bulk Stress (psi)	α, Velocity Factor, Die Land (psi sec/in)	τ_0, Yield Stress (psi)	β, Velocity Factor, Barrel Wall (psi sec/in)
Sodium Caprylate	0.84	6.53	0.73	0.12
Sodium Stearate	1.71	8.82	0.69	0.40
Sodium Stearate/Caprylate	1.41	6.23	0.51	0.54

FORMULATION OF ADDITIVES FOR WATER-BASED TAPE CASTING OF CERAMICS

M. Guiotoku, P. Lemes, C.M. Gomes, C.A. Valente, A.P.N. Oliveira, D. Hotza[*]
Materials Science and Engineering Graduate Program (PGMAT)
Federal University of Santa Catarina (UFSC)
88040-900 Florianópolis, SC, Brazil

ABSTRACT
The effect of additives type and content on rheological properties of water-based suspensions for tape casting was investigated. An aqueous suspension of a ceramic powder (feldspar, $K_2O \cdot Al_2O_3 \cdot 6SiO_2$) was used. Sodium alginate, as a binder, poly(ethyleneglycol), as a plasticizer, and sodium poly(methylmetacrylate), as a dispersant, were added. Viscosity and sedimentation behavior were analyzed for aqueous suspensions as a function of additives. After casting and solvent evaporation, the green densities of the samples were measured. The microstructures of green and sintered samples were evaluated by scanning electron microscopy.

INTRODUCTION
The casting of ceramic tapes is a method of producing flat bodies with controlled thickness and relatively high surface areas, whose application is mainly related with the electronics industry for production of dielectric materials, integrated circuits packaging, piezoelectric materials, superconductors and magnets.

The process consists basically of preparing a suspension of a ceramic powder and casting this suspension on a flat surface. The suspension consists of solvents (water or organic liquids), dispersants, binders and plasticizers. After the evaporation of the solvent, a flexible film remains, which is separated from the surface and can be wound, cut, perforated, printed or laminated. Later, the green material is thermally treated for elimination of the organic substances and sintered.[1] The order of addition of the additives is important since the binder should not get in contact with the powder agglomerates before these are deflocculated using a dispersant, resulting in denser tapes.[2]

The suspension can be classified according to the type of solvent used. The solvent dissolves the organic materials (dispersants, binders and plasticizers) and distributes them uniformly in the suspension. It is the vehicle that maintains the ceramic particles in suspension until it evaporates producing a dense tape.[3] In general, non-aqueous organic solvents such as alcohols, ketones or hydrocarbons are used, because they present low boiling point and avoid hydration of the ceramic powder. However, the use of these solvents needs special care, because they are inflammable and toxic.[4]

The aqueous system is non-toxic, non-inflammable and has low cost[5] and, although of low volatility, it presents a tendency to forming bubbles and making the suspensions susceptible to pH changes.[6] Several formulations using water as solvent have been developed.[7-9]

The rheological behavior of the suspensions depends on the type and concentration of the ceramic powder, solvent, binder, plasticizer, and dispersant. The rheology controls strongly the quality of the final product.[5] A suspension with good rheological characteristics should have low viscosity to facilitate the casting process, but not so low to cause sedimentation. On the other hand, to obtain higher density of green tapes, the solids content (ceramic powder) should be

---

[*] Corresponding author (hotza@enq.ufsc.br).

---

relatively high, in order to reduce the final shrinkage after sintering.[1] A typical solids content for an alumina suspension is in the range from 70 to 80 wt.% of solids.[10]

In this work, the formulation of an aqueous suspension for tape casting was investigated. The rheological behavior of the suspensions was characterized by measuring the viscosities varying the binder and plasticizer concentrations. The densities of green tapes were also measured, and SEM analyses of green and sintered bodies were carried out.

EXPERIMENTAL

As ceramic material a potassium feldspar (Colorminas, mean particle size 27 μm) was used, with the aid of sodium alginate (Sigma, 2 wt.% solution, viscosity 250 mPa·s) and poly(ethileneglycol) (PEG; Synth, molecular weight 400 g/mol) as binder and plasticizer, respectively, and sodium poly(methylmetacrylate) (PMM), as dispersant. Five formulations of 175±2 g were prepared, as shown in Table I, according to compositions found in the literature.[3]

Table I Weight composition of suspensions.

Formulation	Feldspar (wt.%)	Water (wt.%)	PEG (wt.%)	Alginate (wt.%)	PMM (wt.%)
1	56.6	37.9	3.0	1.5	0.9
2	56.9	38.1	2.7	1.3	0.9
3	57.2	38.3	2.3	1.2	0.9
4	57.5	38.5	2.0	1.0	0.9
5	57.8	38.7	1.7	0.8	0.9

The rheological characterization of the suspensions was performed with a viscometer (Haake, VT550), at a shear rate of 20 s[-1]. Suspensions were applied on a flat surface of cellulose acetate. After 12 h at room temperature, the tapes were dried at 110°C for 10 min. Green densities were measured geometrically. Tape samples were sintered in an air oven (EDG, model EDGCON 3P1800) at 1100°C for 60 min.

RESULTS AND DISCUSSION

Rheological behavior

Fig. 1 shows an example of the variation of viscosity as a function of shear rate for the suspension # 1. A characteristic pseudoplastic behavior can be observed.

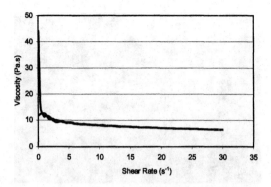

Figure 1. Viscosity as a function of shear rate (suspension #1).

The values of viscosity of suspensions # 1 to 5 at shear rate of 20 s^{-1}, calculated by Herschel Bulkley's model, ranged from 7 to 1 Pa·s, decreasing for lower binder (alginate) + plasticizer (PEG) concentration, Fig. 2. Typically, substances acting as binders are the main responsible for changes in viscosity of tape casting suspensions[3] Increasing plasticizer content (suspension #1) gives flexibility to the tapes, as it can be seen in Fig. 3.

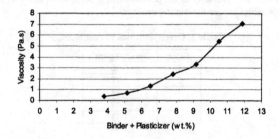

Figure 2. Viscosity as a function of additives concentration.

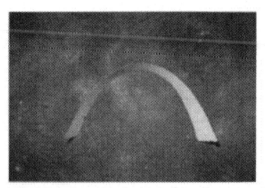

Figure 3. Flexibility of tape given by the plasticizer (from suspension #1).

Casting
    By casting the tapes, it could be noticed that the suspensions with higher binder concentrations (1, 2 and 3) presented appropriate consistency for the application on the substrate, since a low shrinkage of the tapes was observed after casting. After 12 h drying, the tapes were easily taken out of the substrate (cellulose acetate).

Green Density
    For lower additive concentrations, the tape is denser, what corresponds to the higher concentration of ceramic powder, Table 2. These density values are low compared to values found in the literature[11] due probably to the porosities of green bodies, since air bubbles were not removed from suspensions before the casting process.

Table 2. Green densities of cast tapes.

Tapes	Density (g/cm^3)
1	1.36
2	1.37
3	1.38
4	1.38
5	1.39

SEM Analysis
    Fig. 4 presents the microstructural characteristics of a green tape (sample 1). The porosity of such material is relatively high. This implicates a decrease of mechanical strength.
    Fig. 5 presents a typical microstructure of a sintered tape (sample #1). After thermal treatment, the organic additives were burned out. Consequently, a decrease of the porosity can be observed.

216

Figure 4. Micrograph of a green tape (sample #1).

Figure 5. Micrograph of a sintered tape (sample #1).

CONCLUSIONS

Viscosity of suspension is a fundamental parameter of control in the processing of ceramic tapes. In this work, the best results were obtained for suspensions that contained from 2.3 to 3.0 wt.% plasticizer and from 1.2 to 1.5% binder. The optimum amount of dispersant (0.9 wt.%) corresponded to the absence of sediments in periods of more than 24 h.

Alginate, even in small amounts, provides significant viscosity changes to the suspensions, besides giving an appropriate consistency to the tapes after drying. Poly(ethyleneglycol), even in low concentration, gives a good flexibility to the green tapes.

217

ACNOWLEDGEMENT
The financial support of the Brazilian Research Agency (CNPq) is thankfully acknowledged.

REFERENCES

[1] D. Hotza, "Tape Casting of Ceramics," *Cerâmica*, **43**, 159-166 (1997).

[2] J. S. Reed, *Principles of Ceramic Processing*, 2nd ed., Wiley, New York (1995) p. 525.

[3] D. Hotza, and P. Greil,"Review: Aqueous Tape Casting of Ceramic Powders," *Mat. Sci. Eng. A*, **202**, 206-217 (1995).

[4] C. Pagnoux, T. Chartier, M. F. Granja, F. Doreau, J. M. Ferreira, and J. F. Baumard, "Aqueous Suspensions for Tape-Casting Based on Acrylic Binders,"*J. Eur. Ceram. Soc.*, **18**, 241-247 (1998).

[5] B. Bitterlich, C. Lutz, and A. Roosen, "Rheological Characterization of Water-Based Slurries for the Tape Casting Process,"*Ceram. Int.*, **28**, 675-683 (2002).

[6] J.M.F. Ferreira, S. Mei, and M. Guedes, "Aqueous processing of ceramic and glass-ceramic substrates,"*Mat. Sci. Forum*, **442**, 27-36 (2003).

[7] C. Leonelli, L. Barbieri, T. Manfredini, D.S. Blundo, C. Siligard, and A.B. Corradi, "Densification and Properties of CMAS Glass Ceramic Prepared from Compacts of Pressed Powders and of Tape Cast Powder Multilayers," *Brit. Ceram. Trans.*, **95**, 199 (1996).

[8] Y. Zhang, and J. Binner, "Tape Casting Aqueous Alumina Suspensions Containing a Latex]Binder,"*J. Mater. Sci.*, **37**, 1831-1837 (2002).

[9] Y.P. Zeng, D.L. Jiang, and P. Greil, "Tape Casting of Aqueous Al_2O_3 Slurries," *J. Eur. Ceram. Soc.*, **20**, 1691-1697 (2000).

[10] R.E. Mistler, and E.R., Twiname, *Tape Casting: Theory and Practice*, The American Ceramic Society, Westerville, Ohio (2000).

[11] B. Bitterlich, and J.G. Heinrich, "Aqueous Tape Casting of Silicon Nitride," *J. Eur. Ceram. Soc.*, **22**, 2427 (2002).

A NON-LINEAR PROGRAMMING APPROACH FOR FORMULATION OF THREE-COMPONENT CERAMICS AS A FUNCTION OF PHYSICAL AND MECHANICAL PROPERTIES

S.L. Correia, C. M. Gomes, D. Hotza[*]
Materials Science and Engineering Graduate Program (PGMAT)
Federal University of Santa Catarina (UFSC)
88040-900 Florianópolis, SC, Brazil

A.M. Segadães
Department of Ceramics and Glass Engineering (CICECO)
University of Aveiro
3810-193 Aveiro, Portugal

ABSTRACT

The simultaneous effect of raw materials type and content on physical and technological properties of a triaxial ceramics system has been studied in the range of values used in industrial practice. The investigation has been carried out using the statistical design of mixture experiments, a special case in response surface methodologies. Ten formulations of the three raw materials selected were used in the experimental design. Those formulations were processed under conditions similar to those found in the ceramics industry: powder preparation (wet grinding, drying, granulation and humidification), green body preparation (pressing and drying) and firing. A non-linear programming approach was applied to minimize the cost of the three-component ceramics, considering optimum ranges of bulk density, bending strength, water absorption and porosity of sintered bodies. A validation experiment was performed to confirm the optimized composition predicted.

INTRODUCTION

The response surface methodology can be applied to the design of mixture experiments in many cases.[1-6] When a given property depends only on the components fractions in the mixture (x_i, summing up to unity) and not on the mixture amount, the property estimation or response can be determined based on the proportions of the components.

To this aim, a regular array of uniformly spaced points (i.e. a lattice) is first selected. This lattice is referred to as a $\{q, m\}$ simplex lattice, m being the spacing parameter in the lattice and q the number of components. The property is measured for the mixtures corresponding to those simplex points and a regression equation is fitted to the experimental values. The model is considered valid only when the error, i.e. the difference between the experimental and the calculated values is uncorrelated and randomly distributed with a zero mean value and a common variance. Having the regression model, a prediction of the property value can be obtained for any mixture, from the values of the fractions of its components.

The response function f can be expressed in its canonical form, for instance, as a linear, quadratic, or special cubic polynomial.[7-8]

The classification requirements for ceramic floor and wall tiles can be associated with some technological properties.[9] Those properties, such as dimensional changes, water absorption, bulk density, bending strength, and linear thermal expansion, are commonly measured for quality

---

[*] Corresponding author (hotza@enq.ufsc.br).

---

control in industries and laboratories according to international standards.[10-13] Basically, the raw materials mixture and the processing schedule determine changes on the properties. If the processing parameters are kept constant, the properties can be modeled using an optimization methodology applied to the design of mixture experiments.

A mathematical optimization comprises the resolution of a system of simultaneous linear and non-linear equations.[14,15] First, general goals and requirements must be defined, for instance the type of ceramic product to be manufactured. This, in turn, identifies the set of usually inequality constraints corresponding to the desired property value or value range, to meet the specified product type. Finally, one or more optimization variables can be identified as an objective function, to be minimized (such as cost) or maximized (such as robustness to process changes).

The standard form of a cost optimization problem can be then expressed in the following form:[14,15]

Find a vector of optimization variables, $\mathbf{x} = (x_1, x_2, ..., x_n)^T$ in order to minimize the objective function, $f(\mathbf{x})$

subject to

$g_i(x) \leq 0$ (or $g_i(x) \geq 0$), $\quad i = 1, 2, ..., m$, $\quad$ less or higher type inequality constraints

$h_i(x) = 0$ $\quad\quad\quad\quad\quad i = 1, 2, ..., p$ $\quad\quad$ equality constraints

$x_{iL} \leq x_i \leq x_{iU}$ $\quad\quad\quad\quad i = 1, 2, ..., n$ $\quad\quad$ bounds on optimization variables

Thus, for a general optimization problem, p must be less than n. Since the inequality constraints do not represent a specific relationship among optimization variables, there is no limit on the number of inequality constraints in a problem.

This work describes the use of the methodology of mixture design to calculate regression models relating properties of dried ceramic bodies (bending strength), and fired ceramic bodies (bending strength, bulk density, thermal expansion coefficient, linear firing shrinkage and water absorption) with the proportions of clay, feldspar and quartz present in the original raw materials mixture. The regression models so obtained were then simultaneously applied to choose the composition range best suited to produce low-cost ceramic tiles.

EXPERIMENTAL

The raw materials used were: a clay mixture, potash feldspar (99.5 wt.% microcline) and quartz sand (99.5 wt.% α-quartz), all supplied by Colorminas (Criciúma-SC, Brazil). The corresponding chemical compositions were determined by X-Ray Fluorescence (XRF). The crystalline phases present were identified by X-Ray Diffraction (XRD) and quantified by rational analysis.[16]

In the particular case of ceramic mixtures, some lower bound restrictions on the contents of clay, feldspar and quartz must be used, due to processing limitations. A restricted composition triangle has then to be defined.[7,8] Within this pseudocomponents triangle, a simplex-centroid lattice design {3,2}, augmented with interior points, was used to define the mixtures of those raw materials that should be investigated, Figure 1.

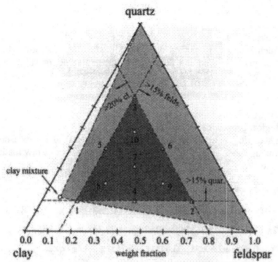

Figure 1. A ternary ceramic system with composition restrictions.

The selected mixtures were processed, following a conventional ceramic tile industrial procedure: wet grinding (residue left in a 325 mesh sieve below 1 wt.%), drying, moisturizing (6.5±0.2 wt.%, dry basis), granulation and uniaxial pressing (Micropressa Gabbrielli, 10 ton hydraulic press). The particle size distribution of the unfired bodies presents a typical bimodal configuration with the first maximum for clay minerals between 2 and 7 μm and the second maximum at about 10-20 μm for quartz and feldspar. The cumulative curves indicate that all the particles are below 35 μm.

To determine the bending strength and the linear firing shrinkage, flat test pieces (50 x 8 x 5 mm3, using 4.0 g of material per test piece) were prepared from each mixture; water absorption was determined on cylindrical test pieces from each mixture (20 x 10 mm3, using 4.5 g of material per test piece). The compaction pressure, in both cases, was 47 MPa. After compaction, the test pieces were oven dried at 110±5 °C until constant weight, fired at 1170 °C for 1 h (heating at 3.20 °C/min up to 600°C, and at 4.75 °C/min from 600 to 1170°C), and naturally cooled.

The linear firing shrinkage (FLS) was calculated from the change in length (measured with Mitutoyo callipers with a resolution of 0.05 mm), upon firing, of the flat test pieces. The modulus of rupture of dried (MoRD) and fired (MoRF) test pieces was determined in three-point bending tests, using a digital Shimadzu Autograph AG-25TA test machine, with a 0.5 mm/min cross-head speed until rupture (in accordance with the standard ISO 10545-4.[12] In all cases, the final test result was taken as the average of the measurements carried out on ten test pieces.

The water absorption (WA) was determined via boiling in water for 2 h, using a Denver DE 100A digital analytical scale with a resolution of 0.1 mg. The bulk density of fired test pieces (BD) was determined using Archimedes' liquid displacement method by immersion in Hg. Both procedures were carried out according to the standard ISO 10545-3.[11] In both cases, the final test result was taken as the average of the measurements carried out on five test pieces.

Those final results were then used to iteratively calculate the coefficients of five regression equations, until statistically relevant models were obtained, relating the MoRD, MoRF, FBD, FLS and WA with the proportions of the clay, feldspar and quartz present in the unfired mixture. Calculations were carried out with Statistica 5.0 (StatSoft, 2000). The type of ceramic product to be manufactured and the desired property value, or value range, to meet the specified product type were then defined and a graphical solution of the simultaneous inequalities was sought.

RESULTS AND DISCUSSION

Models for technological properties

Based on a statistical analysis of significance (not shown) of the measured results of all properties, the following models were selected to describe the effect of raw materials on MoRD, MoRF, FLS, FBD and WA:

$$MoRD = 0.99x_1 - 1.19x_2 - 1.67x_3 + 12.90x_1x_2 + 14.34x_1x_3 + 0.70x_2x_3 \tag{1}$$

$$MoRF = 61.64x_1 + 46.08x_2 + 14.06x_3 + 21.06x_1x_2 - 112.68x_1x_3 - 31.54x_2x_3 + 116681x_1x_2x_3 \tag{2}$$

$$FLS = 14.88x_1 + 6.72x_2 - 11.94x_3 - 3.36x_1x_2 + 20.62x_1x_3 + 77.24x_2x_3 - 134.85x_1x_2x_3 \tag{3}$$

$$FBD = 2.12x_1 + 1.68x_2 + 0.40x_3 + 2.60x_1x_2 + 3.84x_1x_3 + 5.76x_2x_3 - 12.37x_1x_2x_3 \tag{4}$$

$$WA = 0.41x_1 + 7.05x_2 + 44.87x_3 + 1.66x_1x_2 - 55.84x_1x_3 - 99.42x_2x_3 \tag{5}$$

In these Equations, x_1 is the clay fraction, x_2 is the feldspar fraction and x_3 is the quartz fraction.

Low cost optimization

Assuming that the raw materials studied in this work were to be used in the production of porcelain stoneware floor tiles, hence in the AI (extruded) or BI (pressed) groups specified by the European standard EN 87,[9] a set of constraints can be placed on the values of MoRF, FBD and WA:

MoRF $\geq$ 65 MPa (6)

FBD $\geq$ 2.30 g.cm^{-3}, and (7)

WA $\leq$ 0.5 % (8)

Industrial processing imposes extra constraints, namely on the MoRD and the FLS:

MoRD $\geq$ 3.0 MPa, and (9)

FLS $\leq$ 9.5 % (10)

Thus, the selection of a feasible mixture composition range to satisfy all constraints corresponds to finding an intersection of Equations 1 to 5 combined with the inequalities 6 to 10.

This optimization problem aims to find a vector of optimization variables (x_1, x_2 and x_3, which are, respectively, the clay, feldspar and quartz fraction) in order to minimize the objective function, which is the cost:

$$Cost = 9.74x_1 + 77.89x_2 + 32.65x_3 \tag{11}$$

at level of

Cost $\leq$ 34 US$/ton (12)

After all constraint contours are drawn (Equations 1 to 5, and 11), the intersection of adequate sides of all constraints represents the feasible region for the optimization.

Validation of the experiments
The calculated statistical model and the resulting optimization, which gives an estimate for the cost, as a function of the clay, feldspar and quartz contents, subjected to the constraints, was checked using a validation run. Two extra mixtures (11 and 12, Table I) were selected within the feasible region to test the applicability of the statistical and mathematical optimization models. The test compositions were prepared and characterized as described earlier. Table I shows the mixture compositions, the property values and compares the latter with the constraints imposed on the properties.

Table I. Composition of check point mixture and corresponding measured values of MoRD, FLS, MoRF, FBD and WA.

Design mixture	Weight fractions			MoRD (MPa)	FLS (%)	MoRF (MPa)	FBD (g/cm^3)	WA (%)
	x_1	x_2	x_3					
11	0.500	0.260	0.240	3.00	9.11	73.54	2.39	0.10
12	0.560	0.210	0.230	3.00	9.47	68.66	2.36	0.19
Property constraints:				≥ 3.00	≤ 9.50	≥ 65	≥ 2.30	≤ 0.50
Predicted value, mixture 11				3.20	8.97	69.82	2.37	0.12
Predicted value, mixture 12				3.30	9.34	65.42	2.38	0.19

Although quite apart in terms of composition, these mixtures present similar values of MoRF, MoRD, COST, FBD, FLS and WA, all within the specified range of values for the properties. The cost value obtained for mixture 12 was 29.27 US$/ton, which is also comparable to the cost of 29.32 US$/ton, predicted by the relevant model.

CONCLUSIONS
The design of mixture experiments and the use of response surface methodologies enabled the calculation of regression models relating technological properties to the initial combination of raw materials. Furthermore, the use of intersecting surfaces shows that, for the particular raw materials and processing conditions used, an optimized region can be found, in order to minimize materials cost, subjected to restrictions imposed by specification standards.

In this way, a range of compositions can be selected so that the final product has the desired properties and can be successfully manufactured, both from technical and economical points of view.

ACNOWLEDGEMENTS
The authors appreciate the financial support received from the Brazilian Research Agency Capes (Coordenação de Aperfeiçoamento de Pessoal de Nível Superior), and are thankful to Colorminas for providing the raw materials used throughout the work.

REFERENCES
[1] L.B. Hare, "Mixture designs applied to food formulation," *Food Technology*, **28**, 50-62 (1974).

[2] M.J. Anderson, and P.J. Whitcomb, "Optimization of paint formulations made easy with computer-aided design of experiments for mixtures," *Journal of Coatings Technology*, **68**, 71-75 (1996).

[3] G. Piepel, and T. Redgate, "Mixture experiment techniques for reducing the number of components applied for modelling waste glass sodium release," *Journal of the American Ceramic Society*, **80**, 3038-3044 (1997).

[4] S.L. Hung, T.C. Wen, and A. Gopalan, "Application of statistical strategies to optimise the conductivity of electrosynthesized polypirrole," *Materials Letters*, **55**, 165-170 (2002).

[5] L.M. Schabbach, A.P.N. Oliveira, M.C. Fredel, and D. Hotza, "Seven-component lead-free frit formulation," *American Ceramic Society Bulletin*, **82**, 47-50 (2003).

[6] S.L Correia, K.A.S. Curto, D. Hotza, and A.M. Segadães, "Using statistical techniques to model the flexural strength of dried triaxial ceramic bodies," *Journal of the European Ceramic Society*, **24**, 2813-2818 (2004).

[7] R.H. Myers, and D.C. Montgomery, *Response surface methodology: process and product optimization using designed experiments*, John Wiley and Sons, New York, 2002.

[8] J.A. Cornell, *Experiments with mixtures: designs, models and the analysis of mixture data*, John Wiley and Sons, 3rd edition, New York, 2002.

[9] EN 87: Ceramic floor and wall tiles — definitions, classification, characteristics and marking, 1992.

[10] ISO 10545-2: Ceramic tiles — Part 2: Determination of dimensions and surface quality, 1998.

[11] ISO 10545-3: Ceramic tiles — Part 3: Determination of water absorption, apparent porosity, apparent relative density and bulk density, 1998.

[12] ISO 10545-4: Ceramic tiles — Part 3: Determination of modulus of rupture and breaking strength, 1998.

[13] ISO 10545-8: Ceramic tiles — Part 8: Determination of linear thermal expansion, 1998.

[14] M. A. Bhatti, *Practical optimization methods with Mathematica applications*, Springer Telos, New York, 2000.

[15] R. E. Miller, *Optimization foundations and applications*, Wiley, New York, 2000.

[16] C. Coelho, N. Roqueiro, and D. Hotza, "Rational mineralogical analysis of ceramics," *Materials Letters*, **52**, 394-398 (2002).

MICROSTRUCTURAL AND MECHANICAL PROPERTIES OF DIRECTIONALLY SOLIDIFIED CERAMIC IN Al$_2$O$_3$-Al$_2$TiO$_5$ SYSTEM

A. Sayir,[1] M. H. Berger[2] and C. Baudin[3]

[1] NASA-GRC / CWRU, 21000 Brookpark Rd., Cleveland, OH 44135, USA
[2] Ecole des Mines des Paris, 91003 Evry Cedex – FRANCE
[3] Instituto de Cerámica y Vidrio Camino de Valdelatas 28049 –Madrid - SPAIN

ABSTRACT

The mechanical properties of two-phase and poly-phase eutectics can be superior to that of either constituent alone due to the strong constraining effects of the interlocking microstructure. The present work focuses on the solidification characteristics and mechanical properties of Al$_2$O$_3$-Al$_2$TiO$_5$ system. The challenge for the development of Al$_2$O$_3$-Al$_2$TiO$_5$ system is to improve the mechanical strength and toughness concurrently with a high resistance to thermal decomposition. The solidification at the invariant eutectic point and Al$_2$O$_3$ rich region of off-eutectic compositions was studied. Critical to this effort is the correlation of mechanical properties with eutectic growth data. The strength is strongly related to the starting volume fraction of the minor phase. High strength in the order of 340 MPa is associated with high Al$_2$O$_3$ content. Bend tests showed a large displacement for all compositions studied. The off-eutectic composition had Al$_2$O$_3$ phase, new Al$_6$Ti$_2$O$_{13}$ phase as a major phase and Al$_2$TiO$_5$ as a minor phase as determined by the WDX and FEG-TEM-STEM techniques. Samples from eutectic region consisted only Al$_6$Ti$_2$O$_{13}$ and Al$_2$TiO$_5$ phases that bear a resemblance to layered structured materials. The proposed structure of new Al$_6$Ti$_2$O$_{13}$ phase contained one more AlO$_6$ octahedra along [010] direction and the proposed structure was confirmed by the x-ray, HRTEM, STEM and WDX analysis. The spatial arrangement of the new Al$_6$Ti$_2$O$_{13}$ phase and Al$_2$TiO$_5$ phase around the reinforcing Al$_2$O$_3$ dendrites and microcracking were responsible for the improved toughness.

1. INTRODUCTION

Directionally solidified eutectics (DSE) have the potential to be utilized as structural components at elevated temperatures for aerospace applications. The high temperature microstructural stability, large aspect ratio and strong interphase bonding found in many DSE's contribute to their superior creep resistance, compared to conventional ceramics [1]. The eutectic architecture, a continuous reinforcing phase within a higher volume phase or matrix, can be described as a naturally occurring *in-situ* composite. *In-situ* composites exhibit mechanical properties intermediate between monolithic materials and man-made composites [2, 3, 4]. The degree of microstructural anisotropy of DSEs, although a necessary condition in promoting improvements in high temperature mechanical properties, is insufficient to achieve higher toughness. The interfaces between the two phases of eutectic structure typically adopt low-energy orientation relationships during the directional solidification process, this provides strong bonding [5,6, 7] and prevents significant interface debonding. This characteristic of the interface is responsible for the low fracture toughness. The fine eutectic lamellae have little effect in diverting the path of the fracture crack.

It is however conceivable that through incorporation of controlled amounts of primary phase, i.e., the fracture energy of a brittle matrix could be increased. The protruding plates or rods are expected to promote crack deflection at the interfaces between eutectic and primary phase in carefully engineered systems. To examine these postulations, Al_2O_3-Al_2TiO_5 system was selected for directional solidification study. The system was selected, on the basis of previous studies on Al_2O_3-Al_2TiO_5 monolithic composites and layered structures, because of its capacity for debonding and microcracking due to residual thermal stresses [8-9]. The range of structures that can be produced by off-eutectic solidification was examined for the increase in toughness by a controlled amount of primary phase. In addition, the work aims to utilize the anisotropy of the pseudobrookite structure of Al_2TiO_5 phase to promote microcrack formation and to identify toughening mechanisms in the DSE Al_2O_3-Al_2TiO_5 system [10]. The phase diagram of Al_2O_3-Al_2TiO_5 system [11] and solidification reaction at the eutectic invariant point was attractive to produce desirable new phases and microstructures. The fracture paths can be altered through selection of polyphase structures that promotes microcrack formation and debonding. The challenge for the development of Al_2O_3-Al_2TiO_5 is to improve the mechanical strength and toughness concurrently with a high resistance to thermal decomposition.

2. EXPERIMENTAL

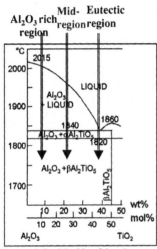

Fig. 1 Al_2O_3 rich side of the phase diagram [9] and studied compositions.

The materials were processed, by the laser heated float zone method [3, 6, 7] using a pulling rate of 5 to 10 mm/min, in the shape of rods of about 4 mm diameter. Source specimens (10 mm in diameter) were fabricated by wet mixing, isostatic pressing into a rod shape and sintering (1450 °C in air). The raw materials were high purity Al_2O_3 (99.999 wt%) and Al_2TiO_5 (99.95 wt%, SiO_2 major impurity). Three different materials were prepared, that will be designated within the region in which their composition is located in the Al_2O_3-Al_2TiO_5 phase diagram: Al_2O_3-rich (80wt% of Al_2O_3-20wt% of Al_2TiO_5), mid-region (50wt% of Al_2O_3-50wt% of Al_2TiO_5) and eutectic region (Fig. 1). The microstructure was examined on cross sections of the directionally solidified rods that were mechanically polished samples using field effect gun – scanning electron microscopy FEG –SEM (LEO Gemini 982). Chemical mappings and profiles were carried out using wavelength dispersive x-ray spectrometry (WDX, CAMECA SX 50). The nanostructure was investigated in transmission electron microscopy using a 200 kV FEG-TEM-STEM (Scanning transmission electron microscopy) equipped with an energy dispersive X-ray spectrometer (FEI Tecnai F20) and 300kV FEG-TEM (Philips CM 300 0ÅM)

3. RESULTS AND DISCUSSION

The microstructures of ceramic that was directionally solidified at the off-eutectic composition of 12m% TiO_2 (Al_2O_3 rich region) are shown in figure 2. The microstructure consists of primary Al_2O_3 dendrites within a continuous bright aluminium titanate phase. The Al_2O_3 dendrites had aspect ratios exceeding a factor of 100 and have a strong tendency to align along the solidification direction. The shape of Al_2O_3 dendrite (the discontinuous phase) was controlled by the imposed solidification conditions and by the anisotropic growth characteristics of the Al_2O_3 structure.

The structure consists of two spatially different microstructural regions, an outer rim and central region. The outer rim region contains closely packed Al_2O_3 dendrites (Fig. 2a) and is bounded by coarse Al_2O_3

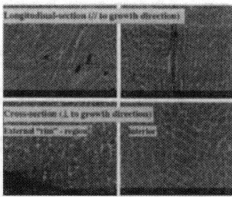

Fig. 2 Microstructure of the directionally solidified Al_2O_3-Al_2TiO_5 eutectic system, 12 m % TiO_2.

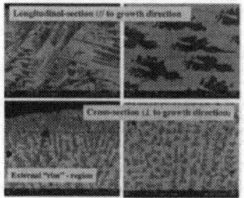

Fig. 3 Microstructure of the directionally solidified Al_2O_3-Al_2TiO_5 eutectic system, 26 m % TiO_2.

dendrites at the interior. In a few cases, the boundary region between outer rim and central region contained porosity. The spatial distribution of the Al_2O_3 dendrites was determined by the three-dimensional heat transfer condition of the molten zone. The rate of heat removal from the molten zone surface at the liquid solid interface was considerably higher than the interior and produced unique microstructure. The solidification reaction of this system can be explained by considering the kinetics of the process and nucleation frequency of the Al_2O_3 phase. The rate of growth of the eutectic is controlled by the temperature of the liquid-solid interface, which in turn is controlled by the extraction of heat from the surface of the molten zone. The volume fraction of the Al_2O_3 dendrites is considerably higher at the rim region than at the interior. The effect of concentration of Al_2O_3 dendrites at the outer rim region to build a tightly skin layer is twofold. First, the large central region is composed of smaller Al_2O_3 dendrites since the overall composition of the liquid is richer in titanium and microstructure adapts to reflect the compositional modification. Hence, any attempt to estimate the volume fraction of phases using limited SEM analysis would lead to incorrect estimates and should require extensive SEM analysis to represent the whole sample. Second, the mechanical properties are strongly affected from the formation of rim region containing primarily Al_2O_3 dendrites as will be discussed.

Figure 3 shows representative microstructural characteristics of the directionally solidified ceramic from off-eutectic composition of 26 m % TiO_2 (mid-region in Fig. 1). The microstructure shows Al_2O_3 dendrite and noticeably larger volume content of titanate phase (bright color). The spatial arrangement of Al_2O_3 dendrites for 26 m % TiO_2 eutectic was similar to 12 m % TiO_2 eutectic with two microstructural regions, an outer rim and central region (Fig. 2 and 3). On the other hand, the outer rim region (~150 μm) of 26 m % TiO_2 composition contained a higher concentration of Al_2O_3 dendrites and also large entrapped pores were observed between rim- and interior regions. The solidification characteristic of Al_2O_3 - Al_2TiO_5 eutectic system for the mid-region (26 m %TiO_2) is controlled more by the efficacy of heterogeneous nucleation centers in the liquid than by any other factors. Under the fast solidification conditions applied in the present investigation, Al_2O_3 dendrite grew ahead of the eutectic, as it is apparent from the formation of well developed Al_2O_3 dendrites (Fig. 3). An increase of TiO_2 content of DSE decreases the extension of dendrites into the liquid as observed from SEM micrographs of

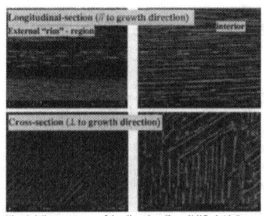

Fig. 4 Microstructure of the directionally solidified Al₂O₃-Al₂TiO₅ at eutectic invariant point 43.9 m % TiO₂.

longitudinal sections. To understand the shortening of the Al_2O_3 dendrite length, it is more pertinent to compare the temperature gradient for the different compositions. An increase of TiO_2 concentration required a larger amount of heat to melt an equivalent amount of molten zone compared to Al_2O_3 rich region. The additional amount of heat requirement for increased TiO_2 content was an indication for the modification of the thermal gradient in radial and axial direction of the molten zone. The alteration of he thermal gradient contributed for the shortening of the Al_2O_3 dendrite length.

The comparison of microstructures obtained at different compositions from off-eutectic region, convey additional information about the Al_2O_3 dendrites. Unlike the process of lamellar growth, the Al_2O_3 dendrite does not grow exactly normal to the moving interface and not exactly parallel to each other. The Al_2O_3 and titanate phases diverge as they solidify as shown in Figs. 2 and 3. If the interphase spacing between Al_2O_3 and titanate phase becomes greater than that of the imposed growth rate, a new Al_2O_3 dendrite has to nucleate in the region between them. Conversely, if two Al_2O_3 dendrites approach each other on the interface, one phase must be terminated due to an insufficient supply of cations in the liquid, for example Al^{3+}. The thickness and lateral spacing of the Al_2O_3 dendrites is determined by the growth rate, and their length is determined by the convergence of two or more phases at the interface. This representation was in agreement with the SEM observations in Figs. 2 and 3, so that solidification was a constantly repeating cycle of nucleation of Al_2O_3 dendrites whenever necessary to keep the interphase spacing at the required value, and their growth was simultaneous with the titanate phase. The epitaxial relationship between Al_2O_3 and titanate phase was confirmed by HRTEM as will be discussed.

The microstructure of directionally solidified materials at the eutectic invariant point (43.9 m % TiO₂) is shown in Fig. 4. The temperature gradient for the eutectic solidification was large enough, so that the lead distance of dendrites become comparable to eutectic and the dendrite formation of Al_2O_3 phase is suppressed even for a very high solidification rate of 10 mm/min. The structure was markedly different than those of off-eutectic compositions, Figs. 2 and 3. Figure 4 exhibits features of a lamellar eutectic structure; both phases were continuous in the growth direction. One of the features of $Al_2O_3 - Al_2TiO_5$ eutectic system was that the titanate phase was discontinuous in the plane normal to the growth direction, as can be seen in Fig. 4c and 4d. The discontinuity of the titanate lamellae was a consequence of 'faulted" [10] regions or "mismatch surfaces" which give the impression that the lamellae have performed a shearing movement normal to their length. The lamellae spacing was approximately 1 to 3 μm which was substantially smaller than the size of Al_2O_3 dendrites observed for the materials solidified at the off-eutectic compositions (Figs. 2 and 3).

The transition from a eutectic lamella structure to a dendritic plus eutectic interface cannot be predicted from the constitutional undercooling arguments alone because constitutional undercooling always exists ahead of the eutectic interface. The understanding and predictions can be made using arguments originally put forward by Tamman and Botschwar [11] which suggest the formation of eutectic or

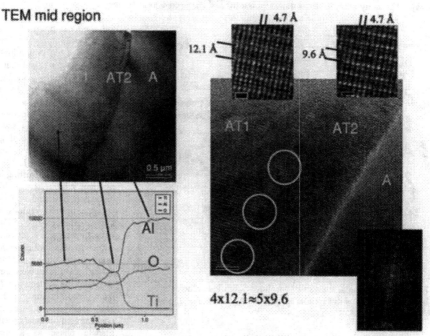

Fig. 5 HRTEM image of polyphase structure obtained from specimen solidified at 26 m % TiO$_2$. The structure and chemical composition of A, AT1 and AT2 phases are discussed in the text.

dendrite plus eutectic, depended on which structure grew faster at given undercooling for the compositions studied (Fig.1). The growth rate for the eutectic composition in the Al$_2$O$_3$ – Al$_2$TiO$_5$ eutectic system increases more readily with undercooling than those of dendrites because of the short diffusion path maintained in the eutectic, see Fig. 4. This conclusion is further supported with high solidification rate experiments (10 mm/min) that also produced lamella structures similar to those shown in Fig. 4. In the absence of information on the growth rate of the primary dendrite phases and eutectic structure, we restrict our statement to qualitative rather than quantitative until a systematic study of the growth rates of each phase is determined.

This discontinuous solidification characteristic of Al$_2$O$_3$ - Al$_2$TiO$_5$ eutectic system offers good explanation for the development of Al$_2$O$_3$ phase but provide very little understanding for the microstructural development of the titanate phase. The Al$_2$O$_3$ dendrite forms as a primary phase in Al$_2$O$_3$-rich and mid-region. The formation of Al$_2$O$_3$ dendrite necessitates the formation of two-phase eutectic structure adjacent to the Al$_2$O$_3$ dendrites. The expected microstructure for both Al$_2$O$_3$-rich and mid-region should consist of dendritic Al$_2$O$_3$ as primary phase plus eutectic structure (Al$_2$O$_3$ and Al$_2$TiO$_5$ phases). The BSE-SEM images shown Figs. 2 and 3, however, did not reveal different phases between the Al$_2$O$_3$ dendrites. The attempts to index the x-ray peaks using Al$_2$TiO$_5$ phase was not successful. Several close forms of orthorhombic Al$_2$TiO$_5$ have been reported in the literature. The orthorhombic forms of Al_2TiO_5 have different group symmetry and are expected to modify the forbidden reflections of x-ray. The x-ray analysis did not corroborate with the expected forbidden reflections to substantiate the orthorhombic phases. From the chemical point of view, the possibility of solid solutions of higher and lower oxidation states exist because samples were strongly colored (blue for 11 m %TiO_2, dark blue for 26 m %TiO_2 and black for 43.9 m%TiO_2). Al_2O_3 is stable and has a cation Al^{3+} a single valence state (a high ionization potential). In contrast, TiO_2 consisting of titanium cation have a low ionization potential and can show extensive regions of non-stoichiometry, TiO_{2-x}, with oxygen pressure. Although the solidification was carried out in air, the solidification rates were very high and the system was far from equilibrium. The analytical efforts concentrated to analyze the titanate phase as a non-stoichiometric and oxygen-deprived phase.

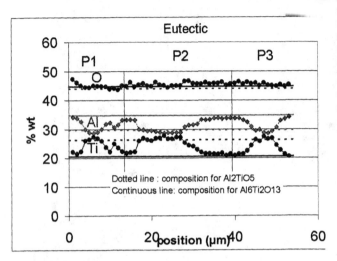

Fig. 6   Chemical analysis of specimen solidified at the eutectic invariant point, 43.9 m % TiO_2.

TEM observations and the SAD patterns shown in Fig. 5 identified the dark phase as α-Al_2O_3 but the pattern of the bright phase of Figs. 2 and 3 could not be assigned to Al_2TiO_5 orthorhombic phase (inter-reticular distances: 1.2 nm and 0.47 nm). The TEM analysis was consistent with EDAX. The x-ray analysis showed that titanate phase was not Al_2TiO_5 alone which agreed with previous observation [12]. Observation of the interface between titanate and alumina phases, Fig. 5, revealed another crystalline interphase layer of a few hundreds nanometers. The STEM-EDX analysis across this interface is also shown in Fig 5. It can be seen that this interphase was another titanate phase, denoted here *AT2*, richer in titanium than *AT1*. High-resolution TEM images of the interface are presented in Figs. 5 together with a Fourier transformation. The inter-reticular distances measured for the *AT2* phase are ~0.47 nm and ~0.97 nm (orthogonal planes). A chemical analysis using STEM-EDX technique has been carried out to investigate the composition across the phases. The concentration measurements were made at over 60 different positions separated by 25 nm intervals using 3 nm probe size along the length scale of 40 to 50 µm. Both plots for mid region and eutectic region are shown in Fig. 6. The chemical analysis of this work showed that *AT1* and *AT2* were corresponding to $Al_6Ti_2O_{13}$ and Al_2TiO_5, respectively. This conclusion corroborate with the x-ray and HRTEM analysis. The 020 and 220 peaks of Al_2TiO_5 are close to $Al_6Ti_2O_{13}$. The indexation of the x-ray data could be attained unequivocally using cell parameters a = 9.439, b = 12.1, c = 3.65 Å for $Al_6Ti_2O_{13}$ and a = 9.439, b = 9.647, c =3.593 Å for Al_2TiO_5 phase. $Al_6Ti_2O_{13}$ phase is constructed of layers that are close to Al_2TiO_5 but had different stacking layers. Based on the STEM-EDX and HRTEM analysis the suggested structure for $Al_6Ti_2O_{13}$ contains one additional

AlO$_6$ octahedron along 010 direction and detailed description of the new Al$_6$Ti$_2$O$_{13}$ structure will be published separately. The HRTEM analysis further revealed epitaxial relationship between Al$_2$O$_3$ and Al$_2$TiO$_5$ phases, [001] Al$_2$TiO$_5$ // [10$\overline{1}$1] Al$_2$O$_3$.

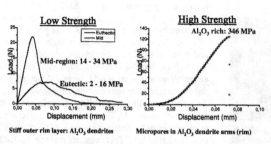

Stiff outer rim layer: Al$_2$O$_3$ dendrites   Micropores in Al$_2$O$_3$ dendrite arms (rim)

Fig. 7 Representative data on the bend strength.

The x-ray analysis of DSE samples solidified at eutectic composition did not show any Al$_2$O$_3$ phase. The absence of Al$_2$O$_3$ phase was also confirmed using SEM, STEM-EDX and HRTEM analytical techniques. This result was contrary to all available phase diagram studies and requires further investigation. These uncertainties to assess the solidification reaction characteristic are compounded when the ideas are applied to Al$_2$O$_3$ - Al$_2$TiO$_5$ eutectic system because the composition of the eutectic point throughout the argument may well be somewhat incorrect as it is the case for many ceramic systems. Nevertheless, we studied structure even though the exact eutectic composition and clear phase diagram may currently not be known. The samples solidified at eutectic composition consist of lamella of Al$_6$Ti$_2$O$_{13}$ and Al$_2$TiO$_5$ phases and had a strong tendency for delamination. The mechanical properties were drastically different for the samples solidified at eutectic composition than those of the samples produced having compositions at the off eutectic region.

The mechanical properties of the specimen produced from the Al$_2$O$_3$-rich region are shown in Fig. 7. The mechanical strength of Al$_2$O$_3$ rich region exceeded 300 MPa and this level of strength is significantly higher than the reported strength values of 15 MPa for the polycrystalline Al$_2$TiO$_5$ system. The non-linear characteristic of the load displacement curves can be explained using spatial distribution of the micropores. The exterior rim region of the specimens contains high density of micropores and fracture typically originates from these pores. The high concentration of the micropores form a complaint outer rim region and is responsible for the nonlinear characteristic of the load displacement curve during the initial loading. The load displacement curves show high strain capability that has not been previously achieved for the polycrystalline Al$_2$O$_3$-Al$_2$TiO$_5$ system or any other oxide DSE. An apparent strain calculated from the displacement measurements were about 0.3 % for maximum load. The fracture surfaces of the specimens were tortuous as shown in Fig. 8. The degree of lifting of Al$_2$O$_3$ dendrite changes from the exterior rim region to the interior due to the changing complaints of the specimen. The Al$_2$O$_3$ dendrite shows pull-out characteristics analogous to the continuous fiber reinforced composites and pull-out lengths are typically 5 to 40 µm. The crack deflection occurs at the boundaries of the Al$_2$O$_3$ dendrites that are microcracked. The arms of the dendrites provide additional interlocking between dendrites to carry the load cooperatively through load sharing. This load sharing between the dendrites were responsible for high strength results achieved for the directionally solidified Al$_2$O$_3$ - Al$_2$TiO$_5$ system even though the material was significantly microcracked. The complex sequence of events including decohesion between the phases, crack deflection and branching were responsible for the non-linear behaviour during loading preceding the fracture.

The mechanical properties of the specimen produced from the mid-region (26 m % TiO$_2$) of the phase diagram exhibited lower strength, 14 to 34 MPa. These low strength results were due to large internal voids. The load displacement curves were distinctly different than the specimen produced from Al_2O_3-rich region. The higher concentration of Al_2O_3 dendrites at the outer rim region increased the stiffness of the specimen and was responsible for the steep and linear portion of the initial loading curve. The lower strength values are due to the increased density of microcracks originating from the presence of the Al_2TiO_5 and $Al_6Ti_2O_{13}$ phases and shortening of Al_2O_3 dendrites. The dendrites which were short and highly interpenetrated by the Al_2O_3-TiO_2 phase in the mid-region (Figs. 1a, 2a) specimens were easily traversed by the cracks whereas those present in the Al_2O_3-rich material, which have high aspect ratio and were much less interpenetrated, deflected the crack (Figs. 1b, 3b). Significantly larger amount of an apparent strain of 0.8 % were calculated from the displacement measurements for maximum load. This increased strain capability is attributed to the formation new $Al_6Ti_2O_{13}$ phase. Extensive microcracks at the interface were observed by SEM and TEM analysis. This suggests that the $Al_6Ti_2O_{13}$ phase have similar thermal expansion characteristic as the Al_2TiO_5 phase. This conclusion can be explained by revisiting the pseudobrookite structure of Al_2TiO_5 phase which has 4 AlO_6 octahedra along [010]. Based on our structural analysis, we proposed that $Al_6Ti_2O_{13}$ phase was built by adding one more AlO_6 octahedra along [010] direction and this proposed structure agrees with previous work [13]. This suggestion will also explain the ratio of 5/4 determined for the different b axis parameter of the $Al_6Ti_2O_{13}$ and Al_2TiO_5 phases. The related structures of Al_2TiO_5 and $Al_6Ti_2O_{13}$ phases were responsible for the strong microcrack formation tendency due to extreme anisotropy of their thermal expansion coefficients. The presence of $Al_6Ti_2O_{13}$ phase increased the tendency for delaminating the phases as observed by fractographic analysis.

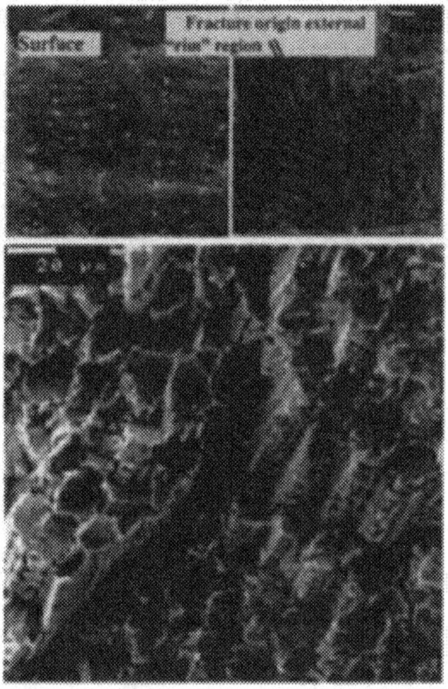

Fig. 8 The exterior rim region contains high density of micropores (top left). Fracture originates from this complaint region (top right). Tortuous fracture characteristics and lifting of Al_2O_3 dendrites (bottom).

The mechanical properties of specimen produced from the eutectic region were significantly different and showed the largest amount of the strain capability and low strength as shown in Fig. 7. The micro crack formation, crack arrest and extensive pull-out along the lamella contributed to the increased toughness. The physical appearance showed distinct cleavage tendency and decohesion characteristics that were similar to the weakly bonded materials like boron nitride or graphite. The epitaxial relationships between Al_2TiO_5 and $Al_6Ti_2O_{13}$ phases and periodically insertion of an additional plane in the $Al_6Ti_2O_{13}$ phase may act as layered structured compound to promote weak layer. The exact effect of the low ionization potential of titanium cations and their distribution at the interface between the Al_2TiO_5 and $Al_6Ti_2O_{13}$ phases require further study.

CONCLUSIONS

The microstructure of the specimens that solidified at Al_2O_3 rich side of the phase diagram was quasi-regular arrays of Al_2O_3 dendrite in a continuous aluminium titanate matrix. A good degree of geometrical relationship between the two phases has been observed even though the solidification rates were extremely high (3 to 10 mm/min). This suggests that a complete variety of microstructures are possible from a single composition depending on the imposed condition of solidification, specifically at different ratios of solidification rate to the thermal gradient at the liquid solid interface. The off-eutectic composition had Al_2O_3 phase, new $Al_6Ti_2O_{13}$ phase as a major phase and Al_2TiO_5 as a minor phase as determined by the WDX and FEG-TEM-STEM techniques. The epitaxial relationship between Al_2O_3 and Al_2TiO_5 phases has been determined using HRTEM analysis. Samples from eutectic region consist only $Al_6Ti_2O_{13}$ and Al_2TiO_5 phases that bear a resemblance to layered structured materials. The proposed structure of new $Al_6Ti_2O_{13}$ phase was built by adding one more AlO_6 octahedra along [010] direction and confirmed by the x-ray, HRTEM, STEM and WDX analysis. The formation of new $Al_6Ti_2O_{13}$ phase is in contradiction to the published phase diagrams. The composition of the eutectic point and the phase diagram may be somewhat incorrect as is the case for many ceramic systems and require further study.

Directionally solidified samples from the Al_2O_3 - Al_2TiO_5 system showed high strength and toughness. The trade-off between strength and toughness is extremely encouraging. The directional solidification processing offer a potential to produce concurrent improvement of strength and toughness. Opportunity for this arises because of the sequence of phase formation such that desirable phases and morphologies can be obtained. So far, the ideas outlined above presents only a qualitative picture of the partitioning of cations in polyphase structures. Much patient work will be necessary to establish details of the modification process and to determine numerical values of the partition coefficients of third elements between the liquid and solid phases, upon which the modification would appear to be dependent.

ACKNOWLEDGEMENTS
This research was funded by NASA cooperative agreement NNC04 AA27A, the European Office of Aerospace Research and Development by AFOSR under Grant No. FA8655-03-1-3040, the Cooperative research CSIC(Spain, 2004FR0024)-CNRS(16212, France) , CAMGRMAT07072004 (Spain) and CICYT, MAT 2003-00836 (Spain).

REFERENCES
1. A. Sayir, in Computer-Aided Design of High-Temperature Materials. Eds. A. Pechenik, R.K. Kalia, P. Vashista, *Oxford University Press (1999) pp.197 – 211.*
2. V. S. Stubican and R. C. Bradt, Ann. *Rev. Mater. Sci., 11 (1981) 267-297.*
3. A. Sayir and S. C. Farmer, *Acta Mat., 48 (2000) 4691 - 4697.*
4. T. A. Parthasarathy, T. Mah, and L.E. Matson, *J. Amer. Ceram. Soc., 76 [1] 29-32 (1993).*
5. A. S. Argon, J. Yi and A. Sayir, *Mat. Sci. and Eng., A319 –321 (2001) 838 - 842.*
6. C. Frazer, E. Dickey and A. Sayir, *J. Crystal Growth, 232 [1-2] (2001) 187- 195.*
7. S. Bueno, R. Moreno, C. Baudín, *J. Eur. Ceram. Soc., 24 (2004) 2785-2791.*
8. S. Bueno, R. Moreno and C. Baudín, *J.Eur. Ceram. Soc., 25 [6] (2005) 847-856.*
9. C. Baudín, A. Sayir and M. H. Berger, *Key Eng. Mater., Trans Tech Publications, Switzerland, 2005.*
10. S. M. Lang, C. L. Fillmore and L. H. Maxwell, *J. Res. Natl. Bur. Stds., 48 (1952) 298-302.*
11. G. A. Chadwick, Solidification, *ASM, Metals Park, Ohio (1971) pp.9-14.*
12. G. Tammann and A. A. Botscwar, *Z. Anorg. Chem., [157] (1926) 26-30.*
13. D. Goldberg, Rev. Int. Hautes Temper et Refract., 5 (1968) 181 –182.
14. M. Gobbels and D. Bostrom, *Ber. DMG, Beih. Eur. J. Min., [9] (1997) 128.*

STUDY OF THE RELATIONSHIP BETWEEN THE YOUNG'S MODULUS AND MICROSTRUCTURE OF VACUUM PLASMA SPRAYED BORON CARBIDE

H. R. Salimi Jazi, Fardad Azarmi, Thomas. W. Coyle and Javad. Mostaghimi
University of Toronto, Centre for Advanced Coating Technologies
40- St George Street, Room- 8292
Toronto, ON, M5S 3G8

ABSTRACT

The influence of the structure of Vacuum Plasma Sprayed (VPS) boron carbide on the elastic modulus of the deposit was studied. Thermal sprayed structures consist of individual splats along with unmelted and partially melted particles, pores, microcracks, and splat boundaries. Thermal sprayed deposits exhibit anisotropic properties parallel and perpendicular to the splat plane. In the present study, micro-indentation hardness tests were performed on the vacuum plasma sprayed B_4C to determine the elastic modulus parallel and perpendicular to the coating surface. Nano-indentation hardness tests were also performed on the cross-sectional microstructure to measure the elastic modulus parallel to the coating surface to compare with the values obtained from micro-indentation tests. A developed object oriented finite-element method also was performed to estimate the elastic modulus. The elastic modulus of the as-sprayed structure decreased significantly compared to the conventionally processed B_4C materials. Due to an anisotropic microstructure, the perpendicular elastic modulus was lower than that parallel to the spray plane. Numerical simulations using finite element method confirmed anisotropic elastic behavior of the structure and highlighted the effect of splat boundaries on the elastic modulus in the as-sprayed structure.

INTRODUCTION

The elastic modulus of thermal sprayed structures is different from the bulk materials due to the unique microstructures which result from the nature of thermal spray processes. The microstructure of thermal sprayed deposits is characterized by the existence of pores, microcracks, splat boundaries, and unmelted and partially melted particles. The anisotropic microstructure of thermal sprayed deposits yields inconsistent results for mechanical properties measured perpendicular and parallel to the coating plane. Many investigations have been performed to measure the effective mechanical properties such as the elastic modulus of thermal sprayed structures and different values have been reported for parallel and perpendicular directions with respect to the coating plane [1-10]. There are a few published experimental and theoretical studies describing empirical relationships between the elastic modulus and microstructures for ceramic coatings fabricated by plasma spraying processes [9-15]. Some of these studies suggested a following relationship between the mechanical properties and porosity [8,9]:

$$S=S_0 \; exp(-bp) \tag{1}$$

where S is a mechanical property of porous materials such as elastic modulus, S_0 is the corresponding individual property of dense materials, p is the volume fraction of porosity, and b is a constant and can have various values, depending on pore shape and distribution. However, there is often a considerable discrepancy between the observed values and the values predicted by the formula and therefore, Equation (1) can not be assumed conclusively for thermal spray deposits [16]. Splat boundaries, pore morphology, and porosity orientation can be considered as important factors which influence the value of elastic modulus in thermal spray structures. Pore morphology is highly anisotropic in thermal spray deposits. The interlamellae types of pores are usually located parallel to the deposition plane and cause reduction in the elastic modulus perpendicular to the splat surface. Li et al. [8,9] suggested that the non-linear elastic behavior of deposits perpendicular to the splat plane is due to opening of interlamellar pores initially clamped together by residual stresses. They proposed a relationship between the elastic modulus perpendicular to the coating plane and the extent of bonding between splats. It has been reported that in the case of bonding ratio larger than 40% the elastic modulus is simply proportional to the bonding ratio. On the other hand, cracks which form perpendicular to the deposition surface reduce the elastic modulus parallel to the splat plane. The higher value of the transverse elastic modulus of as-sprayed structures is attributed to the smaller surface area of cracks compared to the area of the flat pores between splats. H.R. Salimi Jazi et al proposed an analytical model which relates the elastic modulus to the microstructure of thermal sprayed specimens parallel to the coating plane. The suggested relationship between the relative elastic modulus and the fraction of pulled out lamellae, which is a function of the interlamellar bonding ratio is illustrated in Fig. 1. Detailed descriptions of the relationship were given elsewhere [16].

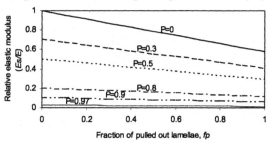

Fig. 1: The relation between relative elastic modulus and fraction of pulled out lamellae. P is the porosity.

Obtaining a pure elastic deformation in thermal spray structures is difficult because of the splat boundary sliding and crack propagation within the microstructure during loading. The elastic modulus of thermal sprayed structures can be determined from a variety of methods such as nano and micro-indentation hardness tests, four-point bending tests, and standard uniaxial tensile tests; however the results vary depending on the method.

Boron carbide is the third hardest material after diamond and boron nitride at room temperature and retains its hardness up to high temperatures. It also has a very low specific weight, a high elastic modulus, and a high resistance to wear and thermal shock. Layered deposition of boron carbide and titanium alloy by vacuum plasma spraying may be attractive for lightweight ballistic protection applications if boron carbide with sufficiently high values of hardness and elastic modulus can be produced. Although boron carbide has been produced by thermal spray processes, low hardness and modulus values limit its applications [11-17].

This work attempts to determine the elastic modulus of vacuum plasma sprayed boron carbide coatings using nano-indentation and micro-indentation tests along with finite element analysis. The results will be used to improve the mechanical properties of the layered deposition of the vacuum plasma sprayed boron carbide.

EXPERIMENTAL PROCEDURE

Boron carbide specimens were manufactured by deposition of commercially available boron carbide feed stock powders on Ti-6Al-4V alloy substrate using vacuum plasma spraying (VPS) by PyroGenesis Inc., Montreal, Canada. Detailed description of the VPS conditions and specimen manufacturing were given in elsewhere [17]. Free standing samples (1×1 cm^2) of the as-sprayed boron carbide with thickness of approximately 500 μm were fabricated by dissolving the titanium alloy substrate using nitric acid. As-sprayed B$_4$C specimens were cross-sectioned with a diamond saw and cold epoxy mounted under low pressure to minimize pull out of particles during preparation. Then, the specimens were polished using No. 800 SiC paper followed by 9, 3, and 1 μm diamond suspensions and finished with an 0.05 μm alumina suspension.

Knoop hardness measurements of the boron carbide deposit were made on cross-sections parallel and perpendicular to the deposition direction using a Zwick microhardness machine under an applied load of 1 kgf for 25 s. The long axis of the Knoop indenter was oriented parallel to the splat spread direction. Nano-indentation experiments were performed using a Shimadzu Dynamic Ultra-micro hardness tester equipped with a Berkovich diamond indenter. The nano-indentation tests were performed by loading to the maximum load (100-190 mN), holding for 20 s, and then complete unloading.

MICROSTRUCTURAL ANALYSES

The lamellar structure is clearly observed in the cross section of the as-sprayed boron carbide shown in Figure 2. The porosity of thermal sprayed coating was determined using image analysis software (Optimas Image Analysis) from the micrographs. Two types of pores can be observed. The first type is rounded, found between the splats, and likely due to gas entrapment during deposition. The second type is characterized by angular edges resulting from the pull out and fragmenting of unmelted particles during sample preparation. The thickness of splats ranges from 1 to 4 μm and the diameter from 10-50 μm. A high density of cracks along splat boundaries can be seen indicating low cohesion between the splats. The grain size ranged from 300 nm to 3 μm [17]. Mixture of amorphous and microcrystalline phases can also be observed. Fine pores, 100-300 nm in diameter, can be seen in the microstructure.

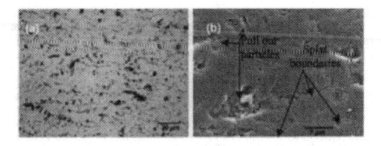

Fig. 2: Optical (a) and SEM (b) micrographs of the polished cross section of the VPS boron carbide.

INDENTATION RESULTS

During loading, a complex elastic and plastic deformation occurs under the indenter. Elastic recovery occurs during unloading in the second half-cycle of indentation. The evaluation of the ratio of hardness to elastic modulus (H/E) is performed by measuring elastic recovery after indentation, following the technique developed by Marshall et al. [18-20]. The ratio of the major and minor diagonals (a/b) of the Knoop indenter geometry is 7.11. The elastic recovery during unloading reduces the length of the minor diagonal and the residual indentation depth. Therefore, the ratio of the major and minor diagonals after unloading (a'/b') would be larger than 7.11. The model developed by Marshal et al. can be expressed as:

$$\frac{b'}{a'} \approx \frac{b'}{a} = \frac{b}{a} - \frac{\alpha H}{E} \qquad (2)$$

The value of α was evaluated experimentally equal to 0.45 by Marshal. S. F. Leigh et al has successfully used the same value for thermal spray structure [19]. H is the Knoop hardness and E is the elastic modulus. The as-sprayed boron carbide has an average hardness of 1417±258 and 1255±150 KHN parallel and perpendicular to the coating surface, respectively. Some cracks along splat boundaries parallel to the large diagonal of the Knoop indenter can be observed (see Fig. 3).

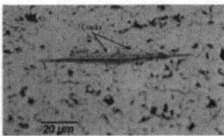

Fig. 3: An optical micrograph of the cross section of VPS boron carbide showing cracks propagation all around the indentation area.

238

Ten indentations were performed on each surface and E and H/E were calculated using Equation (2). Indentations on the cross section of the as-sprayed structure parallel to the substrate show higher value for the elastic modulus than those perpendicular to the surface of the specimen (see Table I). The only explanation is existence of a high density of interlamellar boundaries and pores perpendicular to the loading direction.

Table I: Hardness and related elastic modulus value parallel and perpendicular to the deposition direction.

Indent #	Hardness KHN	$E_\perp$ (MPa)	Indent #	Hardness KHN	$E_\parallel$ (MPa)
1	925	101	1	1315	144
2	1422	156	2	1208	132
3	1354	148	3	1254	137
4	1328	146	4	1219	134
5	988	108	5	1662	182
6	1278	140	6	1858	204
7	1124	123	7	1030	113
8	1254	137	8	1422	156
9	988	108	9	1543	169
10	1341	147	10	1662	182
Average	1200	132	Average	1417	156
STDV	179	19	STDV	258	28

Nano-indentation experiments were performed under load of 100-190 mN on the cross section of the specimen. The measured maximum indentation depth and plastic deformation depth is in the range of 30% to 50% (depending on the applied load and position) of the total depth remains as a residual (plastic) deformation after unloading. The range of the Nano-hardness values was from 41 to 180 GPa, which reflects the in-homogeneity in the structure. Oliver's and Pharr's method has been used for analyzing Nano-indentation load-displacement data [21,22]. This method has been shown to work very well for hard materials such as ceramics. The elastic modulus can be obtained using Equation (3):

$$\frac{1}{E_{eff}} = \frac{1-v^2}{E} + \frac{1-v_i^2}{E_i} \tag{3}$$

where E, v, and E_i, v_i are the elastic modulus and Poisson's ratio for the specimen, and indenter, respectively. The effective elastic modulus, E_{eff}, is derived by:

$$E_{eff} = \frac{\sqrt{\pi}}{2} \frac{S}{\sqrt{A}} \tag{4}$$

where, $S = dP/dh$ is the slope of the upper portion of the unloading curve, and A is the projected area of the elastic contact. We assume that A is equal to the optically measured area of the indentation. Table II shows the nono-indentation depth and hardness followed by measured elastic modulus using Equations (3) and (4).

Table II: The measured elastic modulus respect to the obtained nano-hardness value.

	Position	Max. Depth (nm)	Plastic Depth (nm)	Nano-hardness (GPa)	Effective elastic modulus (N/mm²)	Elastic modulus (N/mm²)
Load 190 mN	1	920	300	43	170	163 ·
	2	668	320	33	150	144
	3	860	270	44	172	165
	4	902	291	39	145	139
	5	863	330	56	208	200
Load 150 mN	1	762	256	44	193	186
	2	785	262	37	179	172
	3	783	260	38	180	173
	4	692	184	61	227	218
	5	706	218	54	216	208
Load 100 mN	1	585	198	51	253	242
	2	618	228	39	228	218
	3	653	249	31	197	189
	4	736	312	19	158	151
	5	634	270	31	197	189

The measured average elastic moduli were 198 ± 34, 191 ± 20, and 162 ± 23 GPa for the applied load of 100, 150, and 190 mN, respectively. These values are higher than those obtained from micro-hardness measurements. It can be due to the wider area of indentation in the micro-hardness test compared to the nano-scale test. In the nano-hardness test the effects of pores and splat boundaries can be ignored due to the small area of contact between indenter and the microstructure while in the microhardness test a bigger indents and area of contacts could encompass all these parameters and affect the results.

FINITE ELEMENT ANALYSIS

An object-Oriented Finite element analysis method (OOF) has been employed to numerically simulate and analyze the elastic modulus of the thermal sprayed deposit. Micrographs of the polished and etched cross-section of the as-sprayed boron carbide have been used in this method. The OOF program is designed to read pixels from the micrographs and group them according to their colors and contrasts [23]. By setting threshold values, the groups are obtained for pores and matrix material. Table III shows the mechanical properties used for the constituents in this analysis.

Table III: Mechanical properties of the VPSF B₄C alloy constituents [24].

Components	Young's modulus (GPa)	Poisson ratio
Boron carbide	289	0.2
Pores	1	0.2

Figure 4 shows the optical microstructure of the as sprayed boron carbide with the converted, meshed, and triangle mode images. All visible cracks and pores are included in the analysis as shown in the converted micrograph (black regions) in Fig. 4b and 4d.

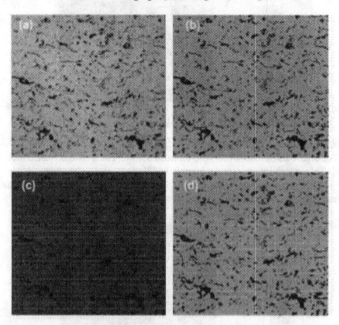

Fig. 4: (a) An optical micrograph from the cross section of the as-sprayed boron carbide, (b) the processed image by *ppm2oof*, (c) the meshed image, and (d) the triangle mode image.

A unixial stress was applied to the cross section parallel to the splat plane in the X-direction (σ_{xx}). A stress map illustrating the stress distribution within the microstructure generated by *oof* is shown in Figure 5. To obtain the average elastic modulus of the structure, several regions were modeled separately. The average result for the transverse elastic modulus is approximately 170 *GPa*. This is approximately 41% less than the fully dense value, while the experimental values from the micro-hardness indentation tests were 55% lower than the fully dense value.

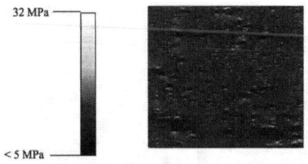

Fig. 5: A stress distribution map parallel to the splat plane.

The same unixial stress was applied perpendicular to the splat plane (σ_{YY}). Figure 6 shows the stress distribution map within the microstructure. The average elastic modulus is equal to 138 *GPa*. This modulus is approximately 52% less than the fully dense value. The experimental values were 68% lower than the fully dense value. All of the elastic modului estimated numerically are significantly higher than the experimental results.

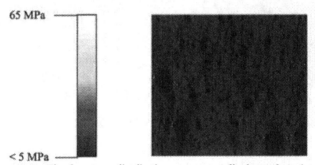

Fig. 6: a stress distribution map perpendicular to the splat plane.

CONCLUSION

In the current study, boron carbide was successfully deposited on Ti-6Al-4V alloy using a VPS process. The deposited structure contains B_4C grains, interlamellae boundaries and pores. The microhardness was measured to be approximately 1200±80 and 1417±250 *KHN* transverse and parallel to the spray direction, respectively. The range in microhardness values was suggested to be due to the inhomogeneous structure. The difference in the hardness values for parallel and perpendicular directions to the substrate surface showed lower cohesion within the lamellae boundaries in the perpendicular direction. The mean elastic modulus, calculated from the Knoop microhardness impressions was approximately 130 ± 19 and 156 ± 28 *GPa* for the perpendicular and parallel directions to the substrate, respectively. The elastic modulus of the as-sprayed structure was more than 50% lower than that for fully dense B_4C. The range of the nano-hardness values was from 41 to 180 *GPa* parallel to the splat plane. The low value of the nano-hardness measurement reflects the effect of the interlamellae boundaries and porosity in the as-sprayed structure. The measured average elastic modulus ranged 162 to 198 GPa, depending on the applied load. The elastic modulus obtained from nano-hardness measurements was higher than those obtained from the micro-hardness measurements. The higher elastic modulus obtained by nano-indentation test is attributed to the smaller number of splat boundaries and other defects encountered compared to the micro-hardness indentation.

The transverse and parallel elastic modului of the as-sprayed structure obtained by the numerical analysis were around 170 and 138 GPa, respectively. The transverse elastic modulus was approximately 40% less than the fully dense structure. All of the values from numerical analysis were slightly higher than those obtained from micro-hardness measurements. The difference may be attributed to the effects of defects whose size was below the resolution of the image, or which were lost after meshing.

REFERENCES

[1]H. R. Salimijazi, T.W. Coyle, J. Mostaghimi, and L. Leblanc, Microstructural Formation of VPSF Ti-6Al-4V Alloys, Thermal Spray 2003: Advancing the Science and Applying the Technology (2003), B.R. Marple and C. Moreau (Eds.), ASM International, Materials Park, OH, pp. 611/616.

[2]S. Kuroda, Properties and Characterization of Thermal Sprayed Coatings – A Review of Recent Research Progress Thermal Spray: Meeting the Challenges of the 21st Century (1998), C. Coddet (Ed.), ASM International, Materials Park, OH, USA, pp. 539/550.

[3]T. McKechine: Near-Net Shape Spray Forming-Metals. Thermal Spray: Surface Engineering via Applied Research (2000), C.C. Berndt (Ed.), ASM International, Materials Park, OH, USA, pp. 1105/16.

[4]H.D. Steffens, M. Dvorak, and K. Nassenstein: Mechanical Properties of Vacuum Plasma Sprayed Titanium and Titanium Alloys. Thermal Spray: International Advances in Coatings Technology (1992), C.C. Berndt (Ed.), ASM International, Materials Park, OH, USA, pp. 369/373.

[5]T.N. McKechine, Y.K. Liaw, F.R. Zimmerman, and R.M. Poorman: Metallurgy and Properties of Plasma Spray Formed Materials. Thermal Spray: International Advances in Coatings Technology (1992), C.C. Berndt (Ed.), ASM International, Materials Park, OH, USA, pp. 839/45.

[6]T. Nguyentat, K.T. Dommer, and K.T. Bowen: Metallurgical Evaluation of Plasma Sprayed Structure Materials for Rocket Engines. Thermal Spray: International Advances in Coatings Technology (1992), C.C. Berndt (Ed.), ASM International, Materials Park, OH, USA, pp. 321/325.

[7]R.T.R. McGrann: Mechanical Fatigue Testing of Thermal Spray Coated Specimens: A Summary of Recent Developments. Thermal Spray 2001: New Surfaces for a New Millennium (2001), C.C. Berndt, K.A. Khor, anf E.F. Lugscheider (Eds.), ASM International, Materials Park, OH, USA, pp. 985/991.

[8]C. J. Li, A. Ohmori, and R. McPherson: The Relationship between Microstructure and Young's Modulus of Thermally Sprayed Ceramic Coatings. J. Materials Science (1997), 32, 997/1004.

[9]C. J. Li, A. Ohmori, and Y. He: Characterization of Structure of Thermally Sprayed Coating. Thermal Spray: Meeting the Challenges of the 21st Century (1998), C. Coddet (Ed.), ASM International, Materials Park, OH, USA, pp. 717/722.

[10]T. Nakamura, G. Qian, and C. Berndt: Effect of Pores on Mechanical Properties of Plasma-Sprayed Ceramic Coatings. J. Am. Ceram. Soc (2000), 83[3], pp. 578/84.

[11]A. Cavasin, T. Brzeinski, S. Grenier, M. Smagorinski, and P. Tsantrizos: "W and B_4C Coatings for Nuclear Fusion Reactors", 15th ITSC (1998), Thermal Spray: Meeting the Challenges of the 21st Century, C. Coddet, (Ed.), ASM International, Materials Park, OH, USA, pp. 957-961, 1998.

[12]L. Bianchi, P. Brelivet, A. Freslon, and C. Cordillot: "Plasma Sprayed Boron Carbide Coatings as First Wall Material for Laser Fusion Target Chamber", 15th ITSC (1998), Thermal Spray: Meeting the Challenges of the 21st Century, C. Coddet, (Ed.), ASM International, Materials Park, OH, USA, pp. 945-950, 1998.

[13]Y. Zeng, C. Ding, and S. Lee: "Young's Modulus and Residual Stress of Plasma-Sprayed Boron Carbide Coatings," J. European Ceramic Society, 21, pp. 87-91, 2001.

[14]J. E. Doring, R. Vaben, J. Linke, and D. Stover: "Properties of Plasma Sprayed Boron Carbide Protective Coatings for the First Wall in Fusion Experiments," J. Nuclear Materials, 307-311 (1), pp. 121-125, 2002.

[15]Y. Zeng, S. W. Lee, and C. Ding: "Study on Plasma Sprayed Boron Carbide Coating," J. Therm. Spray Tech, 11(1), pp. 129-133, 2002.

[16]H. R. Salimijazi, T.W. Coyle, and J. Mostaghimi, Relationship between the Elastic Modulus and Microstructure in Vacuum Plasma Sprayed Structure, Thermal Spray 2005: Advancing the Science and Applying the Technology (2005), B.R. Marple and C. Moreau (Eds.), ASM International, Materials Park, OH.

[17]H. R. Salimijazi, T.W. Coyle, J. Mostaghimi, and L. Leblanc: Microstructr of Vacuum Plasma Sprayed Boron Carbide, J. Thermal Spray Tech., In Press Jan. 2005.

[18]H. J. Kim, Y.G. Kweon, "Elastic Modulus of Plasma Sprayed Coatings Determined by Indentation and Bend Tests", Thin Solid Films, 342, pp. 201-206, 1999.

[19]S.F. Leigh, C.K. Lin, and C. Berndt, "Elastic Response of Thermal Spray Deposits Under Indentation Tests", J. Am. Ceram. Soc., 80 [8], PP. 2093-99, 1997.

[20]J. Li, and C. Ding, "Determining Microhardness and Elastic Modulus of Plasma Sprayed Cr_3C_2-NiCr Coatings Using Knoop Indentation Testing", Surface and Coating Technology, 132, pp. 229-237, 2001.

[21]W.C. Oliver and G.M. Pharr, "An Improved Technique for Determining Hardness and Elastic Modulus Using Load and Displacement Sensing Indentation Experiments", J. Mater. Res., Vol. 7, No. 6, pp. 1564-80, 1992.

[22]A. Bolshakov and G.M. Pharr, "Influences of Pileup on the Measurement of Mechanical Properties by Load and Depth Sensing Indentation Techniques", J. Mater. Res., Vol. 13, No. 4, pp. 1049-58, 1998.

[23]C. Hsueh, J. A. Haynes, Michael. J. Lance, P. F. Becher, M. K. Ferber, E. R. Fuller, S. A. Langer, W. C. Carter, and W. R. Cannon, "Effects of Interface on Residual Stresses in Thermal Barrier Coatings" J. Am. Ceram, Soc., 82 {4} 1073-75 (1999).

[24]CRC Materials Science and Engineering Handbook, p.507.

SCALLOPED MORPHOLOGIES OF ABLATED MATERIALS

Gerard L. Vignoles, Jean-Marc Goyhénèche,
Lab. des Composites ThermoStructuraux (LCTS),
Université Bordeaux 1 – 3, Allée La Boëtie,
F 33600 Pessac, France

Georges Duffa, Anthony Velghe, Ngoc Thanh-Hà Nguyen-Bui,
Commissariat à l'Energie Atomique (CEA)
Centre d'Etudes Scientifiques et Techniques d'Aquitaine (CESTA),
BP2, F 33114 Le Barp, France

Bruno Dubroca,
Mathématiques Appliquées de Bordeaux (MAB)
Université Bordeaux 1 – 351 Cours de la Libération
F 33405 Talence Cedex

Yvan Aspa,
Institut de Mécanique des Fluides de Toulouse (IMFT),
1, Allée du Professeur Camille Soula,
F 31000 Toulouse, France

ABSTRACT

In many cases, ceramic materials undergoing an ablative process (*i.e.* erosion, dissolution, sublimation, oxidation, etching) display a characteristic "scalloped" morphology. The length scales are not related to the ceramic microstructure, and cannot always be related to some flow pattern above the surface.

We propose a simple isothermal model, where diffusion perpendicularly to the average surface and mass loss are in competition. A Hamilton-Jacobi equation is set up, and a steady, non-trivial, analytical solution consists of intersecting circle arcs in 2D. The solution is found to be stable ; the curvature radius is given by the diffusion-to-reaction ratio. A 3-dimensional isothermal simulation confirms the results and the validity of the vertical flux approximation.

Such a model could help to identify intrinsic mass-loss rate constants from morphology examination and the knowledge of gas diffusion coefficients.

INTRODUCTION

Carbon/carbon (C/C) composites are the almost unique class of thermostructural materials that are used as thermal protection of atmospheric reentry bodies. Indeed, the local conditions are dramatically severe : temperatures up to 4500 K and pressures ranging between 0.1 and 100 bars, and a heat flux received by the protection ranging from 0.1 to 500 $MW.m^{-2}$. In such conditions, a non-negligible part of the heat flux is consumed by interfacial mass transfer, which has two principal forms : oxidation (and other chemical reactions) and sublimation. These phenomena are grouped under the generic name of *ablation*[1]. Their particularity is to display a strong coupling with many other physical phenomena, namely fluid flow and heat transfer.

The surface roughness related to the ablation phenomena of the atmospheric reentry body thermal protections has considerable consequences. It is able to promote a laminar-to-turbulent regime transition [2,3], a fact which induces a strong enhancement of heat and mass exchanges between the protection wall and the surrounding environment [4]. The consequences on the surface temperature and heat-affected zone depth are obvious, as well as on the overall ablation velocity.

In turbulent flow conditions, the surface of the material displays a "cellular" roughness (fig. 1) with length scales (here, roughly 1 to 10 μm) that cannot be related to the material heterogeneity nor to turbulent length scales. This kind of "scalloped" surface is strikingly similar to glacier cave walls. Little is known of the creation mechanisms of such a

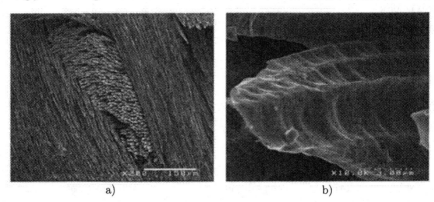

a)   b)

Figure 1: SEM micrographs of an ablated C/C sample (ex-phenolic resin/carbon composite). a) : view at the weave scale, b) : zoom at fiber scale.

surface state, which results from a strong coupling between the material and the surrounding flow. The aim of this work is to set up and use simple models of interaction, involving the least possible phenomena at first, that is, merely the heterogeneous reaction and mass transport from the flowing fluid. The point is to show that roughness scales are not only obtained from intrinsic characteristics of the material, but also from dynamical couplings involving both the material and the surrounding flow. The model presented here is exposed in more details in ref. [5].

MODEL SETUP

Physicochemical description : scales, phenomena

The re-entry body is immersed in a surrounding fluid in which a velocity profile develops itself between the object surface and a detached shock lying in front of it. In this layer, heat production is very strong, and the velocity ranges from moderate close to the walls and at the stagnation point to supersonic. Heat is transferred to the surface, which absorbs it partially by physicochemical phenomena like oxidation, sublimation, and material pyrolysis in the case of carbon/phenolic resin composites. The non-absorbed heat participates to the elevation of the temperature of the body interior, which has to occur with a slow enough rate so that it is kept preserved from damage during reentry. In the case of C/C composites, sublimation is the main phenomenon. It is governed by its intrinsic rate, defined at the wall, and possibly by transport limitations through the surrounding boundary layer, perpendicularly to the surface.

The main idea of roughness set-up is that the wall recession velocity is controlled both by the local reaction rate, which may be a function of the surface location, and by its inclination with respect to the mass flux direction : when a surface is inclined, its vertical recession rate is lower, but if it lies closer to the top of the boundary layer, its rate increases because of a better mass transfer. A dynamical balance between these two factors is able to yield non-flat surfaces, as shown below.

Assumptions and equations

Let us assume that the boundary layer height h_0 is known, and consider a local model placed inside this layer. The domain geometry is illustrated at fig. 2. A 2-dimensional sketch is enough at this point. The model hypotheses are the following : (i) the surface is

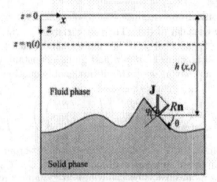

Figure 2: Domain and notations for the model

described by the equation $z = h(x,t)$; (ii) isothermal, isobaric conditions; (iii) pure diffusive mass transfer : no Stefan flux is taken into account; (iv) only one species is tracked : no multicomponent effect is considered; (v) 1st-order reaction rate with unit stoichiometry; (vi) the average height of the boundary layer is fixed, so that the concentration of the fluid is fixed at some height $z = \eta(t)$; (vii) the recession velocity is small with respect to the gas transport velocity, so that diffusion is in quasi-steady state for a given surface profile.

The surface recession is described by the following equations :

$$\partial h / \partial t = v_s R \cos \theta \tag{1}$$

$$\tan \theta = \partial h / \partial x \Longrightarrow \cos \theta = \left(1 + (\partial h / \partial x)^2\right)^{-1/2} \tag{2}$$

$$R = kC(h) \tag{3}$$

where v_s is the solid molar volume, R the molar ablation rate per unit surface, θ is the angle between the surface and the x axis, k is the intrinsic heterogeneous reaction rate in $m.s^{-1}$, and C is the gas-phase species molar concentration. The rate relies on the concentration at the interface, which has to be evaluated. In principle, a complete diffusion equation has to be used with appropriate boundary conditions :

$$\nabla \cdot (\mathbf{J}) = 0 \tag{4}$$

$$\mathbf{J} = -D\nabla C \tag{5}$$

$$\mathbf{J}(z = h) \cdot \mathbf{n} = -R \tag{6}$$

$$C(z = \eta) = C_0 \tag{7}$$

Here the following quantities have been defined : $\mathbf{J}$ is the molar gas-phase species flux, D its diffusion coefficient, $\mathbf{n}$ is the normal vector that points outwards from the surface, and C_0 is the fixed concentration at the boundary layer top.

However, in a 1D vertical-flux approximation, eqs. (4 - 6) reduce to :

$$J_z = -\frac{D}{h - \eta}(C(h) - C_0) \tag{8}$$

At this point, it is convenient to introduce a new variable $h' = h - \eta$ which is the height relative to the top of the boundary layer. The concentration at the interface may be solved by the use of eqs. (3), (6), and (8) . In stationary conditions, it is convenient to use a moving reference frame with origin at $z = \eta$ and global, constant velocity $\mathcal{V} = d\eta/dt$ (*i.e.* $x' = x - \mathcal{V}t; t' = t$), one has the following Hamilton-Jacobi equation :

$$\frac{\partial h'}{\partial t'} = (v_s C_0)\, k\, \left(\frac{Dh'}{k} + \left(1 + \left(\frac{\partial h'}{\partial x'}\right)^2\right)^{-1/2}\right)^{-1} \tag{9}$$

The case of sublimation is exactly similar to the 1st-order reaction case, provided that C is replaced by the saturation difference $C' = (C_0 - C)$. C_0 is now the equilibrium concentration, and k may be evaluated by the Knudsen-Langmuir relationship :

$$k = \frac{\alpha}{4}\sqrt{\frac{8\mathcal{R}T}{\pi \mathcal{M}}} \tag{10}$$

where α is a sticking coefficient.

Analytical solution

An interesting point of the system (9)-(3) is that it has – apart from the trivial flat one – another straightforward steady solution, which is a circle arc, written conveniently in parametric form :

$$\begin{cases} h' &= h_0 + r\,(1 - \cos\theta) \\ x' &= x_0 + r\sin\theta \end{cases} \tag{11}$$

and the radius r is given by the diffusion-to-reaction ratio D/k. The global velocity is given by the same relation as for a flat solution with height h_0 :

$$\mathcal{V} = \frac{1}{(kh_0 D^{-1}) + 1} \cdot (v_s C_0)\,.k \tag{12}$$

Such a solution does not spread over the whole lateral domain, so a complete solution will be made of a collection of circle arcs with radius r and of flat lines. Symmetrical singular points are allowed [5]. Fig. 3 is a sketch of a typical possible solution, with the associated concentration field.

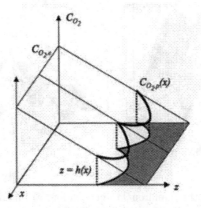

Figure 3: Representation in elevation of the concentration field for a steady, quasi-1D, non-trivial solution of the ablation profile

The stability of the circle arcs with respect to perturbations of concentration and height has been proved [5]. Also, it has been shown that reentrant singular points are not stable, since they tend to transform themselves into smooth circle arcs with upward curvature.

Numerical solutions

In order to check out the model solution and to examine the transient behavior, numerical resolution of equations (4 - 7) and (1-3) has been performed in 3 dimensions. A level-set method has been retained, based on the definition of a function $S(x, y, z, t)$ such that the surface is given by the equation $S(x, y, z, t) = 0$ [6]. Note that the preceding model corresponds in 2D to the choice of $S(x, z, t) = h(x, t) - z$.

The numerical implementation of the transient code is based on a 5-point Weighted Essentially Non-Oscillatory (WENO) scheme [7,8]) and a Lax-Friedrichs scheme for the discretization of the Hamiltonian and a 3rd-order Runge-Kutta scheme for the discretization in time for eqs. (1-3), and a 2nd-order finite-volume scheme for eqs. (4 - 7). The simulation has been started with an initial surface made of reentrant edges and salient curved surfaces, which are anticipated to be completely unstable with respect to the propagation dynamics, illustrated at fig. 4a). The space distribution of the surface extrema is of the same order of magnitude as k/D. After a simulation time equal to 3 times $h_0/\mathcal{V}$, the surface has already transformed itself into a generalization of the analytical solution presented in the preceding section, that is, a set of sphere segments joined together by symmetrical singular edges, which are themselves joined together by symmetrical singular points. The final surface state has acquired the scalloped morphology that comes from diffusion/reaction competition, as seen at figure 4b), with a curvature radius equal to D/k. Figure 5a) is a 2D slice showing the tran-

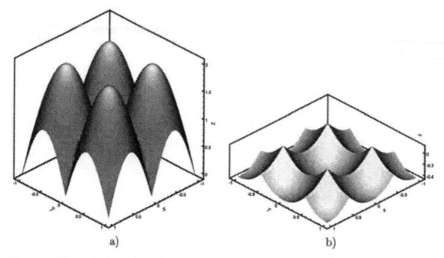

Figure 4: 3D resolution of the Hamilton-Jacobi equation. a) Initial surface, b) Surface after 3 times $h_0/\mathcal{V}$. The space coordinates are in units of h_0

sient behavior of the surface : singular reentrant edges give birth to spherical segments, while salient features end up collapsing together into salient singular edges. Also, a 2D slice of the concentration field is displayed at figure 5b) : it is seen that the vertical flux approximation is quite correct.

PHYSICAL IDENTIFICATION

A numerical evaluation of the curvature radius $r = D/k$ can be made, taking typical values of plasma jet conditions, such that the surface temperature is about 3500 K and the pressure approximately 2.5 MPa. In such conditions, the major phenomenon is carbon sublimation. The value of the sticking coefficient is known experimentally [9,10] : $\alpha \simeq 0.028$. It is considered that, even if the flow is turbulent, there exists a thin layer close to the wall where laminar flow occurs ; this layer has to be traversed by the gaseous species, so the diffusion coefficient in laminar flow is the critical transport property. Its value can be estimated by the classical theory of Chapman-Enskog [11] with typical data [12] : $D \approx 0.01\ cm^2 \cdot s^{-1}$. The resulting value for the curvature radius, about 3 μm, scales reasonably with the features of fig. 1, provided the large uncertainty on the sticking coefficient. In fact, measuring this scale can be a way to have an indirect assessment of the sticking coefficient.

Additionally, the use of eq. 12 with knowledge of the experimental value of the recession velocity (approximately 0.5 $mm.s^{-1}$) allows to evaluate the size of the diffusive boundary layer h_0 to 5.5 μm. It is indeed very small with respect to a thermal diffusion length, so that the isothermal assumption is not so restrictive as it could have been thought. Also, if the heat flux is strongly varied, the fact that sublimation is the main ablation cause renders it "robust", because heat flux variations will primarily lead to changes in sublimation mass flux, but not in surface temperature ; since D/k is mainly a function of T, it will not

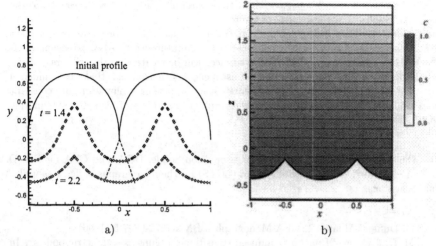

Figure 5: 2D slices of : a) Surface evolution plot, and b) concentration isosurface plot over a stabilized surface

vary strongly when the heat flux varies.

This physical identification provides thus an independent method for the assessment of heterogenous reaction rate or sublimation rate constants from the curvature radius of the scallops and the use of estimates of the diffusion coefficient and of the boundary layer size, which are readily available. Obtaining such a kind of data from geometrical features has already been performed in the very different context of tungsten chemical vapor deposition in a micrometer-size groove[13]. However, inverse methods in the field of surface morphology evolution are not yet very popular.

CONCLUSION AND OUTLOOK

The question of roughness set-up due to ablation in atmospheric reentry has been addressed. Apart from the trivial cause due to the material inhomogeneities, another cause seems to act even in the case of homogeneous materials. It is basically a dynamical effect based on the concurrence between transfer and reaction : when the surface is higher, the consumption rate is higher, but when it is more inclined with respect to the principal transfer direction, the rate is lowered. Thus, points with different altitudes may well have the same recession velocity, leading to a rough but stable surface profile. A simplified isothermal model featuring the interaction of 1D (normal) diffusion with heterogeneous reaction or sublimation has been built, and it shows that in the *intermediate* regime (that is , neither diffusion-limited nor reaction-limited), cellular roughness may appear, under the form of a stable surface made of circle arcs in 2D (and, by extension, of sphere caps in 3D). Such a morphology has been indeed found experimentally in various cases. One of the successes of the presented model is that it helps understand why a characteristic length scale, different from the material-related scales, appears in ablation.

Consequences of such a behavior have to be carefully considered with respect to other experimental results, and the solidity of the model with respect to addition of new physical features, such as the curvature-enhanced chemical reactivity or an explicit coupling between mass, heat and momentum transfers, has to be tested in future work. Also, extensions of the simple model to anisotropic and composite surfaces will be treated in other papers.

Finally, one of the most interesting consequences of this model is that a new, independent method for the assessment of heterogenous reaction rate or sublimation rate constants becomes available. An exciting perspective is to apply it to a variety of cases, like sublimation, etching, dissolution, etc ....

ACKNOWLEDGEMENTS

The authors wish to thank Snecma Propulsion Solide, DGA, and CEA for Ph. D. grants to Y. A. and A. V. and Jean Lachaud (LCTS) for the micrographs of fig. 1 as well as fruitful discussions.

REFERENCES

[1] G. Duffa. "Ablation", CEA Monograph ISBN 2-7272-0207-5 (1996).

[2] M. D. Jackson, "Roughness induced transition on blunt axisymmetric bodies", Interim Report SAMSO-TR-74-86 of Passive Nosetip Technology (PANT) Program, 15, (1974).

[3] D. C. Reda, "Correlation of nosetip boundary-layer transition data measured in ballistic-range experiments", Sandia Report SAND 79-0649 (1979).

[4] M. R. Wool, "Summary of experimental and analytical results", Interim Report SAMSO-TR-74-86 of Passive Nosetip Technology (PANT) Program, 10, (1975).

[5] G. Duffa, G. L. Vignoles, J.-M. Goyhénèche, and Y. Aspa, "Ablation of carbon-based materials : investigation of roughness set-up from heterogeneous reactions", submitted to Int. J. of Heat and Mass Transfer (2005)

[6] I. V. Katardjiev, "A kinematic model of surface evolution during growth and erosion : Numerical analysis", J. Vac. Sci. Technol. A, 7, 3222–3232 (1989).

[7] G.-S. Jiang and D. Peng, "Weighted ENO schemes for Hamilton-Jacobi equations", SIAM J. Sci. Comput., 21, 2126–2143 (2000).

[8] S. Osher and C.-H. Shu, "High-order essentially non-oscillatory schemes for Hamilton-Jacobi equations", SIAM J. Numer. Anal., 28, 907–922 (1991).

[9] H. B. Palmer and M. Shelef, "Vaporization of carbon", Chemistry and Physics of Carbon, 4, 85–135 (1968).

[10] R. L. Baker and M. A. Covington, "The high temperature thermochemical properties of carbon", Interim Report SD-TR-82-19, Office of Naval Research (1982).

[11] J. O. Hirschfelder, C. F. Curtiss, and R. B. Bird, Molecular theory of gases and liquids, John Wiley & sons (1963).

[12] L. Biolsi, J. Fenton, and B. Owenson, "Transport properties associated with carbon-phenolic ablators", Proc. AIAA Conference (1982).

[13] E. J. McInerney and E. Srinivasan and D. C. Smith and G. Ramanath, "Kinetic rate expression for tungsten chemical vapor deposition in different WF_6 flow regimes from step coverage measurements", Z. Metallk., 91, 573–580 (2000).

MECHANICAL AND THERMAL PROPERTIES OF NON-AQUEOUS GEL-CAST CELSIAN AND SIALON CERAMIC MATERIALS

R. Vaidyanathan, T. Phillips
Advanced Ceramics Research, Inc.,
3292 E. Hemisphere Loop, Tucson, AZ 85706

J. Halloran
Advanced Ceramics Manufacturing LLC
Arizona Material Laboratories,
Tucson AZ, 85712

L. Reister, W. Porter, M. Radovic, E. Lara-Curzio, Oak Ridge National Laboratory, P. O. Box 2008, MS 6069, Oak Ridge, TN 37831-6069

ABSTRACT

The mechanical and thermal properties of gelcast, Si_3N_4 reinforced 50% $BaO.Al2O3.2SiO2$ (BAS) – 50% $SrO.Al2O3.2SiO2$ (SAS) and sialon compositions were evaluated from room temperature up to 1300°C. Gelcasting of these compositions was achieved using a non-aqueous monomer system. The mechanical and thermal properties including 4-pt. flexure, bi-axial flexure, Coefficient of thermal expansion (CTE) and thermal conductivity from room temperature to 1300°C for gel-cast celsian and sialon ceramic formulations were analyzed and are presented here.

INTRODUCTION

Ceramics based on Sialons and monoclinic $BaO.Al2O3.2SiO2$ (Celsian BAS), monoclinic $SrO.Al2O3.2SiO2$ (Celsian SAS), or monoclinic solid solutions of BAS and SAS are attracting considerable interest for a variety of high temperature applications due to their unique combination of high refractoriness, low thermal expansion, low dielectric constant and loss tangent both stable over a broad range of temperatures and frequencies. Dielectric ceramics, electronic packaging, structural and wear-resistant ceramics are all possible applications for these ceramic materials [1-3]. In the case of the celsian materials, it is desirable to improve the fracture toughness through the addition of silicon nitride powder (loadings of 20-40%). Unfortunately, expensive hot pressing or hot isostatic pressing is required to produce densified sialons and celsian ceramic materials.

To overcome these processing difficulties, a non-aqueous gelcasting process followed by pressure-less sintering to fabricate near-net shape articles from these ceramic compositions has been developed. The combination of gel-casting and pressure-less sintering would provide a lower cost manufacturing process than is currently available. The non-aqueous nature of the gelcasting process could provide ceramic powders that are not hydrolyzed, leading to a better ceramic material.

A. Celsian and Sialon ceramic materials

Barium aluminosilicate and Strontium aluminosilicate ($BaO.Al2O3.2SiO2$, $SrO.Al2O3.2SiO2$ or Celsian BAS and Celsian SAS) have one of the highest melting temperatures among glass-ceramic materials. The monoclinic forms of these materials have low thermal expansion coefficients of and therefore, are candidate materials for several structural applications [1-4]. However, BAS and other Celsian glass ceramics exhibit poor strength and fracture toughness, limiting their use in structural applications [1-4]. It has been reported that silicon nitride reinforcement, in the form of 20-40% by weight could be used to improve the strength and fracture toughness of these materials. The added advantage of this reinforcement is the ability to use pressure-less sintering techniques to consolidate the ceramic compositions, to prevent the hexagonal form of the composite and to improve the fracture toughness [2].

Another ceramic composition that exhibits enhanced fracture toughness is Sialon [5]. In this case, β'-Sialon that exhibits higher fracture toughness could be consolidated through pressure-less sintering techniques by adding metal oxide additives such as Y_2O_3 [5]. Prior to the consolidation process, parts could be fabricated using standard ceramic processing techniques such as slip casting or gelcasting. However, in slip casting, cracking of the parts could be caused by different rates of shrinkage between the interior and exterior of the casting, especially in complex-geometry parts. The differential shrinkage comes as a result of a moisture gradient throughout the thickness of the casting, a problem enhanced in thin wall components.

B. Gelcasting Process

To overcome fabrication issues with complex-shaped parts, new ceramic processing techniques such as gelcasting have been developed [6]. This is an attractive ceramic forming process for making high-quality complex-shaped ceramic parts. In this process, a slurry made from the ceramic powder and a monomer solution is poured into a mold, polymerized in-situ to immobilize the particles in a gelled part, removed from the mold while still wet, dried and fired. Two variations of this process exist. The first process was developed by Oak Ridge National Laboratory [7], while the second process was developed by Advanced Ceramics Research [8]. The difference in the two processes is based on the nature of the monomer – aqueous or non-aqueous.

Unfortunately, aqueous-based slurries suffer from the drawback that water present in its vehicle causes hydrolytic degradation of the particulate surfaces and forms soluble ionic silicate species within the vehicle [8]. A number of variables magnify this problematic hydrolysis effect including, allowing the slurry to stand for prolonged time periods and exposure to elevated temperatures, or formulating the slurry under alkaline pH conditions. Slurry particle hydrolysis is undesirable for several reasons.

First, the formation of soluble silicate hydrolysis by-products often changes slurry rheology by increasing its viscosity when subjected to low shear (as encountered when pouring the slurry). This is undesirable since many green ceramic forming methods rely upon slurries that have predictable and controllable viscosities that do not change upon aging. Second, the soluble silicates may polymerize and induce slurry gelation. Further, particulate hydrolysis may change the overall slurry chemical composition by increasing the amount of oxygen (in the form of silica) within the ceramic formulation. This is particularly undesirable for silicon carbide (SiC) and silicon nitride (Si_3N_4) slurries since the properties of sintered ceramics made from these materials are highly sensitive to small changes in chemical composition. An increased amount of silica in an aged SiC and Si_3N_4 slurry may manifest itself as a change in the composition and properties of the intergranular glass phase responsible for binding the individual SiC or Si_3N_4 grains together within the sintered ceramic body. Elevated silica levels in intergranular glasses, for example, have been shown to decrease the high temperature creep resistance of sintered Si_3N_4 ceramics [8]. Consequently, there is a significant need for methods and formulations for siliceous ceramic slurries that do not have the disadvantages of conventional methods and formulations to fabricate components for various high-temperature structural applications.

Recently, a non-aqueous based gelcasting process that does not suffer from the issues of hydrolysis during long-term storage has been developed [8]. The major advantage of this process is the ability to retain its viscosity even after long-term storage. Figure 1 shows the viscosity plot of a non-aqueous silicon nitride slurry as a function of storage time. Even though the data is not added here, it was observed that the viscosity did not vary as a function of shelf-life greater than 3 months. This suggested that a non-aqueous gel-casting process would be a suitable fabrication technique for new ceramic formulations such as celsians or sialons. A schematic of the gelcasting process is shown in Figure 2.

The advantages of the non-aqueous gelcasting process are as follows:

- The gelcasting slurry viscosity is low enough to be capable of being injected into molds or fiber performs
- The low viscosity also allows the injection into complex-shaped molds
- The components solidify without any appreciable shrinkage during the gelling process
- The gelcasting process can produce near-net shape components with good surface finish

254

- The gelcasting slurry has a long shelf life

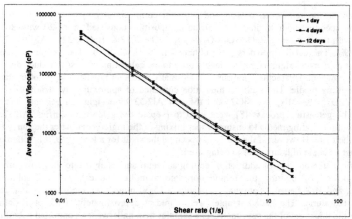

Figure 1. Non-aqueous gelcasting slurry viscosity as a function of shelf life

Gel Casting Process Schematic

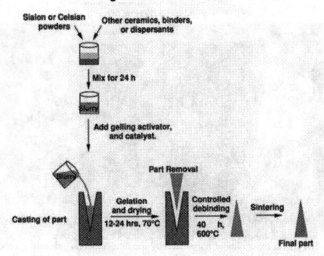

Figure 2. A schematic of the non-aqueous gelcasting process

The objective of this study was to develop a fabrication process for celsian and sialon type of ceramic formulations based on a non-aqueous gelcasting process and to characterize the thermo-mechanical properties of the materials at temperatures ranging from room to 1300°C. Four-point flexural strength, biaxial flexural strength, coefficient of thermal expansion, as well as thermal conductivity for both the silicon nitride reinforced celsian and sialon compositions were evaluated and are reported here.

EXPERIMENTAL PROCEDURES

A. Materials and slurry preparation

Silicon nitride reinforced celsian and a sialon composition prepared in-house was used for fabricating the specimens for testing. The celsian composition consisted of 13-20 wt% silicon nitride (UBE E-05) with the rest being BaO-SrO-SiO$_2$-Al$_2$O$_3$ (SrO = 7.13%, BaO = 32%, Al$_2$O$_3$ = 27.86%, and SiO$_2$ = 33.01%). The celsian powders were obtained from H. C. Stark. The celsian composition was milled for four days using alumina milling media prior to the gelcasting process to reduce the particle size of the celsian powders. Diffraction analysis showed that the celsian powders were not contaminated by the alumina milling media. The sialon composition consisted of approximately 70% UBE E-05 silicon nitride, 1% Y2O3, 1% AlN, 1% SiO2, and the rest Al2O3 by weight. Dispersants and binders as described in the gelcasting process [8] were used to prepare the gelcasting slurries. All slurries were milled with alumina milling media to ensure proper mixing of the ingredients. After the slurry is prepared, it has a long shelf life. However, if the slurry has been on the shelf for a long time, it is rolled on a rolling mill for at least a couple of hours prior to being used.

The final step consists of addition of gelling activator and catalyst into the formulation at 5µL/g of slurry, followed by pouring the formulation into a mold. The catalyst is a solution of 90% butyrolactone, and 10% benzoyl peroxide. Typically, a 15-30 minute working time is available prior to hardening of the slurry. The gelled samples were cured at approximately 70°C for 12-24h, the time depending on the thickness of the specimens. The samples then underwent a controlled debinding process from room temperature up to 600°C in air. The celsian compositions were subsequently sintered at approximately 1600°C for the celsian samples while the sialon samples were sintered at 1750°C. Flat plates and sub-scale parts were fabricated using the gelcasting process. Some typical examples are shown in Figure 3.

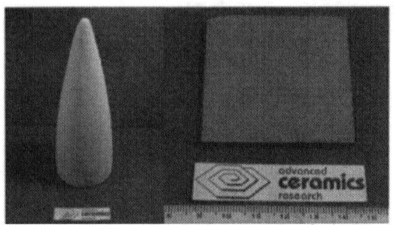

Figure 3. Typical examples of parts built by the non-aqueous gelcasting process

B. Mechanical Properties

4-point flexure testing was performed from room temperature up to 1300°C for both the materials. However, since the celsian materials exhibited a loss in strength and stiffness beyond 1100°C, they were not tested beyond this temperature. Flexure tests were performed on 4.0 mm x 3.0 mm x 50.0 mm flexure bars. These tests were performed at Oak Ridge National Laboratory in test frames with pneumatic actuators with a loading capacity of 2.2 kN. The test frames are capable of testing 3 samples simultaneously up to 1500°C.

Biaxial testing was performed using a concentric ring-on-ring configuration at room temperature and 900°C [10]. For this test, disk-shaped test specimens with a diameter ≈ 25 mm were placed concentrically between a loading ring of diameter (D_l) 5.5 mm and a supporting ring of diameter (D_s) 20 mm. The loading and unloading rings were made from alumina. Load was applied to the samples at a constant cross-head displacement rate of 1 mm/min. The equibiaxial strength was calculated using the following equation:

$$\sigma_f = \frac{3 \cdot F}{2\pi h^2}\left[(1-\nu)\frac{D_s^2 - D_l^2}{2D^2} + (1+\nu)\ln\frac{D_s}{D_l}\right] \tag{1}$$

where F, h, and ν are breaking load, sample thickness and Poisson's ratio, respectively. The Poisson's ratio for the celsian and sialon materials was estimated to be 0.23 and 0.28 respectively [11, 12]. Generally, it is expected that the stronger and harder the ceramic, the higher will be its Poisson's ratio.

C. Thermal Properties

The thermal expansion coefficient, thermal diffusivity and thermal conductivity for both the gelcast celsian and sialon materials were measured at Oak Ridge National Laboratory. The thermal expansion of bulk celsian and sialon samples was measured under flowing He at a flow rate of 5 cc/min. A temperature range of 25-1200°C was used for the celsian materials, while a temperature range of 25-1400°C was used for the sialon materials. The measurements were carried out on heating and cooling using a ramp and soak cycle in a dilatometer (Unitherm, Anter Corp, Pitt., PA). The heating rate was 3°C/min, and at every 200°C on both heating and cooling the sample was allowed to equilibrate with the furnace for 15 min before resuming the heating. The data was collected every minute. The dilatometer was calibrated using a single crystal sapphire rod.

Thermal diffusivity, α, was measured by the laser flash technique at the Oak Ridge National Laboratory. These were performed on a Flashline 5000 Thermal Diffusivity System (Anter Corporation, Pittsburgh, PA). The room temperature diffusivity was measured using a xenon flash technique. The measurements were made on samples sputter coated with a thin coating of gold or gold-palladium, followed by a thin coating of colloidal graphite. Laser flash measurements were performed between 100 and 1200°C, in 100 degree intervals, in nitrogen. Specific heat capacities, C_p, were determined in the temperature range of 100-1200°C by differential scanning calorimetry on an Omnitherm DSC 1500 using sapphire as the baseline standard. The thermal diffusivity, specific heat and density were used to calculate the thermal conductivity κ between 100-1200°C according to,

$$\kappa = \alpha \, \rho \, C_p \tag{2}$$

RESULTS AND DISCUSSION

A. Mechanical Properties

Table I shows the average mechanical properties of the celsian and sialon compositions ranging from room temperature up to 1300°C. This is also plotted in Figure 4, which shows a straight line connecting the average values (based on 5 samples each) at each test temperature. It can be seen that the mechanical properties generally decrease with an increase in the testing temperature. This became apparent when typical flexural stress-strain curves for both the materials were plotted as a function of test temperature. Figure 5 and 6 show the typical stress-strain curves for the gelcast celsian and sialon compositions respectively. It is also clear that the stiffness of both the materials decrease as a function of test temperature.

257

Table I. Average 4-point flexural strength as a function of test temperature for gelcast celsian and sialon compositions

Temperature	Celsian Average	SD-Celsian	Sialon average	SD-Sialon
25	155	10	297	7
800	96	22	315	52
1125	107	39		
1200			263	30
1300			251	4

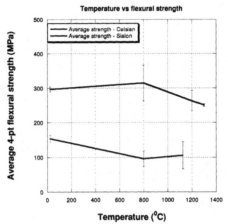

Figure 4. Average 4-point flexural strength as a function of test temperature for gelcast celsian and sialon compositions

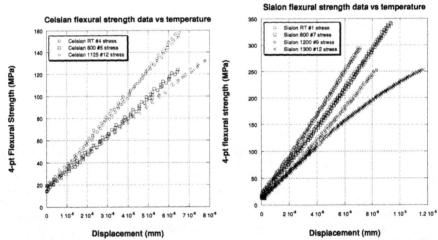

Figure 5. Typical stress-strain curves for celsian Compositions as a function of temperature

Figure 6. Typical stress-strain curves for sialon compositions as a function of temp.

Table II shows the average biaxial strength of the gelcast celsian and sialon compositions. Figure 7 and 8 show the typical biaxial stress-strain curves for the celsian and sialon compositions respectively. Biaxial strength values are typically more useful to designers compared to flexural strength values, since they are more conservative. The maximum test temperature was 900°C, due to the capability of the test fixture and grips. It can be seen that the biaxial strength was lower than the 4-pt flexural strength for both the ceramic compositions. It was also seen that the biaxial strength and modulus values did not change up to 900°C.

Table II. Average biaxial strength results for gelcast celsian and sialon compositions

Material	RT	SD	900 deg C	SD
Celsian	104	8	119	15
Sialon	160	16	184	20

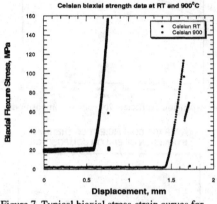

Figure 7. Typical biaxial stress-strain curves for Celsian compositions at RT and 900°C

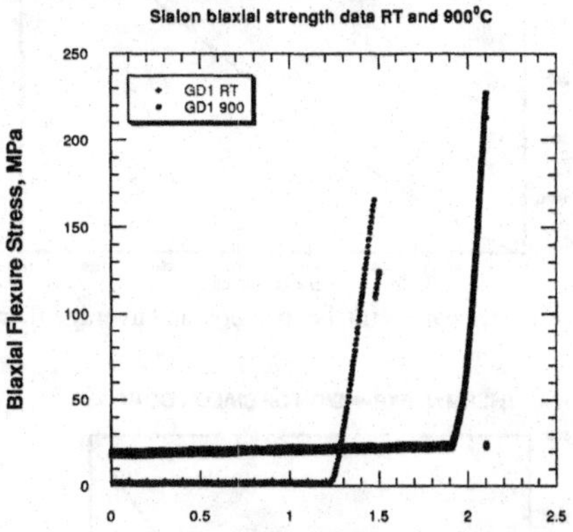

Figure 8. Typical biaxial stress-strain curves for sialon compositions at RT and 900°C

Figure 9 and 10 show the optical micrographs of typical flexural test samples. Both the room temperature and elevated temperature samples showed typical mirror, mist, and hackle features. Most fractures were observed to have originated from surface flaws, including porosities. This suggested that further processing improvements are required to obtain increased mechanical properties.

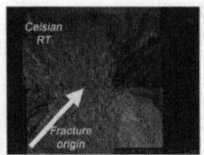

Figure 9. Typical fracture surface of celsian Samples tested at RT

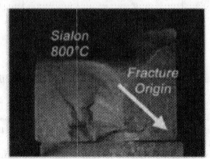

Figure 10. Typical fracture surface of sialon samples tested at 800°C

B. Thermal Properties

Figure 11 shows the linear and average CTE for the Celsian composition up to 1200°C. It can be observed that the CTE values are consistent during heat up and cooling and that there is no hysterisis associated with the material composition. It can also be seen that the average CTE ranges from 2.5 to 4.5 ppm between room temperature and 1200°C. The linear thermal expansion data and average thermal expansion values for the sialon materials are shown in Figure 12. It was also observed that the average CTE varied from 1 to 4 ppm, which is lower than that of the celsian composition.

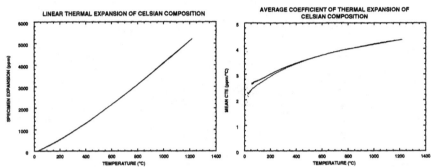

Figure 11. Linear thermal expansion and average thermal expansion coefficient for celsian compositions

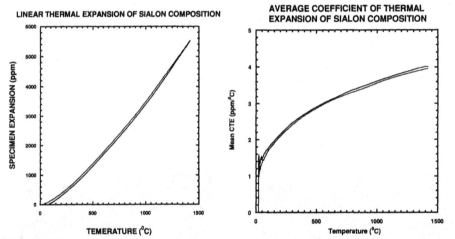

Figure 12. Linear thermal expansion and average thermal expansion coefficient for sialon compositions

The thermal conductivity measurements were made in the following manner. The thermal diffusivity data was measured first for both the materials, followed by the specific heat measurements. The specific heat was measured both during heat up as well as cool down, starting from approximately 140°C.

Table II lists the specific heat, thermal diffusivity and the calculated thermal conductivity at each temperature for both the sialon and celsian compositions. The thermal conductivity is calculated to be the product of specific heat, thermal diffusivity as well as the density of the material, for each temperature.

260

The measured specific heat capacity for the celsian and sialon materials is shown in Figure 13, while the obtained thermal conductivity values are shown in Figure 14. It can be seen that the thermal conductivity values for the celsian materials varied from approximately 1.8 W/m-K at room temperature to 1.5 W/m-K at 1200°C while that for sialon varied from 6.4 to 4.2 W/m-K at room temperature to 1200°C.

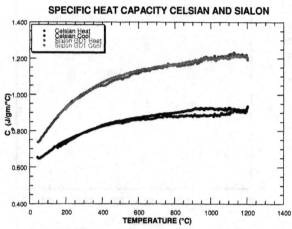

Figure 13. Specific heat capacity for gelcast celsian and sialon compositions

Table II. Thermal diffusivity, specific heat and thermal conductivity values for celsian and sialon compositions

Temperature	Thermal diffusivity (cm2/s)		Specific heat (J/gm/deg.C)		Density (g/cc)		Thermal conductivity (W/m.K)	
(deg. C)	Celsian	Sialon	Celsian	Sialon	Celsian	Sialon	Celsian	Sialon
148	0.00894	0.0247	0.71337	0.87818	2.89	2.93	1.84	6.36
201	0.00858	0.0234	0.74467	0.93957	2.89	2.93	1.85	6.44
306	0.00784	0.0211	0.79304	1.00726	2.89	2.93	1.80	6.23
407	0.00735	0.0198	0.8271	1.04853	2.89	2.93	1.76	6.08
505	0.00702	0.0182	0.84469	1.08369	2.89	2.93	1.71	5.78
653	0.0066	0.0164	0.86465	1.13103	2.89	2.93	1.65	5.43
704	0.0063	0.0158	0.87389	1.14471	2.89	2.93	1.59	5.30
805	0.0063	0.0148	0.87767	1.16192	2.89	2.93	1.60	5.04
904	0.006	0.0142	0.8839	1.1753	2.89	2.93	1.53	4.89
1003	0.006	0.0138	0.8878	1.19768	2.89	2.93	1.54	4.84
1102	0.0054	0.0118	0.89638	1.21339	2.89	2.93	1.40	4.20
1201	0.0054	0.0102	0.90835	1.21901	2.89	2.93	1.42	3.64

CONCLUSIONS

The feasibility of the non-aqueous gelcasting process to fabricate different ceramic compositions has been demonstrated under this program for celsian and sialon compositions. The non-aqueous gelcasting process has the potential to reduce manufacturing costs, is applicable to several ceramic compositions and could reduce the turn around time for prototype components. The mechanical and thermal properties of gelcast celsian and sialon compositions were evaluated. It was observed that the flexural strength was fairly stable up to 1000°C and softened beyond this temperature. The average CTE of these materials up to 1200°C was measured to be between 2.5 – 4.5 ppm for celsian materials and 1 – 4 ppm for the sialon materials. The average thermal conductivity of celsian and sialon materials was observed to be 1 – 2 W/m.K and 4 – 6 W/m.K respectively.

Celsian and Sialon thermal conductivity data

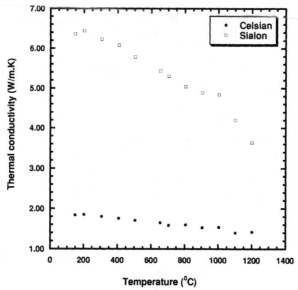

Figure 14. Calculated celsian and sialon thermal conductivity data from room temperature to 1200°C

ACKNOWLEDGMENTS

The authors gratefully acknowledge the financial support from Naval Surface Warfare Center through a phase I SBIR contract # N00167-04-C-0010. The authors would also like to thank Dr. Inna Talmy and Mr. Curtis Martin at NSWC for their continued support and encouragement. Research also sponsored by the Assistant Secretary for Energy Efficiency and Renewable Energy, Office of FreedomCAR and Vehicle Technologies, as part of the High Temperature Materials Laboratory User Program, Oak Ridge National Laboratory, managed by UT-Battelle, LLC, for the U.S. Department of Energy under contract number DE-AC05-00OR22725.

REFERENCES

1. Talmy, I. G., Zaykoski, J. A., *U. S. Patent* No. 5, 538, 925, (1996).
2. Talmy, I. G., Zaykoski, J. A., *U. S. Patent* No. 5, 578, 534, (1996).
3. Talmy, I. G., Zaykoski, J. A., *U. S. Patent* No. 5, 641, 440, (1997).
4. F. Ye, J. C. Gu, Y. Zhou, and M. Iwasa, "Synthesis of BaAl2Si2O8 glass-ceramic by a sol-gel method and the fabrication of SiCpl/BaAl2Si2O8 composites," *Journal of the European Ceramic Society* **23** (2003) 2203–2209.
5. F. L. Riley, "Silicon nitride and related materials," *J. Am. Ceram. Soc.*, 83 [2] 245–65 (2000).
6. M. A. Janney, S. D. Nunn, C. A. Walls, O. O. Omatete, R. B. Ogle, G. H. Kirby and A. D. McMillan, "Gelcasting," found at http://www.ms.ornl.gov/researchgroups/process/cpg/gelpubs/GelChap.pdf
7. M. A. Janney, C. Walls, D. M. Kupp, and K. W. Kirby, "Gelcasting Sialon radomes," *Am. Cer. Soc. Bull.*, 83(7), pp. 9201-9206, (2004).
8. J. L. Lombardi, G. Artz, K. Johnson, D. Dent, "Dispersant system and process for formulating non-

aqueous siliceous particulate slurries," U. S. Patent No. 6, 221, 921, April 24, 2001.

9. K. W. Kirby, A. Jankiewicz, M. Janney, C. Walls, D. Kupp, "Gelcasting of GD-1 ceramic radomes," Proceedings of the 8th DoD Electromag. Windows Symposium, pgs. 287-295 (2000).

10. M. Radovic and E. Lara-Curzio, "Mechanical properties of tape cast nickel-based anode materials for solid oxide fuel cells before and after reduction in hydrogen," *Acta Materialia*, 52, pp. 5747-5756, (2004).

11. B. Bergman, T. Ekström, A. Micski, "The Si-Al-O-N system at temperatures of 1700-1775°C," *Journal of the European Ceramic Society* Vol. 8, pp. 141-151 (1991).

12. C. M. Sheppard, K. J. D. MacKenzie and M. J. Ryan, "The Physical Properties of Sintered X-phase Sialon Prepared by Silicothermal Reaction Bonding," Journal *of the European Ceramic Society*, Vol. 18, pp. 185-191, (1998).

263

BEND STRESS RELAXATION CREEP OF CVD SILICON CARBIDE

Yutai Katoh and Lance L. Snead
Oak Ridge National Laboratory
P.O.Box 2008 / MS-6138, 1 Bethel Valley Road
Oak Ridge, TN, 37831-6138

ABSTRACT

Bend stress relaxation (BSR) creep of two forms of chemically vapor-deposited beta phase silicon carbide, namely polycrystalline and single-crystalline, was studied. The experiment was primarily oriented to demonstrate the applicability of BSR technique to irradiation-induced / enhanced creep behavior of silicon carbide in nuclear environments. It was demonstrated that thin strip samples with sufficient strength for BSR experiment could be machined and the small creep strains occurred in those samples could be measured to sufficient accuracy.

The thermal creep experiment was conducted at 1573 – 1773K in argon to maximum hold time of 10 hours. Both materials exhibited similar primary creep deformation at the initial stresses of 65 – 100 MPa. The relative stress relaxation determined in the present experiment appeared significantly smaller than those reported for a commercial CVD SiC fiber at given temperature, implying a significant effect of the initial material conditions on the relaxation behavior. The analysis based on the relaxation time – temperature relationship gave an activation energy of ~850 kJ/mol for the primary responsible process in CVD SiC.

INTRODUCTION

Creep property is among the major potential lifetime-limiting factors for high temperature materials, including silicon carbide (SiC) ceramics and SiC-based ceramic composites. SiC-based ceramics and composites are considered for application in advanced fission and fusion power systems[1,2]. In nuclear environments, irradiation-induced / enhanced creep ('irradiation creep') is added to thermally-activated creep deformation. In many cases, irradiation creep is caused by preferred absorption of supersaturated point defects at edge dislocations in favor of stress relaxation, and hence generally dominates at relatively low temperatures where thermal creep is not of concern[3]. Integrity of gas reactor fuel particles will be affected by creep of SiC shell as the primary fission gas container. Lifetime of SiC-based structural composites in fusion system will be potentially limited by irradiation creep[4].

Irradiation creep data for ceramics are extremely limited because of difficulty in applying conventional pressurized tube technique[5] or other external loading techniques to ceramic samples in nuclear reactors. Bend stress relaxation (BSR), developed for evaluation of creep properties of ceramic fibers[6], is a technique that is easily applicable to irradiation creep studies. BSR technique has been applied to irradiation creep studies on SiC-based ceramics in forms of thin fibers[7]. However, creep properties of bulk SiC (monolithic SiC or matrix material of ceramic composites) might differ significantly from those of SiC-based fibers, because of significant differences in grain size, micro/nano-structures and chemistry. Also, the stress range of concern for bulk SiC is significantly lower than that for fibers in many cases.

In this work, BSR creep experiment was performed using thin strip specimens machined out of chemically vapor deposited (CVD) SiC in two different material classes, in a stress range of general interest for structural ceramics and composites. The primary objective of the experiment was to demonstrate the applicability of BSR technique to the (irradiation) creep studies of bulk SiC which are not made into a fiber form. Additionally, it was attempted to help understanding the high temperature deformation mechanism for high purity and stoichiometric SiC using the limited data obtained.

EXPERIMENTAL

Materials

The materials used were CVD-produced polycrystalline and single-crystalline beta-phase SiC. The polycrystalline material was the standard resistivity grade of 'CVD-SiC' produced by Rohm and Haas Advanced Materials (Woburn, MA), with manufacturer-claimed purity of >99.9995%. The crystal grains of CVD-SiC are highly elongated along the growth direction and have a bi-modal size distribution of large (10~50μm in column width) and small (typically 1~several μm) grains. The crystal grains are heavily faulted and have a preferred crystallographic orientation of <111> directions in parallel to the CVD-growth direction but randomly oriented in the normal plane. All the flexural specimens were machined with longitudinal directions normal to the growth direction.

The single-crystalline (referred to as 'SC-SiC' hereafter) material was 3C-SiC {100} surface orientation wafer produced by Hoya Advanced Semiconductor Technologies Co., Ltd. (Tokyo, Japan)[ ]. The wafer was nitrogen-doped n-type with a carrier density of ~1x10^{19} cm^{-3}, and the thickness of 250μm. The wafer reportedly contains stacking faults with a mean interspacing of ~2μm as the dominating defects[ ].

The creep specimens were machined into thin strips with dimensions of 25mm x 1mm x 0.05mm. Both faces were polished with 0.06μm diamond powders in order not to spoil strength. The surfaces of the CVD-SiC strips were normal to the CVD-growth direction. The SC-SiC strips were machined so that the all sides are normal to <100> orientations.

Fixture and Strain Measurement

Figure 1 shows a schematic illustration of the fixture designed for the SiC creep study. The thin strip specimens are fixed within a narrow gap between curved surfaces of a pair of the loading plates. This configuration was preferred to conventional four-point flexural configuration, in which the specimen temperature may unacceptably deviate from the designated temperature during in-reactor experiments, which relies on volumetric nuclear heating for controlling

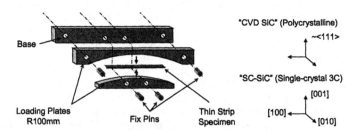

Figure 1. Illustration of the bend stress relaxation creep fixture designed for in-reactor experiment. All the fixture parts are made of CVD SiC.

266

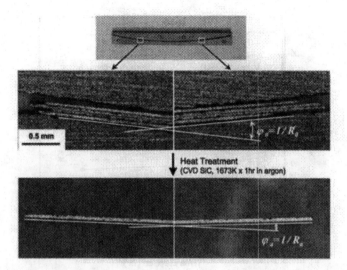

Figure 2. Measurement of bend radius of constrained and relaxed thin strip samples by optical microscopy.

temperature. Outer dimensions of the assembled fixture were designed to be 40mm x 4mm x 2mm so that it fits into slots for the standard miniature composite tensile specimens in irradiation vehicles. All the fixture parts were made of CVD-SiC to avoid potential chemical reactions.

Figure 2 shows a photograph of the assembled fixture, and optical micrographs of constrained and the relaxed specimens. The bend radius of constrained specimens varied in a range of 120 – 160mm, depending on gap width in the assembled fixture. Bend radius of the specimens was smallest in the center and largest at the both ends, but the variation was within ~10% over the specimen length except for the very end regions. The bend radius averaged over the entire specimen length was used to determine the residual stress after relaxation (σ_a):

$$\sigma_a = \frac{Et(\varphi_0 - \varphi_a)}{2L} \tag{1}$$

where E is the Young's modulus, t the specimen thickness, L the specimen length, φ_0 the initial bend angle, and φ_a the residual bend angle in the freed sample. Thus, the BSR ratio (m) was determined by the equation below:

$$m = \frac{\sigma_a}{\sigma_0} = 1 - \frac{\varphi_a}{\varphi_0} \tag{2}$$

where σ_0 is the initial stresses. The accuracy in bend angle determination by digital optical microscopy was <0.1°. This gives the potential error of ~1% in BSR ratio (m) determination regardless of m value in the present experimental configuration.

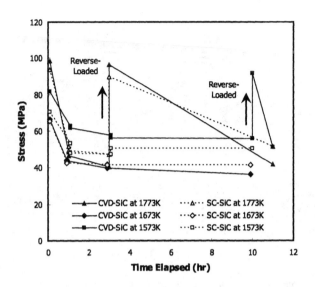

Figure 3. Relaxation behavior of the maximum surface stress in polycrystal (CVD-SiC) and single-crystal (SC-SiC) beta-SiC samples during BSR experiment. Reverse loading was applied to some specimens by putting back into the fixture after flipping.

BSR Conditions

The stress relaxation experiment was performed in a flow of commercial ultra-high purity argon at temperatures of 1573, 1673, and 1773K. The maximum cumulative hold time at the designated temperature was 10 hours. The heating rate at the end of heating sequence was ~15K/min. The cooling rate was not controlled but higher than the heating rate. The specimens were measured for creep strain after heat treatment for 1 and 3 hours and then put back to the fixture for further heat treatment. In some occasions, specimens were flipped before additional heat treatment in order to examine the effect of reverse loading. The initial flexural stress levels were estimated to be 65 – 100 MPa, assuming the room temperature Young's modulus of ~450GPa[10] and ~94, ~93, and ~92% retention of the room temperature modulus at 1573, 1673, and 1773K, respectively[11-14].

RESULT AND DISCUSSION

The relaxed bars were curved generally smooth over the length. There was no sign of significantly localized deformation noticed, in a macroscopic scale, in either the relaxed single- and poly-crystalline samples.

Flexural stress in the constrained specimens, calculated from the constrained and unconstrained bend radii, is plotted in Fig. 3 as a function of cumulative hold time at the designated temperature. In all cases, the stress exhibited a steep initial drop during the first hour, followed by the periods of much slower relaxation process. Despite of the somewhat varied

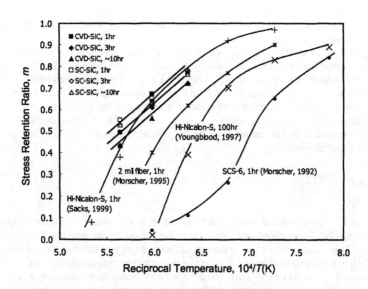

Figure 4. Bend stress relaxation ratio (m) for in polycrystal (CVD-SiC) and single-crystal (SC-SiC) beta-SiC samples plotted against reciprocal temperature.

initial stresses, higher relaxation rates were generally noted at higher temperatures, indicating the responsible mechanism is thermally activated. The residual stress approached to an asymptotic value, which varied but did not appear to systematically depend on materials or temperature. The observed stress relaxation behavior is believed to be due to primary creep, since SiC is known to exhibit only primary creep in the temperature range of this work[15].

CVD-SiC and SC-SiC exhibited similar stress relaxation behavior during the first hour at all temperatures, although the absolute relaxation rate of SC-SiC might be slightly lower than that of CVD-SiC. The residual stress in SC-SiC remained almost unchanged after the first hour of heat treatment, whereas CVD-SiC continued to reduce in stress between 1 and ~10 hours. The minimum stresses achieved were higher for SC-SiC than for CVD-SiC at 1673 and 1773K, while they appeared opposite at 1573K, where the initial stress in the CVD-SiC sample was significantly higher than in the SC-SiC sample. Despite of the differences noted above, the similarity in stress relaxation behavior between CVD-SiC and SC-SiC during the first hour suggests that the identical operating mechanism is responsible for the initial stage of the primary creep, eliminating the possibility of major contribution of the grain boundary diffusion and/or sliding[16,17].

Carter, et al., reported a lack of compressive stress creep deformation in CVD-SiC at temperatures <1923K when the specimens were loaded along the CVD-growth direction, and attributed it to the dislocation motion on {111} slip planes as the responsible creep mechanism[18]. If we assume the dislocation glide along <110> directions on {111} planes as the primary operating mechanism, the loading orientation in SC-SiC specimens should give significantly higher average Schmid factor than in CVD-SiC specimens. The possibly lower stress relaxation rate and the higher minimum stress in SC-SiC observed in this work indicate that the operation of such mechanism is unlikely.

Figure 4 plots the BSR ratio, m, against the reciprocal temperature. The m values for CVD-SiC exhibited the expected hold time-dependence, whereas those for SC-SiC did not due to the very small deformation during the period beyond 1 hour. The stress relaxation in the SC-SiC samples needs to be measured after shorter hold time, or preferably be measured by an in-situ type experiment, in order to determine the effect of hold time on the m values.

Assuming a thermally activated relaxation mechanism, the activation energy (Q) for the responsible process can be determined from the temperature dependence of m values at different hold times by the relationship,

$$Q = R \cdot \frac{\ln(t_2/t_1)}{1/T_2 - 1/T_1} \tag{3}$$

where T_1 and T_2 are the temperatures at which m is equal to a constant value after hold times of t_1 and t_2, respectively[6]. From the present experiment, one can derive the activation energy of ~850kJ/mol. This value is very close to the self diffusion energy of either carbon or silicon in beta-SiC[19,21], implying that the bulk diffusion is a potential controlling mechanism.

However, it should be noted that this analysis may incorporate potentially a large error in estimating the activation energy because of the following reasons. First, the initial stress was not constant but significantly varied in the runs at different temperatures, as seen in Fig. 3. For these materials in the stress range tested, a linear stress dependence of the strain rate has not been demonstrated. Second, the asymptotic values of the remaining stress seem to be dependent on temperature. The asymptotic stress at 1573K and 1673K probably corresponds to m=~0.5 in this experiment, so the activation energy would be most accurately determined at the constant value of m at significantly higher than 0.5. However, m=~0.5 has already achieved at 1773K after the shortest hold time. Therefore, the optimum m value at which the activation energy could be estimated in a sufficiently credible manner was not identified in the present experiment.

Plotted together in Fig.4 are the published bend stress relaxation ratio data for SCS-6™ CVD SiC fiber[8], Hi-Nicalon™ Type-S near-stoichiometric polymer-derived SiC fiber[7,22], and a developmental "2 mil" CVD SiC fiber that consists of the core and inner layers for SCS-6 fiber[15]. Data points from the present experiment appear at significantly higher temperatures than those for the CVD SiC fiber for the same m value, implying the smaller extent of stress relaxation in the CVD SiC studied than in the fiber. It is believed that this difference is primarily due to the presence of excess silicon in the outer shell of the SCS-6 CVD SiC fiber. Morscher and DiCarlo have derived an activation energy of ~560kJ/mol for the SCS-6 fiber at $m = 0.5$ and attributed the deformation to anelastic grain boundary sliding controlled by the grain boundary diffusion of excess silicon[8]. On the other hand, m values obtained here are close to those for the Hi-Nicalon™ Type-S fiber, which does not include excess silicon but is slightly rich in carbon. Another potential reason for the different relaxation behavior is the initial strain; in the present experiment it was ~0.02%, while it was 0.1~0.3% in the referred work by Morscher, et al[8], and Youngblood, et al[7]. The initial stress or strain in the experiment by Sacks is not reported[22].

Specimen reversed in bending direction after experiencing significant stress relaxation exhibited a large initial stress drop similar to that in virgin samples. The residual stress levels reached after heat treatment under the reverse loading were similar to those in non-flipped specimens. These features indicate that the microstructural defect motion responsible for the primary creep deformation is reversible. Further study is necessary to identify the responsible operating mechanism.

CONCLUSIONS

BSR creep of polycrystalline and single-crystalline chemically vapor deposited beta phase SiC was studied. The applicability of BSR technique to creep study of bulk ceramics was demonstrated; thin strip samples with sufficient strength were successfully machined and the small creep strains occurred in those samples were measured to sufficient accuracy.

The thermal creep experiment was conducted at 1573 – 1773K in argon to maximum hold time of 10 hours. Both materials exhibited similar primary creep deformation at the initial stresses of 65 – 100 MPa. The stress retention ratios determined in the present experiment appeared significantly higher than those reported for SiC fibers at a given temperature, implying a significant effect of the initial material condition on the relaxation behavior. The preliminary analysis based on the relaxation time – temperature relationship gave an activation energy of ~850 kJ/mol for the primary responsible process in CVD SiC.

REFERENCES

[1] R.H. Jones, et al., "Promise and Challenges of SiC$_f$/SiC Composites for Fusion Energy Applications," *Journal of Nuclear Materials*, **307-311**, 1057-72 (2002).

[2] G.O. Hayner, et al., *Next Generation Nuclear Plant Materials Research and Development Program Plan*, INEEL/EXT-04-02347 Revision 1, Idaho National Engineering and Environmental Laboratory, Idaho Falls (2004).

[3] J.L. Straalsund, "Radiation Effects in Breeder Reactor Structural Materials," Metallurgical Society of American Institute of Mining, Metallurgical and Petroleum Engineers, New York (1977).

[4] R. Scholz and G.E. Youngblood, "Irradiation Creep of Advanced Silicon Carbide Fibers," *J. Nucl. Mater.*, **283-287**, 372-75 (2000).

[5] G.W. Lewthwaite, "Irradiation Creep during Void Production," *J. Nucl. Mater.*, **46**, 324-28 (1973).

[6] G.N. Morscher and J.A. DiCarlo, "A Simple Test for Thermomechanical Evaluation of Ceramic Fibers," *J. Am. Ceram. Soc.*, **75**, 136-40 (1992).

[7] G.E. Youngblood, R.H. Jones, G.N. Morscher, and A. Kohyama, "Creep Behavior for Advanced Polycrystalline SiC Fibers," *Fusion Reactor Materials Semiannual Progress Report*, DOE/ER-0313/22, 81-86 (1997).

[8] H. Nagasawa, K. Yagi, and T. Kawahara, "3C-SiC Hetero-Epitaxial Growth on Undulant Si(001) Substrate," *J. Crystal Growth*, **237-239**, 1244-49 (2002).

[9] E. Polychroniadis, M. Syvajarvi, R. Yakimova, and J. Stoemenos, "Microstructural Characterization of Very Thick Freestanding 3C-SiC Wafers," *J. Crystal Growth*, **263**, 68-75 (2004).

[10] Y. Katoh and L.L. Snead, "Mechanical Properties of Cubic Silicon Carbide after Neutron Irradiation at Elevated Temperatures," *Journal of ASTM International*, in press.

[11] J.R. Hellmann, D.J. Green, and M.F. Modest, "Physical Property Measurements of High Temperature Composites," *Projects Within the Center for Advanced Materials*, J.R. Hellmann and B.K. Kennedy, Eds., 95-114 (1990).

[12] W.S. Coblenz, "Elastic Moduli of Boron-Doped Silicon Carbide," *Journal of the American Ceramic Society*, 58, 530-531 (1975).

[13] R.G. Munro, "Material Properties of a Sintered -SiC," *Journal of Physical and Chemical Reference Data,* 26, 1195-1203 (1997).

[14] J. Kubler, "Weibull Characterization of Four Hipped/Posthipped Engineering Ceramics Between Room Temperature and 1500 °C," *Mechanische Charakterisierung von Hochleistungskeramik Festigkeitsunte,* 1-88 (1992), EMPA Swiss Federal Laboratories for Materials Testing and Research.

[15] G.N. Morscher, C.A. Lewinsohn, C.E. Bakis, R.E. Tressler, and T. Wagner, "Comparison of Bend Stress Relaxation and Tensile Creep of CVD SiC Fibers," *J. Am. Ceram. Soc.,* **78**, 3244-52 (1995).

[16] C.A. Lewinsohn, L.A. Giannuzzi, C.E. Bakis, and R.E. Tressler, "High-Temperature Creep and Microstructural Evolution of Chemically Vapor-Deposited Silicon Carbide Fibers," *J. Am. Ceram. Soc.,* **82**, 407-13 (1999).

[17] J.A. DiCarlo, "Creep of Chemically Vapour Deposited SiC Fibers," *J. Mater. Sci.*, 21, 217-224 (1986).

[18] C.H. Carter, Jr., R.F. Davis, and J. Bentley, "Kinetics and Mechanisms of High-Temperature Creep in Silicon Carbide: II, Chemically Vapor Deposited," *J. Am. Ceram. Soc.,* **67**, 732-40 (1984).

[19] M.H. Hon, R.F. Davis, and D.E. Newbury, *J. Mater. Sci.*, 15, 2073 (1980).

[20] J. Li, L. Porter, and S. Yip, "Atomistic Modeling of Finite-Temperature Properteis of Crystalline Beta-SiC, II. Thermal Conductivity and Effects of Point Defects," *J. Nucl. Mater.*, 255, 139-152 (1998).

[21] F. Gao, W.J. Weber, M. Posselt, and V. Belko, "Atomistic Study of Intrinsic Defect Migration in 3C-SiC," *Phys. Rev.*, B 69, 245205 (2004).

[22] M.D. Sacks, "Effect of Composition and Heat Treatment Conditions on the Tensile Strength and Creep Resistance of SiC-based Fibers," *J. Eu. Ceram. Soc.,* **19**, 2305-15 (1999).

Hardness and Wear Resistance of Ceramics

MECHANICAL RESPONSES OF SILICON NITRIDES UNDER DYNAMIC INDENTATION

Hong Wang and Andrew A. Wereszczak

Metals and Ceramics Division
Oak Ridge National Laboratory
Oak Ridge, TN 37831-6068

ABSTRACT

Mechanical responses of five silicon nitrides have been experimentally studied by using dynamic and standard hardness tests. The indentation loads ranged from several newtons to several hundred newtons. In the dynamic test, a load pulse with ~ 100 µs rise time was achieved. Indentation size effect (ISE) and loading rate effects on the Vickers hardness were observed for the silicon nitrides. The ISE was further characterized by using the Meyer law and the energy-balance or proportional specimen resistance (PSR) model. The conventional brittleness approach is not satisfactory in characterizing transition between plasticity and fracture, and a new brittleness concept is proposed based on the PSR model that appears to capture the dynamic indentation responses of the materials.

INTRODUCTION

Silicon nitride is used as rolling contact elements in diesel engines and is attracting increasing attention for use in hybrid bearings due to its unique strength and fracture toughness properties [1]. To ensure the surface integrity of components, an effective and efficient machining (grinding) method is essential and that, in turn, requires a thorough understanding of material removal mechanisms and their effects on the subsurface damage [2, 3]. Among other methods, indentation simulates the abrasive (grit)-workpiece interaction by using an idealized indenter and emerges as an effective approach to investigating the grinding responses of ceramics including silicon nitride [4]. The residual impression of indentation and the underlining plastic zone are believed to be representative of the surface grinding groove and subsurface grinding-affected-zone such as residual stress layer. On the other hand, the median/radial crack is mainly responsible for the strength degradation, whilst the lateral crack is related to the chip formation and grinding material removal. Therefore, an appropriate characterization and understanding of indentation deformation and fracture should provide insightful information for the design and control of the grinding process (parameters), for example, grit depth-of-cut, single-grit normal grinding force and wheel speed, etc.

When using Vickers hardness to quantify the plastic deformation of silicon nitrides, one finds that plasticity is affected not only by the porosity, microstructure, phases and additives present [5-8], but also by the surface condition, temperature and load level [9-13]. Measured hardness values increase at low load levels, and this effect is called the indentation size effect (ISE). Under conditions where the ISE exists, the traditional hardness can not fully capture the plastic deformation behavior of the material. As a result, a number of approaches such as power law functions, polynomial series and reduced polynomial series have been proposed to relate the indent load to indent size [14-

17]. At the same time, by using in-situ optical microscopic observations, acoustic emission monitoring, observations of cross section of indentations, different crack systems have been revealed including radial, median and lateral crack. However, it remains uncertain as to how the crack development is partitioned between the loading and unloading cycles, especially for the radial/median crack [18-22]. Numerous equations have been proposed empirically or theoretically to relate the indent load to crack size to different degrees of satisfactions [23-25].

Indent diagonal and crack length are two cross-correlated quantities with the indentation process [18]. It is hypothesized that the indentation is dominated by plasticity at low loads and by fracture at high loads [26]. The investigation of ISE should facilitate the development of appropriate grinding conditions in the respective regimes (e. g., brittle or ductile). However, most previous research efforts pertained to quasi-static conditions. Considering that the surface of ceramic components is generally subjected to dynamic loading during machining and in many other applications, it becomes imperative to understand how the ceramic responds under the dynamic indentation. In this paper, we address this important issue and present results obtained on five different silicon nitrides by using dynamic hardness testing. Specifically, the ISE and loading rate effects are focused on. It should be emphasized that the present study concerns residual impression measurements and not in-situ measurements of indenter load-displacements.

EXPERIMENTAL TECHNIQUES

The dynamic indentation system (developed at Michigan Tech, Houghton, MI [27, 28]) is shown in Fig. 1. The major part of the system is an incident bar, which is similar to that used in a Split-Hopkinson Pressure Bar system (marage steel, ϕ6.4 x 1,130 mm). A Vickers indenter is attached to the right end of the incidental bar extending from the indenter house. The specimen is sandwiched between the indenter and a load cell (Kistler 9213, 200 KHz, Amherst, NY) that is fixed to a rigid stopper and aligned horizontally with the loading axis. A solenoid is used to trigger the gas gun. After launching the projectile (ϕ6.4 x 250 mm), a compressive stress wave is generated, which is split into two parts. One part proceeds to load the specimen directly, and the other is transformed into a tensile stress wave through a series of reflections to withdraw the indenter, resulting in a single loading pulse as desired [27, 28]. For the low range of dynamic load ($\leq$20-30 N), a short steel bar (ϕ6 x 75 mm) is used to drive the system by manually tapping the incident bar. The force signal was wired to a charge amplifier (Kistler 5010B, Amherst, NY), and then to the computer interfaced with a digitizer (National Instruments NI5102, 2 Channels, 15 MHz, 8-Bit, Austin, TX), and finally acquired via a LabVIEW program. Under current configuration, a load pulse with a ~ 100 μs rise time can be achieved in the respective load ranges with two driving methods. On the other hand, the standard indentation was conducted by using micro-hardness tester (Instron Wilson-Wolpert Tukon 2100B, Canton, MA), which serves as a base line for comparative study. Typical loads ranged from 1 to 400 N with the load time of 3 seconds. An optical microscope (Nikon Nomarski Measure Scope MM-11, Tokyo, Japan) was used for the high-load indent measurements (> 20 -30 N) made immediately after each test, while a SEM (Hitachi S4100 field emission scanning electron microscope, San Jose, CA) was used for the small indent size in order to obtain a better resolution.

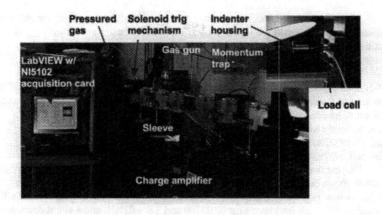

Fig. 1 Experimental setup of dynamic indentation system.

Table 1 Tested silicon nitrides with physical and mechanical properties

Silicon nitrides	$\rho*$ (g/cc)	σ_f* (MPa)	$E^\dagger$ (GPa)	$\nu^\dagger$	$H_{True}^\ddagger$ (GPa)	$K_{IC}*$ (MPa√m)	Specimen preparation
Ceralloy SN147-31N, GPSRB, Ceradyne	3.21	800	326.0	0.258	15.44	6.0	From ASTM C1161 bars, 4x4x3 mm, surface ground at least 30 μm and polished to 0.5 μm
CerBec NBD 200, HIPed, Saint Gobain	3.16	900	322.0	0.271	15.77	5.5	From φ6.35mm balls, ground into two parallel flats with 3 mm height, polished to 0.5 μm
CerBec SN101C, HIPed, Saint Gobain	3.21	1000	307.0	0.270	16.09	6.5	From φ6.35mm balls, ground into two parallel flats with 3 mm height, polished to 0.5 μm
TSN03-NH, HIPed, Toshiba	3.22	900	309.0	0.278	15.98	6.7	From φ11.11mm balls, ground into two parallel flats with 3 mm height, polished to 0.5 μm
NC132, Hot-pressed, Norton	3.18[§]	710[§]	316.7	0.297	16.77	4.6	In 4x4x3 mm, surface polished to 0.5 μm.

* Based on manufacturer's data sheet, except specified otherwise.
† Based on RUS measurements [29].
‡ Based on quasi-static case, Fig. 4(b).
§ Larsen, D. C. and Adams, J. W., Property screening and evaluation of ceramic turbine materials, AFWAL Report TR-83-4141, 1984, Nat'l Tech Information Service.

MATERIALS

Five silicon nitrides, as given in Table 1, were selected for the current study either because of their widespread use in bearing applications or as a standard reference material. The physical and mechanical properties are listed in the above for reference. These ceramics represent three of the typical methods of material processing (e.g., gas pressure sintering, hot pressing and hot isostatic pressing.

EXPERIMENTAL RESULTS

A plot of H_V versus indent load for Ceralloy 147-31N is shown in Fig. 2(a). While somewhat scattered, the data reveal a consistent trend with respect to ISE and loading rate effect. The H_V in the static case shows a small increase from 14 to 15 GPa with decreasing load, which is more pronounced in the dynamic loading case where H_V reaches 18-19 GPa. Plots for the other silicon nitrides show very similar trends.

Overall, the surface radial crack size as a function of indent diagonal data for dynamic loading does not vary significantly from that for static loading for Ceralloy 147-31N as shown in Fig. 2(b). The similar observation holds for other silicon nitrides. The dynamical effect on the crack sizes of these ceramics is generally hardly appreciable with a little increase in high load levels for NBD200 and TSN-03NH, and a slight decrease for SN101C.

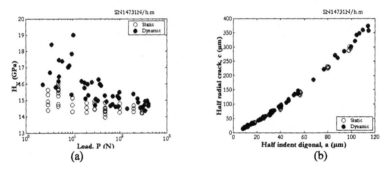

Fig 2 Results of Ceralloy 147-31N with respect to (a) hardness and (b) indentation crack.

ANALYSES AND DISCUSSIONS

The generic form proposed by Meyer states that the indent load P can be related to the indent diagonal d in a power law form [14],

$$P = Ad^n \tag{1}$$

in which A and n are two constants determined by log-log curve-fitting. A number of studies have shown that the Meyer model fits the data of ceramics quite well [11, 13, 17], and that n is related to microstructure of material, increasing with increases in grain size and relative density [7]. As a result, there was interest in the applicability of the Meyer law for the dynamic, as well as static, loading conditions. Our resulting analysis showed that the P-d data points actually fall on a straight line with a high correlation coefficient R when plotted in a log-log scale. Effects of static and dynamic loading were revealed on both the Meyer parameters and the correlation coefficient (Fig. 3(a)). Values of n were less than 2 and were consistently slightly lower for dynamic as compared to static loading, while the coefficient A exhibited the opposite trend. The A-n plot for both static and dynamic loading as shown in Fig. 3(b) depicts the evolution of the Meyer parameter. Note that it is the deviation of n value from 2 that determines the ISE. So the decreasing n indicates a larger deviation and higher ISE can be obtained in dynamic loading, as observed in Fig. 2(a). Considering the above-mentioned effect of microstructure on n [7], it seems that the rising loading rate is equivalent to reducing grain size, and similarly makes the material appear to be stronger.

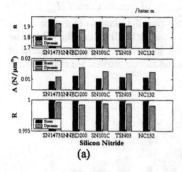 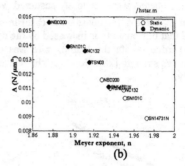

(a)   (b)

Fig. 3 Variations of Meyer model parameters.

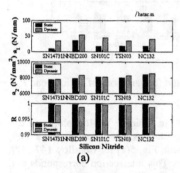 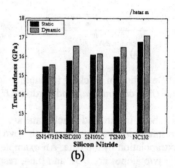

(a)   (b)

Fig. 4 Variations of (a) energy-balance or PSR model parameters and (b) hardness values.

Another popular method of interpretation is the energy-balance or proportional specimen resistance (PSR) model. In this model, indent load P can be related to diagonal d in a reduced series [11, 15-17],

$$P = a_1 d + a_2 d^2 \qquad (2)$$

where the $a_1 d$ term is related to the contribution from the surface generation or specimen resistance (between indenter and specimen), and the $a_2 d^2$ term relates to volume plastic deformation. The parameters a_1 and a_2 can be determined by linear regression of P/d against d. An alternative form of PSR model is

$$P/d^2 = a_1/d + a_2 \qquad (3)$$

which explicitly demonstrates how a_1 measures the extra d-dependent component and a_2 defines the so-called true hardness. The study showed that the current data could be fit as well by the PSR model as illustrated in Fig. 4(a), in which a_1 is more sensitive than a_2, being raised significantly by the dynamic loading. On the other hand, the modified PSR model [13] could not give an effective characterization of indent-load relation when the related parameters were randomly scattered. With the use of true hardness ($H_{True} = 2a_2$, as given in Table 1 for quasi-static case) in characterizing the plastic deformation, one can find that this quantity is increased by the dynamic loading in all cases (Fig. 4(b)). The rise in hardness with the loading rate was observed to be related to the increase of compressive strength [28]. However, the relevant data on silicon nitrides are not available currently.

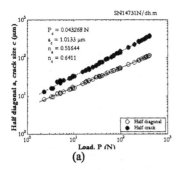

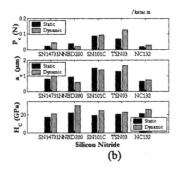

(a)   (b)

Fig. 5 (a) Definition of cracking indentation load Pc and (b) variations of critical load P_c, indent size (half diagonal) a_c, and hardness H_c.

When both indent and crack data are presented in a single log-log plot, one can define an intersection point between the two resulting plots, which have different slopes, by the extrapolation of the data. An example is given in Fig. 5(a) for Ceralloy 147-31N, where the two slopes are 0.52 and 0.64, respectively. This intersection corresponds to a point where indent size equals crack size, and defines a critical load, P_c, which should signify the indentation fracture initiation as suggested by Lawn et al [26]. It is interesting to find that P_c of TSN03-NH is significantly higher, especially in dynamic case, in comparison with other materials (Fig. 5(b)). These values of P_c fall within the range of Bolse and Peteves' observation on the hot-pressed silicon nitride [9]. At the same time, the critical hardness ($H_C = P_C/2a_C^2$) shows a more appreciable increase under the dynamic loading for each material as compared to the hardness in Fig. 4(b).

It should be pointed out that the relation of indentation dimension to the indenter load in terms of residual diagonal and crack sizes as given in Fig. 5(a) reveals the essence of ISE. At loads less than P_c, the plastic deformation dominates the response of material,

280

while, at the level higher than P_c, fracture participates and dominates the process. Lawn et al [26] have theoretically deduced that the half diagonal at the point of fracture initiation, a_c, is directly related to the square of the fracture toughness to hardness ratio. A smaller value of a_c manifests a stronger propensity to cracking, and then represents a more brittle behavior. With a unit of μm^{-1}, the quantity $(H/K_{IC})^2$, which is denoted as B, and is inversely proportional to a_c, should thus depict the brittleness of material in some senses [26]. Considering the fracture toughness of silicon nitride is quite insensitive to loading rate [30], the K_{IC} values in Table 1 can be used into above B formula, which immediately shows that the formula merely reflects the rate sensitivity of the square of the hardness. In other words, the brittleness of all the silicon nitrides tested is promoted by dynamic loading, which conflicts with the observations in Fig. 5(b) where a_c does not decrease universally. To characterize the dynamic response of studied silicon nitride more effectively, a new brittleness B is introduced,

$$B = a_2 / a_1 \qquad (4)$$

by re-examining Eq. (2). This definition is a natural extension of the brittleness concept elaborated by Quinn and Quinn [11] because a_2 is related to plastic deformation and a_1 is to the generation of new surface (fracture actively involved). Fig. 6 shows that the proposed brittleness has actually been suppressed substantially under dynamic loading for each material, suggesting that this concept is much more capable of capturing the dynamic responses of materials.

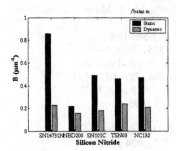

ig. 5 Variation of brittleness B in Eq. (4).

While the hardness is relatively insensitive to loading rate as seen in Fig. 4(b) (< 5% change), the inspection of Fig. 5(b) and Fig. 6 reveals that the critical hardness and brittleness responses to the loading rate are quite impressive. Ceralloy 147-31N (GPS RBSN) shows the strongest rate-effect on brittleness (~ 74% reduction) and notable effect on critical hardness (~ 21% increase). On the other hand, SN101C (HIPed), TSN03-NH (HIPed) and NC132 (HPSN) display relatively higher rate effects on brittleness (48 to 63% reductions) and critical hardness (11 to 27% increases), whilst NBD200 (HIPed) exhibits the smallest (~29%) reduction in B and the largest (~36 %) increase in H_c. Observations of these rate effects should be very helpful in explaining controlling mechanism of plastic and fracture responses and the loading rate sensitivities of ceramics.

CONCLUSIONS

- Vickers hardness of all the silicon nitrides exhibits the indentation size effect (ISE) and dynamic loading enhances the hardness as well as the ISE.

- TSN-03NH exhibits the highest dynamic fracture-initiation load (P_c) as defined by the intersection of extrapolation of the indent and crack size against load plots.
- Brittleness derived from the energy-balance/PSR model can capture the plastic-fracture competition during the dynamic indentation effectively.
- Dynamic loading significantly raised the critical hardness and reduced the brittleness of the silicon nitrides studied.

ACKNOWLEDGMENTS

This research was sponsored by the U.S. DOE, Office of Transportation Technologies, as a part of the Heavy Vehicle Propulsion System Materials Program, under contract DE-AC05-00OR22725 with UT-Battelle, LLC. The work was supported in part by an appointment to the ORNL Postdoctoral Research Associates Program, sponsored by the U.S. DOE and administered by the ORISE. Authors would like to thank Drs. M. K. Ferber, P. J. Blau and P. F. Becher of ORNL for their time on experimental system set-up and manuscript review. Particularly, authors are grateful to G. Subhash of MTU, G. D. Quinn of NIST, and R. C. Bradt at U. of Alabama for their valuable suggestions and inspiring discussions on experimental issues as well as ISE and indentation fracture.

REFERENCES

[1]   Mikijelj, B., Mangels, J., Belfield, E., Inst Mech Eng Fuel Injection Systems Conference, Nov 26-28, 2002, London, UK
[2]   Malkin, S., Ritter, J. E., J. Eng. Ind.-Trans. ASME, 111: 167-174 1989
[3]   Zhang, B., Zheng, X.L., Tokura, H., Yoshikawa, M., J. Mater. Process. Tech., 132: 353-364 10 2003
[4]   Malkin, S., Hwang, T. W., Annuals CIRP, 45: 569-580 1996
[5]   Chakraborty, D., Mukerji, J., J. Mater. Sci., 15: 3051-3056 1980
[6]   Greskovich, C., Gazza, G. E., J. Mater. Sci. Lett., 4: 195-196 1985
[7]   Babini, G. N., Bellosi, A., Galassi, C., J. Mater. Sci., 22: 1687-1693 1987
[8]   Berriche, R., Holt, R. T., J. Am. Ceram. Soc., 76: 1602-1604 1993
[9]   Bolse, W., Peteves, S. D., J. Test. Eval., 23: 27-32 1995
[10]   Ekstrom, T., J Hard Mater, 4:77-95 1993
[11]   Quinn, J. B., Quinn, G. D., J. Mater. Sci., 32: 4331-4346 1997
[12]   Dusza, J., Steen, M., J. Am. Ceram. Soc., 81: 3022-3024 1998
[13]   Gong, J. H., Wu, J. J., Guan Z. D., J. Eur. Ceram. Soc., 19: 2625-2631 1999
[14]   Meyer, E., Phys. Z. 9: 66 1908
[15]   Bernhardt, E. O., Z. Metallk. 33: 135-144 1941
[16]   Frohlich, F., Grau, P. and Grellmann, W., Phys. Stat. Sol. 42: 79-89 1977
[17]   Li, H. and Bradt, R. C., J. Mater. Sci., 28: 917-926 1993
[18]   Evans, A. G., Wilshaw, T. R., Acta Metal., 24: 939-956 1976
[19]   Niihara, K., Morena, R., Hasselman, D. P. H., J. Mater. Sci. Lett., 1: 13-16 1982
[20]   Shetty, D. K., Wright, I. G., Mincer, P. N., Clauer, A. H., J. Mater. Sci., 20: 1873-1882 1985
[21]   Cook, R. F., Pharr, G. M., Am. Ceram. Soc., 73: 787-817 1990
[22]   Lube, T., J. Eur. Ceram. Soc., 21 (2): 211-218 2001
[23]   Ponton, C. B., Rawlings, R. D., Mater. Sci. Tech., 5: 865-872 1989

[24]   Li, Z., Ghosh, A., Kobayashi, A. S., Bradt, R. C., J. Am. Ceram. Soc., 72: 904-911 1989
[25]   Liang, K. M., Orange, G., Fantozzi, G., J. of Mater. Sci., 25: 207-214 1990.
[26]   Lawn, B. R., Jensen, T., Arora, A., J. Mater. Sci., 11 (3): 573-575 1976
[27]   Koeppel, B. J., Subhash, G., Exper. Tech., 21: 16-18 1997
[28]   Anton, R. J., Subhash, G., Wear, 239: 27-35 2000.
[29]   Wereszczak, A. A. Ceramic Eng. & Sci. Proc., in press, Vol. 26 2005.
[30]   Kishi, T., Takeda, N. Kim, B. N., Ceram. Eng. Sci. Proc., 11: 650-664 1990.

SCRATCH RESISTANCE AND RESIDUAL STRESSES IN LONGITUDINALLY AND TRANSVERSELY GROUND SILICON NITRIDE

G. Subhash*, M.A. Marszalek
Department of Mechanical Engineering-Engineering Mechanics,
Michigan Technological University, Houghton, MI 49931 USA

A.A. Wereszczak, M.J. Lance
Ceramic and Technology Group
Oak Ridge National Laboratory, Oak Ridge, TN USA

ABSTRACT

An instrumented scratch tester is utilized to investigate the scratch resistance of surface ground silicon nitride. Single-grit scratches were conducted, both in longitudinal and transverse direction to grinding. The volume of material removed during the scratch event and the measured force were utilized to quantify the scratch resistance and then estimate the bounds for ductile to brittle transition limits for material removal. Residual stress measurements and the scratch resistance were rationalized based on the induced damage features.

*Corresponding author: subhash@mtu.edu, Ph: (906) 487-3161, Fax: (906) 487-2822

INTRODUCTION

Grinding is the most commonly employed machining operation to fabricate components from structural ceramics. However, the grinding operation has been known to introduce significant surface and sub-surface damage[1,2]. It has been shown that the grit size of the diamond wheel used during the grinding process directly correlates to the amount of sub-surface damage depth induced in the material[2,3]. It was also shown that the bend strength of transversely machined specimens was affected at grinding grit sizes of 600 or coarser and the strength of the longitudinally machined specimens was not affected down to a grit size of 320[2]. The transversely machined bend bars have exhibited a bend strength that is 18% lower than that of the longitudinally machined specimens. It has been concluded that grinding direction has the strongest influence on the characteristic strength of the final component[4].

In other studies, it has been shown that there is an inverse relationship between depth of cut (DOC) during grinding and the specific energy absorbed[5,6,7]. A smaller DOC during machining leads to larger values of specific energy being absorbed by the material. This phenomenon has been attributed to ductile regime grinding in ceramics. On the other hand, larger depths of cut are known to induce a greater sub-surface damage, dominated by brittle fracture.

In the past, scratch hardness has been used to assess the scratch resistance of ceramics[5,8]. Subhash and co-workers[9,10] have utilized an instrumented scratch tester to develop a scratch resistance measure that quantifies the susceptibility of a ceramic to brittle fracture. The current study further extends the above technique on longitudinally and transversely machined ceramics and attempts to correlate the scratch depth to the ductile or brittle mode of material removal.

MATERIAL

Reaction bonded silicon nitride (Si$_3$N$_4$) specimens of 3 mm x 4 mm x 12 mm ground longitudinally or transversely to the length axis as per ASTM C1161B-1994, were used in this study. The average tensile strength of longitudinally and transversely ground Si$_3$N$_4$ specimens are 816 and 650 MPa, respectively.

EXPERIMENTAL PROCEDURE

The experiments were performed using a custom built instrumented scratch tester[9,10]. The scratch tool, a single-point 90° conical diamond dressing tool, is used for scratching. Typical scratch velocities are on the order of 1 m/s. After the test, the loose debris in the groove was forced out by blowing compressed air and then scanned using a laser profilometer at a resolution of 0.1 microns. The scratches were then viewed under a scanning electron microscope (SEM) to analyze the induced damage features. A select number of scratched specimens were analyzed using Raman spectroscopy to measure residual stress within and surrounding the scratches. The residual stress was calculated using the method described by Tochino and Pezzotti[11].

EXPERIMENTAL RESULTS

A typical normal force profile, the associated SEM image and the laser profilometer scan of the scratch on a transversely machined Si$_3$N$_4$ specimen are shown in Fig. 1. The fluctuations in the measured force signal correspond to random cracking surrounding and beneath the scratch.

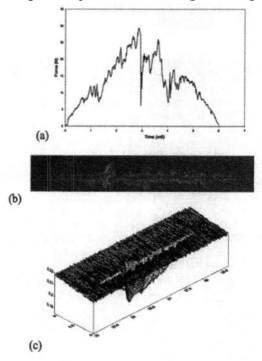

Fig. 1 (a) Measured force profile during single-grit scratching on transversely machined Si$_3$N$_4$, (b) corresponding scanning electron micrograph and (c) 3-D view of scratch.

It can be seen in the above figure that there is minimal cracking when the DOC is low and large amounts of lateral cracking when DOC is high. The 3-D view shows a large volume of material removal below the scratch trace in the mid-section of the scratch. The scratch resistance measure[10] defined by $\dfrac{V * \sigma_f}{F * B * l}$ where V is the volume of material removed, $_f$ is the tensile strength of the material, F is the mean normal force, B is the brittleness measure as previously defined by Zhang and Subhash[8] and l is the length of the scratch was used to reveal the susceptibility of a ceramic to scratching. The values are given in Table I.

Table I: Data on various scratch parameters machined Si_3N_4 specimens

Longitudinal

Length (mm)	Volume (mm^3)	Mean Force (N)	Scratch Resistance Measure $V*\sigma_f/(F*B*l)$	Scratch energy (J)	Specific Energy (J/mm^3)	Residual Stress (GPa)
1.03	0.00002	3.2051	257	0.00329	164.40	
1.11	0.00008	5.2465	487	0.00584	72.96	
1.33	0.00016	4.2182	1299	0.00561	35.07	
2.05	0.00103	8.3901	2379	0.01723	16.73	
2.28	0.00065	6.9004	1717	0.01573	24.20	2.493
2.43	0.00177	9.3561	2973	0.02278	12.87	
2.45	0.00128	8.2210	2451	0.02014	15.74	
2.68	0.00121	10.7846	1617	0.02886	23.85	
2.96	0.00226	10.0668	3592	0.02980	13.18	
3.07	0.00354	14.5402	3040	0.04469	12.62	
3.1	0.00484	21.1990	3433	0.06572	13.58	

Transverse

Length (mm)	Volume (mm^3)	Mean Force (N)	Scratch Resistance Measure $V* \sigma_f/(F*B*l)$	Scratch energy (J)	Specific Energy (J/mm^3)	Residual Stress (GPa)
1.03	0.00002	3.4151	175	0.00350	175.17	3.1944
1.23	0.00007	3.3176	532	0.00409	58.48	
1.30	0.0001	6.7021	363	0.00870	87.03	
1.57	0.00049	6.3292	1771	0.00996	20.32	
1.93	0.00062	9.5977	1045	0.01852	29.88	
2.20	0.00215	17.0729	1899	0.03756	17.47	
2.26	0.00123	14.3235	1228	0.03237	26.32	
2.30	0.00101	10.5593	1368	0.02429	24.05	4.3375
2.44	0.00278	10.7950	2741	0.02634	9.47	
2.50	0.0016	17.2377	1244	0.04309	26.93	
2.66	0.00229	9.9205	3156	0.02634	12.60	
2.77	0.00399	24.8047	1852	0.06871	17.22	
3.02	0.00425	16.0833	2966	0.04857	11.43	1.775
3.18	0.00394	15.7643	2580	0.05013	12.72	

The above table shows a good correlation between the scratch lengths and the various measured and calculated parameters. As expected, increasing scratch length corresponds to an increasing trend in volume of material removed and the mean force required to create the scratch. Fig. 2 shows the volume of material removed plotted against the measured mean force. The plot indicates that for small values of force, both longitudinal and transverse scratches have low values of volume of material removed and the rate of increase in the volume removed is low. However, beyond a value of 7 N, the rate of volume removed increases rapidly with a majority of the longitudinal scratches generating more material removal volume than transverse scratches for similar force levels. This trend can be rationalized as more extended cracking in the former than in the latter as will be substantiated in the section on 'Microscopic Observations'.

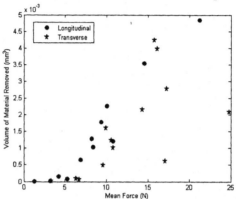

Fig. 2: Volume of material removed for scratches in longitudinal and transverse directions.

Fig. 3 shows the plot of scratch resistance measure vs. volume of material removed by the scratch. Once again, note that at low volume, the scratch resistance measure is low but increases rapidly at first and then appears to stabilize as the volume increases. This behavior can be attributed to different modes of material removal depending on the DOC. At low DOC, cracks do not extend beyond the scratch groove and the material removal process is dominated by the

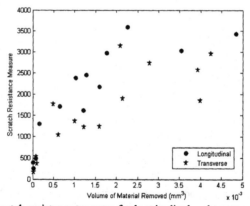

Fig. 3: Scratch resistance measure for longitudinal and transverse scratches.

ductile mode of deformation. As the DOC increases, cracking starts to occur around the scratch groove and thus the material removal is dominated by brittle fracture. Also, at large DOC, the proposed measure appears to reach a steady value, reflecting the material removal process to be mostly by brittle fracture.

The specific energy is an important parameter to consider while characterizing the mode of material removal in ceramics. It has been shown[7] that the specific energy has an exponential relationship with volume of material removed during the grinding process. For our scratch experiments, a similar trend is observed as seen in Fig. 4. It is clear that when the volume of material removed is small or the DOC is low, the specific energy is high. This relationship has been attributed to ductile material removal during grinding. It is also noted that the specific energy is slightly higher for transverse scratches than for longitudinal scratches, especially at low volume of material removed. This behavior further confirms dominance of ductile mode of material removal at low DOC in transverse scratches and dominance of brittle fracture at high DOC for both scratch orientations.

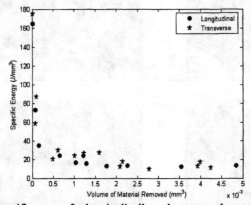

Fig. 4: Specific energy for longitudinally and transversely scratched Si_3N_4

While there are residual stresses present on the ground ceramics before scratching, additional residual stresses are induced in the material after the scratch event. Depending on the imposed DOC, the induced damage surrounding the groove and the resulting residual stresses will vary. As will be seen later in the section on 'Microscopic Observations', when DOC is small, cracking does not extend beyond the groove and the material removal occurs mostly via ductile mode of fracture. Therefore, residual stresses are expected to be high. On the other hand, as the DOC increases, cracking begins to occur around and beneath the scratch groove. These cracks relieve the residual stress present in the material as they propagate. The measured residual stress values for scratches on transversely machined specimen given in Table I supports this trend. More measurements on longitudinally machined specimens are being conducted to further validate the trends.

MICROSCOPIC OBSERVATIONS

The objective here is to identify the mode of material removal and the extent of cracking as a function of DOC due to scratching in both the grinding directions. Micrographs of scratch induced damage are presented for low, medium and high DOC (or length) scratches. Figs. 5 and 6 show the central segments of two scratches conducted parallel to and across the grinding direction, respectively. Both scratches are of equal length and the force measured during the scratch was also of the same magnitude. It is seen in both scratches that the extent of lateral cracking and the associated damage are minimal. However, in scratches conducted transverse to the grinding marks, the generated debris size was of the order of grain size, see Fig. 6 (b). These scratch edges are relatively smooth and do not show significant damage. On the contrary, when the scratch is conducted parallel to the grinding marks such fine debris is not created and is swept off completely by the scratch tool, see Fig. 5 (a). The edges of the groove are wavy due to the induced damage as shown in Fig. 5 (b). At higher force levels (or DOC), this damage may

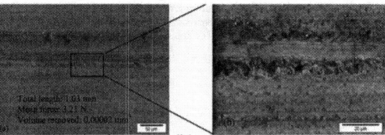

Fig. 5 (a) Mid-section of a scratch parallel to the grinding direction and (b) associated damage.

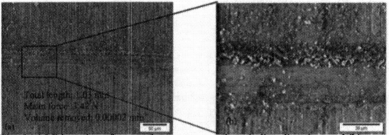

Fig. 6 (a) Mid-section of a scratch transverse to the grinding direction and (b) scratch boundary

extend as lateral cracking and contribute to large volumes of material removed. Thus, it appears as though the limit for large lateral cracking or damage initiation is lower when the scratch is conducted parallel to the grinding marks than transverse scratching. This inference can also be extracted from Figs. 2-4, where the volume removed and the scratch resistance measure are higher, and the specific energy is lower, for longitudinal scratches, compared to transverse.

Figs. 7 (a) and (b) show a comparison between two scratches of similar lengths, but with large differing mean force values. The scratch conducted longitudinal to the grinding marks, Fig. 7 (a), reveals lateral cracks in the form of lobes along the boundary of the scratch. The cracks are circular in shape and are almost evenly spaced. Fig. 7 (b) shows a scratch transverse to the grinding marks, of equal length. The mean force required to create the scratch was over twice that shown in Fig. 7 (a) and therefore, the extent of lateral cracking is more severe. Scratches conducted longitudinal and transverse to the grinding marks, with similar measured force, but differing lengths, are shown in Figs. 8 (a) and (b), respectively. Although the force levels and the associated severity of cracking appear to be the same in both the scratches, the DOC is lower in the longitudinal scratch. The sensitivity to the dominance of brittle cracking seems to originate at lower DOC, or force levels, in longitudinal scratches than in transverse.

 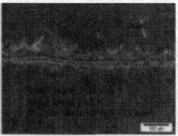

Fig. 7: Mid-sections of a scratch (a) parallel and (b) transverse to the grinding direction

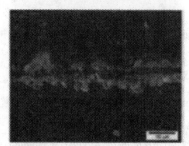 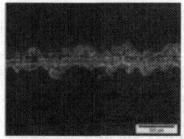

Fig. 8: Mid-sections of a scratch (a) parallel and (b) transverse to the grinding direction

The above inference has been consistently noted in the data presented in Figs. 2-4 as well as in the micrographs provided in Figs. 5-8. This damage sensitivity to the direction of scratching with respect to grinding marks can be attributed to relative activation of pre-existing damage in the machined ceramic. As stated in the Introduction, the longitudinally ground tensile specimens have 25% higher tensile strength than transversely ground tensile specimens. This means that cracks propagate at lower stress levels along the grinding marks, assuming maximum normal stress criterion for failure in brittle materials, when a tensile load is applied transverse to the grinding marks. In other words, the residual damage parallel to transverse grinding is more sensitive to tensile loading than residual damage perpendicular to longitudinal grinding. Thus, in a scratch process, as the tool traverses parallel to longitudinal grinding marks, it initiates cracking at lower stress levels then for transverse scratching as observed in our current experiments. This explanation is also consistent with the results by Quinn et al.[2] stated in the Introduction. However, the experimental data available thus far is limited, and more data needs to be generated to further validate the above conjuncture.

CONCLUSIONS

Based on the analysis of single-grit scratch experiments on machined ceramic specimens and the microscopic observations of the induced damage, the following conclusions can be drawn.

- Scratches conducted in both longitudinal and transverse directions to grinding reveal dominance of ductile mode of material removal at low DOC and of brittle mode of material removal at high DOC.
- In general, for the same DOC or the same imposed force, scratches parallel to the grinding direction result in greater damage than scratches conducted transverse to the grinding direction.
- The residual stress measurements in transverse scratches by Raman spectroscopy also support the above conclusion by revealing higher stress levels in low DOC scratches and low residual stress levels in high DOC scratches.

ACKNOWLEDGEMENTS

This research was supported by Oak Ridge National Laboratory, Oak Ridge, TN and sponsored by the U.S. Department of Energy, Office of Transportation Technologies, as part of the Heavy Vehicle Propulsion System Materials Program, under Contract DE-AC05-00or22725 with UT-Battele, LLC.

REFERENCES

[1] K. Li, and W.T. Liao, "Surface/subsurface damage and the fracture strength of ground ceramics" *Journal of Materials Processing Technology,* 57 [3-4] 207-220 (1996).

[2] G.D Quinn, L.K. Ives and S. Jahanmir, "On the fractographic analysis of machining cracks in ground ceramics: A case study on silicon nitride" *NIST Special Publication 996,* (2003).

[3] L.K. Ives, C.J. Evans, S. Jahnmir, R.S. Polvani, T.J. Strakna and M.L. McGlauflin, "Effect of ductile-regime grinding on the strength of hot-isostatically-pressed silicon nitride" *NIST Special Publication 847,* 341-352 (1993).

[4] T.J. Strakna, S. Jahanmir, R.L. Allor and K.V. Kumar, "Influence of grinding direction on fracture strength of silicon nitride" *American Society of Mechanical Engineers, Applied Mechanics Division, AMD, Machining of Advanced Materials,* 208 53-64 (1995).

[5] J.B. Quinn and G.D. Quinn, "Indentation brittleness of ceramics: a fresh approach" *Journal of Material Sciences,* 32 [16] 4331-4346 (1997).

[6] T.W. Hwang, C.J. Evans and S. Malkin, "Size effect for specific energy in grinding of silicon nitride" *Wear,* 225 862-867 (1999).

[7] T.W. Hwang and S. Malkin, "Grinding mechanisms and energy balance for ceramics" *Journal of Manufacturing Science and Engineering,* 121 623-631 (1999).

[8] W. Zhang and G. Subhash, "An elastic-plastic cracking model for finite element analysis of indentation cracking in brittle materials" *International Jounral of Solids and Structures,* 38 [34-35] 5893-5913 (2001).

[9] G. Subhash, J.E. Loukus and S.M. Pandit, "Application of data dependant systems approach for evaluation of fracture modes during a single-grit scratching" *Mechanics of Materials,* 34 [1] 25-42 (2002).

[10] G. Subhash and R. Bandyo, "A new scratch resistance measure for structural ceramics" *Journal of the American Ceramic Society,* (in press 2005).

[11] S. Tochino and G. Pezzotti, "Micromechanical analysis of silicon nitride: a comparative study by fracture mechanics and Raman microprobe spectroscopy" *Journal of Raman Spectroscopy,* 33 709-714 (2002).

IMPROVING THE DAMAGE-RESISTANCE OF Si_3N_4-BASED BALL AND ROLLER BEARING COMPONENTS AND THEIR NON-DESTRUCTIVE EVALUATION

Rolf Wagner, Gerhard Wötting
H.C. Starck Ceramics GmbH & Co. KG
Lorenz- Hutschenreuther- Str. 81
D-95100 Selb

ABSTRACT

Silicon nitride (Si_3N_4) is the material of choice to combine mechanical and physical properties for use in high performance bearing devices. A critical drawback, however, is its limited impact strength. During manufacturing and precision hard machining impact forces may result in damage in the form of so-called C-cracks. To reduce this effect, an improved new silicon nitride material was developed with increased impact strength. In order to test this property, special equipment was constructed based on a pendulum test. It reveals significant differences between various silicon nitride grades as well as other ceramics. Based on these results, the material was also adapted to produce roller bearing components.

In order to use silicon nitride balls or rollers in safety-relevant applications like aircraft engines a non-destructive evaluation is necessary to assure there are no critical defects. The appropriately required equipment was developed and improved over several years. By now, automatic apparatuses for ultrasonic testing of bearing balls and rollers are available and will be introduced, In particular a new prototype for rapid evaluation of up to 2000 balls per day allows for cost-effective large scale testing.

INTRODUCTION

There is increasing demand for bearings to be operated at higher speed, elevated temperature and higher power density. These applications require improved materials as well as improved system reliability. In recent years, so-called hybrid bearings with steel races and ceramic balls and rollers have performed well and are now used in a number of technical applications. As ceramic bearing components, more or less only silicon nitride has shown the performance necessary. In particular, this statement holds for the hot-isostatically pressed (HIPed) grades.

Only materials and components grades processed this way – without pores or other defects – guarantee maximum security under extreme operating conditions. A typical microstructure of such a silicon nitride is shown in Figure 1. It is preferably a fine-grained material with a narrow grain size distribution and columnar grain morphologies that are surrounded by light glassy phases. The average visible length of the columnar grains is 2-3 μm; its thickness is less than 1 μm.

Although such silicon nitride materials show outstanding properties, there is still a problem

Figure 1: Microstructure of bearing grade silicon nitride

with limited impact strength, which can cause defects during fabrication or when used as high precision bearing components[1,2]. Impact forces may result in damage in form of so-called C-cracks (see Figure 2). These cracks grow under load and can result in flakes and, at the very least, bearing damage.

Due to a favorable combination of properties in comparison with other ceramic materials, dense, pore- and defect-free silicon nitride generally shows high impact strength. Hadfield[2] and Cundill[3], however, demonstrated that C-cracks can be found in commercial hot-pressed silicon nitride. Mechanical contact stresses are causing these defects. A high load between inflexible materials can result in surface cracks, which extend to the subsurface. These defects

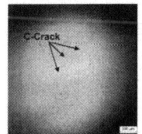

Figure 2: C-crack on a silicon nitride bearing ball

occur when the tensile load of the material is higher than the tensile strength. This value depends on material and microstructure properties like Young and Shear Modulus, fracture toughness, tensile and shear strength and microstructural defects and pores.

EXPERIMENTAL PROCEDURE FOR DETERMINING AND IMPROVING IMPACT STRENGTH

For determining and improving the impact strength, various silicon nitride materials were prepared with a focus on varying Young's modulus, hardness and Poisson ratio. This was achieved by using slightly different processing parameters, which resulted in different sizes and amounts of residual C-particles within the microstructure. From these base materials ball blanks for finished balls with a diameter of 12.7 mm were prepared by cold-isostatic pressing (CIP), consolidated by hot-isostatic pressing (HIP, SN A-C) or by gas-pressure sintering (GPS; 100bar, SN-D) and ground and polished to fulfill requirements for DIN-ISO ball-grade G5. Material SN-E is a commercially available SN ball grade. The properties of the various materials are summarized in Table 1.

	Density	Bending Strenght	K_{Ic}	HV	Young's Modulus	v	C-Particles	C-Content	C-Cracks	
	g/cm³	MPa	MPa m$^{1/2}$	GPa	GPa		[µm]	[%]		%
SN-A	3,238	1050	6,5	15,2	297	0,274	3,600	0,30	2	17
SN-B	3,237	1020	6,6	15,0	294	0,273	4,700	0,90	0	0
SN-C	3,238	900	5,8	14,8	295	0,275	4,200	0,40	5	42
SN-D	3,230	965	6,5	14,8	302	0,274	2,700	0,20	4	33
SN-E	3,164	980	6,0	16,2	310	0,269	0	<0,05	12	100

Table 1: Properties of various silicon nitride compositions

A pendulum-like machine was used to test the impact strength. The length of the pendulum was l=156 mm, using a weight for the hammer of m=360 g. A schematic drawing of the impact testing machine is shown in Figure 3. Two balls with a diameter of 12.7 mm are fixed in a ball take-up of the pendulum and the end trestle. Before using the balls, they were examined for

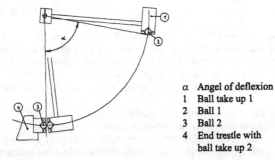

α   Angel of deflexion
1   Ball take up 1
2   Ball 1
3   Ball 2
4   End trestle with
      ball take up 2

Figure 3: Schematic drawing of the impact testing machine

defects using dye penetrant inspection. By changing the angle of deflection, the impact velocity and energy can be varied. The position of the balls was changed by 90° after every impact. In this way, six possible defects could be created on every ball. After this test, the balls were again examined by dye penetrant inspection, and the number of impact cracks was determined.

Using this pendulum test, the impact energy W is determined by the angle of deflection, the impact mass m_a, pendulum length l and acceleration of gravity g according to:

$$W = g(1 - \cos\alpha)m_a l \qquad (1)$$

Mass m_a and pendulum length are combined to a machine constant K for the parameters used

$$K = 0{,}1562 \ kgm \qquad (2)$$

Therefore the critical angle of deflection α_c for the critical impact energy can be calculated based on:

$$\cos\alpha_c = 1 - (32Ea_c^5 / 15R^2 gK) \qquad (3)$$

with E = Young's modulus.

In experimental trials, an angle of 40° was found to be useful to differentiate the impact strength of various ball compositions. All tests were carried out with six impacts on the materials shown in Table 1.

RESULTS AND DISCUSSION

In addition to Table 1, the results of the impact test with the pendulum for the different SN

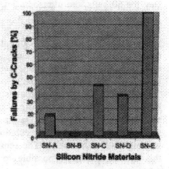

Figure 4: Results of the impact tests

materials are shown in Figure 4. The maximum damage resulting from 6 impacts (6 per ball=2x6=12) was set as 100% in this graph.

The materials clearly show different sensitivities for damage as result of the impacts. Macroscopic defects and microstructural inhomogeneities are, of course, detrimental and have to be avoided. Besides this, a combination of material properties seems to be responsible for an improved impact resistance, comprising relatively low values for the hardness and the elastic constants. All these data should not succeed certain values, which obviously provides some elasticity.

To establish these material properties, the formation of some residual carbon turned out to be successful without negative impact on mechanical properties and performance. In this respect it is important that the size and amount of C-particles are within a relatively small range. This can be achieved by selecting the appropriate processing parameters and does not necessitate to admix C-particles to the starting composition.

ULTRASONIC EVALUATION OF BEARING BALLS

Ultrasonic immersion testing has been evaluated for the bearing ball and has previously been shown to give full coverage of all areas of the ball and in the flaw size range needed[4,5]. This testing method achieved all of the significant challenges of the flaw size and orientation requirements, thus enabling a reliable inspection. Ultrasonic inspection transducers of 25 MHz are used for testing by a pulse echo method. A transducer is adjusted in such a way that Transversal and Rayleigh waves are created. In this way both, surface and subsurface areas can be inspected (Figure 5).

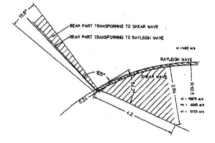

Figure 5: Ultrasonic beam directions for surface and subsurface testing

The ultrasonic inspection can be performed using self-constructed equipment for diameters between 15 and 32 mm. Using a tray for 60 balls, the ultrasonic inspection is fully automatic. A master ball with a defined defect (laser notch 100x25x50 μm^3 length x width x depth) is used for calibration. This master defect results in a clear and high ultrasonic response. Using this signal, it is possible to calculate a possible response of a 50 μm defect. To guarantee the inspection of the whole ball surface, the balls are manipulated six times. As result of these scans, Figure 6 shows the orientations of the Rayleigh wave directions. The sensitivity of this system makes it possible to find C-

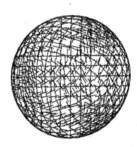

Figure 6: Rayleigh wave directions after six scans

296

Figure 7: New prototype ultrasonic inspection system for rapid
evaluation of up to 2000 balls per day

Cracks (see Figure 1) of > 50 μm. Using the C-scans of every measured direction, it is possible to detect the position of a defect on the surface of a ball. Information about the kind of defect (scratch or C-crack) and their depth, however are not achievable.

Major disadvantage of this ultrasonic testing procedure is the inspection time which requires about eight minutes per ball. For this reason, new equipment was enhanced (Figure 7) employing the principal of the ultrasonic technique of the system described above. This equipment was constructed for testing silicon nitride balls of between 6.35 and 18 mm in diameter. In comparison with the first generation of ultrasonic testing system where the identity of every ball is guaranteed and information can be obtained on the location of the defect, the new system was developed for rapid testing, which means the identity of the balls is lost after inspection. The inspected balls can be selected to a maximum of four classes. These classes are defined individually. Quantitatively, up to 2000 balls (based on 14.28 mm balls) can be evaluated ultrasonically per day. Both systems are offered as services.

ULTRASONIC EVALUATION OF BEARING ROLLERS

Based on the results of the impact test, the experience gained with the material was also used for bearing rollers. Their performance was analyzed by very intensive test rig evaluations[6].

Such finished rollers with dimensions of 8 x 8 mm and 18 x 15 mm were also ultrasonically inspected, using a similar technique as the one used for the balls. To do this, a special machine, which was originally constructed to

Figure 8: Ultrasonic inspection system for rollers

inspect silicon nitride rollers for the space shuttle program, was adapted (Figure 8). To be sure to detect all kinds of defects, the transducer is moved in seven directions. For a reliable inspection, master rollers with laser notches were used (Figure 9). By using these master rollers, the ultrasonic system could be calibrated, and it is possible to determine the size of defects on the rollers.

After the ultrasonic inspection, the rollers without any defect signals were tested in a test rig for aircraft engines showing highest performance without damaging[6].

These results indicate that an ultrasonic inspection technique is indispensable for a reliable inspection of rollers or balls.

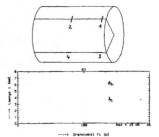

Figure 9: C scan of the master roller (Laser Notches 50 x 25 µm)

CONCLUSION

A method for analysis of the impact strength, based on the work of Cundill[2] was enhanced and used for the work presented here. It was demonstrated that the impact strength of silicon nitride could be improved by the formation of fine dispersed carbon particles in the microstructure due to appropriate selection of processing conditions. These particles obviously generate favorable hardness data and elastic properties, leading way to an improved impact resistance of SN bearing components such as balls and rollers.

While a dye penetrant inspection was used to determine the C-cracks created by the impact test, ultrasonic inspection is indispensable for a reliable analysis confirming the absence of critical surface- and sub-surface defects of the ceramic or metal parts. For this, a special system was developed which runs inspections automatically. For rapid inspection of balls with diameters between 6.35 and 18 mm, a prototype machine was established with a capacity of up to 2000 balls per day.

In conclusion, improved materials as well as ultrasonic inspection systems are available on an industrial scale for bearing components. The new material as well as the introduced full prove inspection are designed to guarantee reliability in all safety-relevant applications such aircraft engines.

REFERENCES

[1]J. Wemhöner, E. Streit, T. Preusser, G. Wötting,R. Westerheide, "Quality safe and reliable production of ceramic rolling elements for roller bearings", Proc. "Materials Week"2000, Munich, published by DGM

[2]Hadfield, "Failure of Silicon Nitride Rolling Elements with Ring Crack Defects", Ceramics International 24, 379-386 (1998).

[3]Cundill, "Impact Resistance of Silicon Nitride Balls", Proc. 6th Int. Symp. on Ceramics Materials and Components for Engines, Arita, Japan, 556-561 (1997).

[4]G. Ojard, P. Komater, W. Mowrer, J. Loftis, K. Clodfelter, K. Woodis, R. Neuschaefer, Procceedings of the 1994 Advanced Earth to Orbit Propulsion Technology Conference, NASA Conference Publication 3282, Vol II., 331-338.

[5]G. Ojard, F. Reed, P. Komater, W. Mowrer, J. Loftis, E. Donahue, D. Ehlert, K. Woodis, B. Neuschaefer, Proceedings of the 1994 JANNAF Conference, Hill Air Force Base, UT, October 24-27 (1994).

[6]G. Wötting, R.Wagner, E. Streit, G. Martin, "Qualitätssichere und zuverlässige Siliciumnitrid-Wälzkörper für hochbelastete Wälzlager", to be published 2005

299

Properties of
Fiber-Reinforced Composites

COMPARISON OF ELEVATED TEMPERATURE TENSILE PROPERTIES AND FATIGUE BEHAVIOR OF TWO VARIANTS OF A WOVEN SiC/SiC COMPOSITE

Sreeramesh Kalluri
Ohio Aerospace Institute
NASA Glenn Research Center
21000 Brookpark Road M/S 49-7
Brook Park, OH 44135

Anthony M. Calomino
NASA Glenn Research Center
21000 Brookpark Road M/S 49-7
Brook Park, OH 44135

David N. Brewer
US Army Research Laboratory
NASA Glenn Research Center
21000 Brookpark Road M/S 49-7
Brook Park, OH 44135

ABSTRACT
        Tensile properties (elastic modulus, proportional limit strength, in-plane tensile strength, and strain at failure) of two variants of a woven SiC/SiC composite, manufactured during two separate time periods (9/99 and 1/01), were determined at 1038 and 1204°C by conducting tensile tests on specimens machined from plates. Continuous cycling fatigue tests (R = 0.05 and 20 cpm) were also conducted at the same two temperatures on specimens from both composites. In this study, average tensile properties, 95% confidence intervals associated with the tensile properties, and geometric mean fatigue lives of both composite materials are compared. The observed similarities and differences in the tensile properties are highlighted and an attempt is made to understand the relationship, if any, between the tensile properties and the fatigue behaviors of the two woven composites.

INTRODUCTION
        Advanced gas turbine engines, which are typically designed to operate at augmented efficiencies, require higher operating temperatures. In order to design for such an environment, materials that can withstand these higher operating temperatures as well as provide sustained performance are required. Ceramic matrix composites (CMCs) are strongly preferred for these tasks because of their high temperature capability and high specific strength. A woven SiC/SiC composite (manufactured by a slurry-cast, melt-infiltration process) has been identified as a potential combustor liner material for the advanced aircraft gas turbine engines[1]. For design purposes, both the reliability and reproducibility of this woven CMC need to be established and suitable tensile property and fatigue life data bases are required.
        The two woven composites evaluated in this study were manufactured by General Electric Power Systems Composites, LLC during September 1999 (9/99) and January 2001 (1/01) with the Chemical Vapor Infiltrated (CVI) SiC/slurry-cast/melt-infiltration process. Tensile properties and fatigue behavior (both without and with hold-time) at 1038 and 1204°C

for the 9/99 composite were previously reported[2]. Tensile properties of both the 9/99 and 1/01 composites at 816 and 1204°C and the variability exhibited by each of those properties were also documented earlier[3]. In this investigation, additional tensile tests at 1038°C and fatigue tests at 1038 and 1204°C were conducted on the 1/01 composite material. The tensile properties and fatigue lives from the 9/99 and 1/01 woven SiC/SiC composites generated at 1038 and 1204°C were compared to identify similarities as well as differences. In particular, the dependency of fatigue lives on the tensile properties of the woven composites was explored.

EXPERIMENTAL DETAILS

Sylramic[TM] fiber in a 5HS weave, 20 EPI configuration with a [0/90]$_{4s}$ lay-up constituted the fiber pre-form for both the 9/99 and 1/01 composites. In the case of the 1/01 composite, before matrix densification, the fiber pre-forms were subjected to an in-situ BN (iBN) heat treatment. Chemical vapor infiltrated, silicon-doped BN interphase and a CVI SiC matrix further subjected to a slurry-cast, melt infiltration process (MI) were utilized for both the composites. Rectangular composite plates (229 mm length, 152 mm width, and 2 mm thickness) were produced by the manufacturing process and test specimens for both tensile and fatigue tests were machined from these plates. The test section within these specimens was 10.2 mm wide and 28 mm in length. Additional details on the test specimen geometry and test system, including test specimen gripping method, test frame alignment procedure, specimen heating and temperature measurement techniques, strain measurement method, and test procedures, were reported previously[2,4,5]. In fatigue tests, triangular waveform with a frequency of 0.33 Hz (20 cpm) and an R-ratio (minimum load to maximum load) of R = 0.05 were used. In all the fatigue tests, failure was defined as separation of the test specimen into two pieces.

TENSILE PROPERTIES

Tensile properties investigated in this study are the following: elastic modulus, E; proportional limit strength (0.005% offset), PLS; in-plane tensile strength, ITS; and strain at failure, SF. Mean values and associated standard deviations of these four tensile properties are shown in Tables I and II, respectively, for the 9/99 and 1/01 composites. A total of 24 tensile tests were conducted at 1204°C for both the 9/99 and 1/01 composites. At 1038°C, due to a limited number of specimens, only six and eight tests were conducted for the 9/99 and 1/01 composites, respectively. Mean ITS and SF values of the 1/01 composite were significantly higher than the corresponding values for the 9/99 composite. No clear cut differences were observed among the mean E and PLS values between the two composites. In general, at both temperatures, standard deviations of the 1/01 material were higher than those exhibited by the 9/99 material with two exceptions (PLS at 1204°C and ITS at 1038°C).

Table I. Means and Standard Deviations of Tensile Properties for 9/99 MI SiC/SiC Composite

Temperature [°C]	Number of Tests [n]	E [GPa]	PLS [MPa]	ITS [MPa]	SF [%]
1038	6	209 {15}*	168 {20}	325 {29}	0.44 {0.06}
1204	24	182 {14}	166 {28}	307 {21}	0.46 {0.07}

* {} Denotes Standard Deviation

Table II. Means and Standard Deviations of Tensile Properties for 1/01 MI SiC/SiC Composite

Temperature [°C]	Number of Tests [n]	E [GPa]	PLS [MPa]	ITS [MPa]	SF [%]
1038	8	185 {27}*	162 {35}	426 {19}	0.60 {0.11}
1204	24	184 {25}	155 {28}	399 {37}	0.57 {0.12}

* {} Denotes Standard Deviation

Average tensile properties and corresponding 95% confidence intervals (error bars) at 1038 and 1204°C for the two composites are compared directly with bar charts in Figs. 1 to 4. The confidence intervals were estimated as twice the standard deviation for each tensile property. Student's t-test with a risk level, $\alpha = 0.02$, and a two-tailed distribution was used to compare mean values of all the tensile properties for both composites. At the investigated temperatures, no statistically significant differences were found to exist between the two composites for the mean values of E and PLS. Differences in the mean ITS and SF values were found to be statistically significant between the two composites. Higher scatter exhibited by the mean tensile properties of the 1/01 composite in comparison to the 9/99 composite and the statistically significant differences observed among the mean ITS and SF values of both composites might be due to the iBN heat treatment given to the fiber pre-forms of the 1/01 composite.

FATIGUE BEHAVIOR

Fatigue tests were conducted with a maximum stress, σ_{max}, of 179 MPa and an R-ratio, R, of 0.05 on specimens from both composites. The selected maximum stress was above the average PLS values of the two composites at both temperatures (Tables I and II). Three tests were conducted at each temperature to obtain a representative fatigue life. The observed geometric mean fatigue lives (equivalent to arithmetic mean fatigue lives in log space) are listed in Table III. For the 9/99 and 1/01 composites, the geometric mean fatigue life decreased as temperature increased from 1038 to 1204°C. At both temperatures, geometric mean fatigue lives of the 1/01 composite were higher than those of the 9/99 composite. Since average PLS values of both composites were very similar, the higher fatigue lives of 1/01 composite could have resulted from the higher average in-plane tensile strengths of this composite.

Table III. Geometric Mean Fatigue Lives of MI SiC/SiC Composite ($\sigma_{max} = 179$ MPa; R = 0.05)

Temperature [°C]	Number of Tests [n]	9/99 [Cycles]	1/01 [Cycles]
1038	3	17 173	32 028
1204	3	4 093	20 405

Specimens fatigued to failure at both 1038 and 1204°C are shown in Figs. 5 and 6. In general, a majority of the specimens failed in the test section with only two specimens failing in the transition region between the gripping area and the test section (Fig. 5). Fatigue life data from these two specimens were included in calculating the geometric mean fatigue life because of the following reasons: 1) failure locations were closer to the test section than the grip section indicating that temperatures at these locations were very close to the test temperatures and 2) the large transition radius used in the specimen design, between the grip and test sections, did not significantly increase the dimensions at the failure locations. Arithmetic mean logarithmic cyclic lives and associated minimum and maximum values are plotted in Fig. 7 for both composites. For the limited number of fatigue tests conducted, fatigue lives of 1/01 composite exhibited much less scatter at both temperatures than those from the 9/99 composite. This observed trend in fatigue lives was opposite to that exhibited by the tensile properties for both the composites. Note that more tensile tests than fatigue tests were conducted on each composite and the scatter in the fatigue lives of the 1/01 composite might eventually increase if more fatigue tests were conducted.

DISCUSSION

The primary difference in processing between the two composites was the iBN heat treatment provided for the fiber pre-forms of the 1/01 composite before matrix densification. Similar treatment was not given to the fiber pre-forms of the 9/99 composite. As evidenced by the Student's t-test results of mean tensile properties of both composites, the iBN treatment seems to improve the mean ITS and SF values while having no significant influence on the mean E and PLS values. Moreover, the 1/01 composite, which received this treatment, exhibited much larger variation in all four tensile properties than the 9/99 composite. The higher geometric mean fatigue lives observed for the 1/01 composite as compared to the corresponding lives of the 9/99 composite correlated well with higher mean ITS values exhibited by the 1/01 composite. The reason for lower scatter observed in the fatigue lives of the 1/01 composite, which is opposite to the trend observed in the tensile properties, is not very clear due to the limited number of fatigue tests conducted in this study. A systematic fatigue investigation that includes at least six to ten specimens per test condition is required to quantify scatter in the fatigue data in a reliable manner.

CONCLUDING REMARKS

Tensile properties and fatigue behavior of two variants of a woven SiC/SiC composite manufactured during two separated time periods (9/99 and 1/01) were investigated at 1038 and 1204°C. Fiber pre-forms of the 1/01 composite were iBN heat-treated prior to matrix densification by slurry-cast, melt-infiltration process. No such treatment was given to the fiber pre-forms of the 9/99 composite. Average tensile properties and associated standard deviations were reported and compared for both composites. Geometric mean fatigue lives of both the composites are presented along with the associated scatter bands. The 1/01 composite exhibited higher average values of in-plane tensile strength and strain at failure in the tensile tests and higher geometric mean fatigue lives compared to the 9/99 composite. The 1/01 composite exhibited more variation in the tensile properties than the 9/99 composite, whereas the opposite was true for the scatter observed in the fatigue lives.

ACKNOWLEDGEMENTS

Financial support for this work was obtained from NASA Glenn Research Center, Brook Park, Ohio under cooperative agreement NCC-3-1041 through the Ultra Efficient Engine Technology Program. The authors are grateful to Mr. John D. Zima and Mr. William L. Brown for conducting the tensile and fatigue tests at NASA Glenn Research Center.

306

REFERENCES

[1]D. Brewer, "HSR/EPM Combustor Materials Development Program," *Materials Science and Engineering*, **A261**, pp. 284-291, 1999.

[2]S. Kalluri, A. M. Calomino, and D. N. Brewer, "High Temperature Tensile Properties and Fatigue Behavior of a Melt-Infiltrated SiC/SiC Composite," *Fatigue 2002*, Proc. of the 8th Intl. Fatigue Congress, Vol. 3/5, A. F. Blom, Ed., Stockholm, Sweden, pp. 1965-1972, 2002.

[3]S. Kalluri, A. M. Calomino, and D. N. Brewer, "An Assessment of Variability in the Average Tensile Properties of a Melt-Infiltrated SiC/SiC Composite," *Ceramic Engineering and Science Proceedings*, Vol. 25, Issue 4, 28th International Conference on Advanced Ceramics and Composites: B, Edgar Lara-Curzio and Michael J. Readey, Eds., pp. 79-86, 2004.

[4]M. J. Verrilli, A. M. Calomino, and D. N. Brewer, "Creep-Rupture Behavior of a Nicalon/SiC Composite," *Thermal and Mechanical Test Methods and Behavior of Continuous-Fiber Ceramic Composites, ASTM STP 1309*, M. G. Jenkins, S. T. Gonczy, E. Lara-Curzio, N. E. Ashbaugh, and L. P. Zawada, Eds., American Society for Testing Materials, pp. 158-175, 1997.

[5]M. J. Verrilli, A. Calomino, and D. J. Thomas, "Stress/Life Behavior of a C/SiC Composite in a Low Partial Pressure of Oxygen Environment, Part I: Static Strength and Stress Rupture Database," 26th Annual Conf. on Composites, Advanced Ceramics, Materials, and Structures: A, *Ceramic Engineering and Science Proceedings*, Vol. 23, Issue 3, H.-T. Lin and M. Singh, Eds., pp. 435-442, 2002.

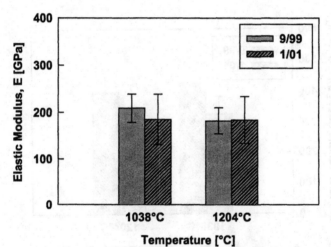

Figure 1: Average Elastic Moduli and 95% Confidence Intervals for Two Variants of a Woven SiC/SiC Composite

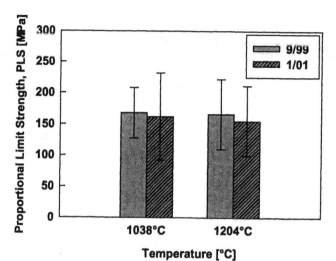

Figure 2: Average Proportional Limit Strengths and 95% Confidence Intervals for Two Variants of a Woven SiC/SiC Composite

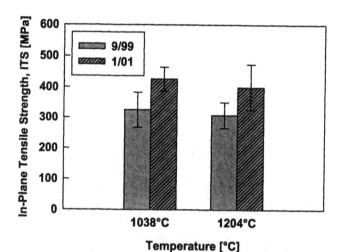

Figure 3: Average In-Plane Tensile Strengths and 95% Confidence Intervals for Two Variants of a Woven SiC/SiC Composite

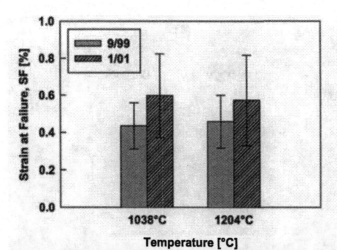

Figure 4: Average Strains at Failure and 95% Confidence Intervals for
Two Variants of a Woven SiC/SiC Composite

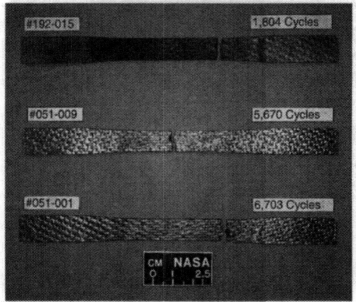

Figure 5: Specimens Tested in Fatigue at 1204°C:
9/99 SiC/SiC Composite

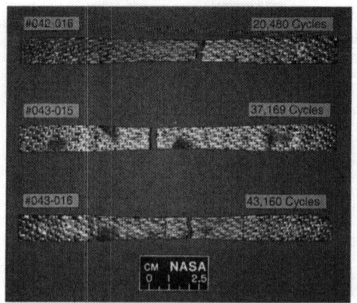

Figure 6: Specimens Tested in Fatigue at 1038°C:
1/01 SiC/SiC Composite

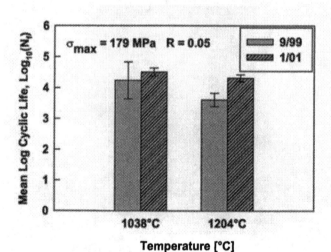

Figure 7: Arithmetic Mean Logarithmic Fatigue Lives and Extreme Values for
Two Variants of a Woven SiC/SiC Composite

TENSILE AND THERMAL PROPERTIES OF CHEMICALLY VAPOR-INFILTRATED SILICON CARBIDE COMPOSITES OF VARIOUS HIGH-MODULUS FIBER REINFORCEMENTS

Takashi Nozawa, Yutai Katoh, Lance L. Snead
Metals and Ceramics Division, Oak Ridge National Laboratory
Oak Ridge, TN 37831-6151

Tatsuya Hinoki, Akira Kohyama
Institute of Advanced Energy, Kyoto University
Gokasho, Uji, Kyoto 611-0011, Japan

ABSTRACT

Chemically vapor-infiltrated (CVI) silicon carbide (SiC) matrix composites are candidate structural materials for proposed nuclear fusion and advanced fission applications due to their high temperature stability under neutron irradiation. To optimize the thermal stress properties for nuclear applications, CVI-SiC matrix composites were produced with three-dimensional (3D) fiber architectures with varied Z-fiber content, using the highly-crystalline and near-stoichiometric SiC fiber Tyranno™-SA. In addition, hybrid SiC/SiC composites incorporating carbon fibers were fabricated to improve thermal conductivity. The purpose of this work is to obtain thermal and mechanical properties data on these developmental composites. Results show that the addition of small amount (>10 %) of Tyranno™-SA fiber remarkably increases the composite thermal conductivity parallel to the fiber longitudinal direction, in particular the through-thickness thermal conductivity in the orthogonal three-dimensional composite system due to the excellent thermal conductivity of Tyranno™-SA fiber itself. On the other hand, tensile properties were significantly dependent on the axial fiber volume fraction; 3D SiC/SiC composites with in-plane fiber content <15 % exhibited lower tensile strength and proportional limit failure stress. Results show that the composites with axial fiber volume >20 % exhibit improved axial strength. The carbon fiber was, in general, beneficial to obtain high thermal conductivity. However matrix cracks induced due to the mismatch of coefficients of thermal expansion (CTE) restricted heat transfer via matrix, limiting the improvement of thermal conductivity and reducing tensile proportional limit stress.

INTRODUCTION

Silicon carbide fiber reinforced silicon carbide (SiC/SiC) composites are candidate materials for nuclear fusion and advanced fission reactors because of elevated-temperature mechanical capability, low induced-radioactivity, and after-heat [1]. The latest composites fabricated from high modulus SiC fibers, i.e., highly-crystalline and near-stoichiometric SiC fibers such as Tyranno™-SA and Hi-Nicalon™ Type-S, and β-SiC matrix provide good geometrical stability and strength retention after neutron irradiation [2-4]. Also, enhanced thermal and thermo-mechanical properties of the highly crystalline SiC composites have the added advantage to providing higher system efficiency. In addition, composites with higher thermal conductivity and strength exhibit much better resistance to thermal shock.

Constituent materials are one important factor to maximize the thermal properties of composites. Among various processing techniques, chemical vapor infiltration is regarded as the technique that produces the highest crystallinity of SiC with inherently high thermal conductivity [5]. Tyranno™-SA and Hi-Nicalon™ Type-S are also beneficial to use due to their crystalline structure. Tyranno™-SA fiber exhibits the thermal conductivity of 65 W/m-K, while 18 W/m-K for Hi-Nicalon™ Type-S. Further thermal conductivity improvements have been proposed whereby hybrid composite concepts using carbon fibers as reinforcements mixed with SiC fibers (SiC-C/SiC composites) are utilized [6]. Specifically, pitch-based carbon fibers possess much higher thermal conductivity (22~1000 W/m-K), as compared to SiC or other graphite fibers. For many nuclear applications, heat transport in the direction orthogonal to the plane of the primary stress is required. For this reason, a Z-stitch of high conductivity graphite fiber into an X-Y weave of SiC fibers has been considered. Matrix densification is also important to keep good heat transfer via matrix. The through-thickness thermal conductivity of ~70 W/m-K, which is higher than that of conventional composites (~20 W/m-K), has been reported for two dimensional (2D) SiC/SiC composite with the density of ~3.1 g/cm^3, fabricated by nano-infiltration transient eutectic phase sintering (NITE) process [7].

The objective in this study is to evaluate the thermal and mechanical properties of various architecture types of SiC/SiC and hybrid SiC-C/SiC composites reinforced by the high-modulus SiC and carbon fibers, designed to provide high thermal conductivity.

Table I. SiC/SiC and hybrid SiC-C/SiC composites under investigation.

Composite ID	Fiber	Architecture	X- (and Y-) fiber volume fraction [%]	Z-fiber volume fraction [%]	Density [Mg/m³]	Porosity [%]
P/W SA	SA[*1]	P/W	20	-	2.51	20
S/W HNLS	HNLS[*2]	5-harness S/W	22	-	2.52	19
3D(1:1:4) SA	SA[*1]	3D(X:Y:Z=1:1:4)	10	40	2.76	10
3D(1:1:1) SA	SA[*1]	3D(X:Y:Z=1:1:1)	15	15	2.78	11
Hybrid 2D	SA[*1]+C[*3]	P/W	SA[*1]:8, C[*3]:18	-	2.25	19
Hybrid 3D	SA[*1], C[*3]	3D(X,Y:SA[*1], Z:C[*3])	SA[*1]:12	C[*3]:21	2.19	25

[*1]SA: Tyranno™-SA Grade 3 fiber, [*2]HNLS: Hi-Nicalon™ Type-S fiber, [*3]C: P120S fiber

EXPERIMENTAL

The materials were CVI-SiC matrix composites with the highly crystalline and near-stoichiometric SiC fibers: Tyranno™-SA Grade-3 and Hi-Nicalon™ Type-S. Also hybrid CVI-SiC matrix composites with Tyranno™-SA Grade-3 and pitch-based carbon (P120S: ~640 W/m-K in longitudinal, ~2.4 W/m-K in radial) fibers were prepared. All materials were produced by isothermal/isobaric CVI process. Two types of 2D SiC/SiC composites with plain-weave (P/W) Tyranno™-SA and 5-harness satin-weave (S/W) Hi-Nicalon™ Type-S, two types of orthogonal 3D SiC/SiC composites with differed through-thickness (Z-direction) fiber content: X: Y: Z = 1: 1: 1 and 1: 1: 4, and two types of hybrid composites with a P/W and an orthogonal 3D architecture composed of Tyranno™-SA and P120S fibers were fabricated, respectively (Table I). It is noted that the Z-direction fibers of both 3D SiC/SiC composites were made into SA grade (Si-Al-C) from AM grade (Si-Al-C-O) Tyranno™ fiber at the temperature of ~2073 K in inert environment after weaving. All composites had 150 nm thick pyrolytic carbon (PyC) as fiber/matrix (F/M) interphase.

Tensile specimens were machined from the composite plates so that the longitudinal direction was parallel to either of X or Y fiber directions. Miniature tensile geometry that had been developed for neutron irradiation studies on ceramic composites [8] was employed. The rectangular geometries (length × width × thickness) of 15.0 mm × 4.0 mm × 2.3 mm for the 2D SiC/SiC and/or the 2D and 3D hybrid SiC-C/SiC composites, and 20.0 mm × 6.0 mm × 2.5 mm for the 3D SiC/SiC composites are used. Tensile tests were conducted following the general guidelines of ASTM standard C1275. The tensile test incorporated several unloading/reloading sequences in order to allow hysteresis analysis. For the testing at room temperature, specimens were clamped by wedge grips with aluminum end tabs on both faces of the gripping sections. The strain was determined by averaging the readings of strain gauges bonded to both faces of the center gauge section. The crosshead displacement rate was 0.5 mm/min for all tests.

Room temperature thermal diffusivity was measured using a xenon flash technique and thermal conductivity was calculated using the measured composite density and specific heat. Specific heat of composites was calculated assuming the rule of mixtures. Microstructures and tensile fracture surfaces to selected samples were examined with an optical microscopy and a field emission scanning electron microscopy (FE-SEM).

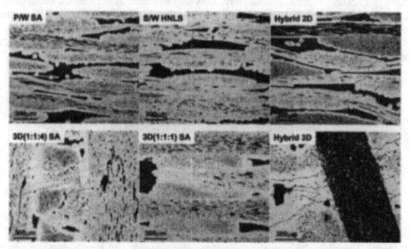

Figure 1. Cross-sectional images of as-received SiC/SiC and hybrid SiC-C/SiC composites. Both hybrid SiC-C/SiC composites underwent matrix cracking due to CTE-mismatch.

RESULTS AND DISCUSSION
Microstructure

Figure 1 shows the typical cross-sectional images of as-received composites. Two dimensional SiC/SiC composites had large pores in most cases in weaving cross-sectional pockets, resulting in a density of ~2.5 g/cm^3. By contrast, both 3D SiC/SiC composites were well-densified even in the pocket regions, yielding a density of ~2.7 g/cm^3. Both 2D and 3D hybrid composites had large pores in pocket regions. Therefore less densification of the matrix yielded a lower density (~2.2 g/cm^3) than that of ideally densified composites (~2.8 g/cm^3). In addition, the hybrid composites contained transverse matrix cracks around every carbon fiber bundle. Specifically major cracks propagated within laminated plies for some 2D hybrid composites. This is attributed to the large CTE mismatch between the SiC matrix and the carbon fiber.

Thermal Conductivity

Table II lists a relationship between the axial fiber content and the thermal conductivity. The 3D SiC/SiC composites with the Tyranno™-SA fiber exhibited the highest improvement of the thermal conductivity in Z-direction. The in-plane thermal conductivity was also high (~50 W/m-K) similar to the through-thickness thermal conductivity due to the presence of continuous X- or Y-direction fibers. In contrast, the SiC/SiC composite of Hi-Nicalon™ Type-S with relatively lower thermal conductivity exhibited less improvement of the in-plane thermal conductivity, though high axial (in-plane) fiber volume fraction. Similarly, the carbon fiber with higher thermal conductivity of 640 W/m-K is, in general, considered effective to provide the high through-thickness thermal conductivity. Less improvement of the thermal conductivity of the 3D hybrid composite was primarily due to large porosity. However, many CTE mismatch induced cracks might prevent thermal diffusion via matrix of carbon fiber containing composites. In addition, inherently lower radial thermal conductivity of the carbon fiber is also disadvantage to obtain the good in-plane thermal conductivity.

Table II. In-plane and through-thickness thermal conductivity.

Composite ID	In-plane		Through-thickness	
	Fiber volume fraction [%]	Thermal conductivity [W/m-K]	Fiber volume fraction [%]	Thermal conductivity [W/m-K]
P/W SA	20	47.8 (1.3)	0	18.9 (0.3)
S/W HNLS	22	34.7 (1.3)	0	18.1 (0.3)
3D(1:1:4) SA	10	46.2 (1.4)	40	58.2 (1.8)
3D(1:1:1) SA	15	47.4 (1.0)	15	46.1 (0.9)
Hybrid 3D	12	32.6 (0.5)	21	53.0 (13.5)

*The numbers in parenthesis indicate standard deviation.

Tensile Properties

Figure 2 exhibits the typical tensile stress-strain curves and tensile data are shown in Table III. Each of the curves for 2D SiC/SiC composites comprises an initial proportional segment that corresponds to the elastic deformation, followed by a non-linear portion due to domination of the matrix cracks and the fiber failures. The highest tensile strength of about 200 MPa for 2D SiC/SiC composites is similar to that obtained from conventional 2D CVI-SiC/SiC composites [5]. The higher tensile Young's modulus of ~270 GPa of the 2D SiC/SiC composites is attributed to the use of highly-crystalline SiC fibers and matrix. The large scatters were due to varied pore content. In contrast, the dense 3D SiC/SiC composites exhibited high Young's modulus, though significantly lower tensile strength than the 2D CVI SiC/SiC composites. They are roughly distinguished into two types of fracture patterns: brittle and quasi-ductile with non-linear portion, although both fracture behaviors exhibited limited short fiber pullout (Fig. 3). Most of the brittle specimens failed just after the proportional limit stress (PLS). The brittle specimens of 3D(1:1:1) therefore exhibited very lower tensile strength (~65 MPa). However, there is no significant difference between the brittle and the non-brittle composites for 3D(1:1:4). Young's moduli for both the brittle and the non-brittle composites were still in same range.

A feature of the hybrid composites is significantly lower Young's moduli, ~140 GPa for 2D and ~100 GPa for 3D, respectively. In contrast, the estimated Young's moduli by parallel-serial approach based on the rule of mixtures [9] yielded ~270 GPa for 2D and ~100 GPa for 3D. It is noted that the Young's modulus of the P/W hybrid composites was roughly estimated by

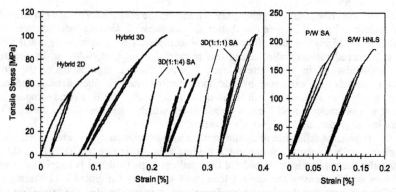

Figure 2. Typical tensile stress-strain curves of SiC/SiC and hybrid SiC-C/SiC composites.

Table III. Tensile data of various SiC/SiC and hybrid SiC-C/SiC composites.

Composite ID	Young's modulus [GPa]	Proportional limit stress [MPa]	Tensile strength [MPa]	Misfit stress [MPa]
P/W SA	273(23)	103(21)	199(2)	11(11)
S/W HNLS	244(6)	101(9)	226(50)	9(7)
3D(1:1:4) SA - brittle	213(26)	64	64	-
3D(1:1:4) SA	224(5)	35(2)	59(13)	7(2)
3D(1:1:1) SA - brittle	264(18)	60(4)	65(4)	-
3D(1:1:1) SA	277	53	102	-7(8)
Hybrid 2D	138(28)	8(1)	94(28)	32
Hybrid 3D	95(10)	47(1)	98(3)	18(4)

*The numbers in parenthesis indicate standard deviation.

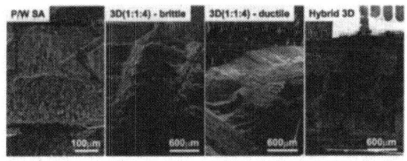

Figure 3. Typical tensile fracture surfaces of SiC/SiC and hybrid SiC-C/SiC composites.

replacing with the orthogonal 2D architecture. Therefore, the estimated Young's modulus should be significantly influenced by pore distribution and waviness. Processing-induced matrix cracks perpendicular to the loading axis of the 2D hybrid composites could reduce the initial Young's modulus. Damaged matrices also caused less proportional segment, followed by progressive damage accumulations in lower stress level. However, severe degradation of the Young's modulus might also be explained by additional factors such as fiber damage induced by preparing a hybrid weaves, although it is still unclear. Contrarily, the 3D hybrid composites with transverse matrix cracks around Z-stitch carbon fibers, which existed parallel to the axial fibers, exhibited less degradation of the Young's modulus.

The significant difference of tensile strength in the composite types is attributed primarily to the effective fiber volume content in the loading axis. According to Curtin [10], the intact fibers transfer the applied load beyond the fiber failure. In particular, it is simply assumed that intact fibers equally convey the applied load, well known as the global load sharing (GLS) theory. This indicates that the ultimate tensile strength is proportional to the axial fiber volume fraction. Composites tested in this study were designed with varied fiber volume content. As plotted in Fig. 4(a), the large differences in axial fiber content provided the differed tensile strength in each composite. Both 3D SiC/SiC composites with large structural unit, ~3.0 mm, are subjected to change in fiber volume fraction in cutting position. The bare fibers located at the edges of the specimen are mostly damaged, i.e. discontinuous along the gauge length, and those are ineffective in load transfer. In this case, less ability to transfer the applied load resulted in the quasi-brittle failure at the PLS. Even in the non-brittle case, very low axial fiber content leaded to low composite tensile strength. Additionally, the possible slight processing damage on reinforcing fibers of 3D SiC/SiC composites during heating up to 2073 K may have degraded the composite tensile strength.

As with the tensile strength, the difference of the PLS in composite types may be attributed primarily to the axial fiber volume content (Figs. 4(b)). In particular, the 2D hybrid composites with severe processing-induced matrix damages had much lower PLS.

The residual thermal stress is inevitable when the coefficients of thermal expansion are different between the fibers and the matrix. In the Tyranno™-SA or Hi-Nicalon™ Type-S/PyC/CVI-SiC system, the residual stress should be minimal, because the fibers consist primarily of cubic (beta-phase) SiC, i.e. equivalent to the matrix constituent. The thermal residual stress estimated by the method proposed by Vagaggini, *et al.* [11] exhibits nearly zero for 3D SiC/SiC composites and residual tension of ~10 MPa for 2D SiC/SiC composites (Table III), while ~100 MPa in a ceramic grade Nicalon™/PyC/CVI-SiC composite [12]. The slightly larger residual stress of 2D SiC/SiC composites might be due to restriction by weaving structure. The hybrid SiC-C/SiC composites exhibited relatively higher tensile residual stress due to large CTE mismatch of the SiC and the carbon.

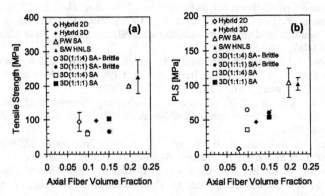

Figure 4. Relationships among tensile properties and the axial fiber volume fraction.
(a) Tensile strength and (b) proportional limit stress

CONCLUSIONS

This work aims to obtain thermal and mechanical property data for advanced CVI silicon carbide matrix composites using high-modulus SiC and carbon fibers. The major conclusions drawn from this study are summarized as follows.

1. The 3D SiC/SiC composites with Tyranno™-SA fibers exhibited the highest thermal conductivity, due to excellent thermal conductivity of Tyranno™-SA fiber itself

2. Tensile properties of SiC/SiC composites were significantly dependent on the axial fiber volume fraction; 3D SiC/SiC composites with less in-plane fiber content exhibited lower tensile strength and proportional limit stress. The high volume fraction of the axial fibers, >20%, is required to obtain good tensile properties.

3. The carbon fiber was, in general, beneficial to obtain high thermal conductivity. However matrix cracks due to the mismatch of coefficients of thermal expansion restricted heat transport via matrix, limiting the improvement of thermal conductivity, and significantly reducing tensile properties.

ACKNOWLEDGEMENTS

The authors would like to thank Dr. Edgar Lara-Curzio and Dr. Hsin Wang at Oak Ridge National Laboratory for mechanical testing and thermal conductivity measurement, respectively.

This research was sponsored by the Office of Fusion Energy Sciences, US Department of Energy under contract DE-AC05-00OR22725 with UT-Battelle, LLC and 'JUPITER-II' US-Department of Energy/Japanese Ministry of Education, Culture, Sports, Science and Technology (MEXT) collaboration for fusion material system research.

REFERENCES

[1] A.R. Raffray, R. Jones, G. Aiello, M. Billone, L. Giancarli, H. Golfier, A. Hasegawa, Y. Katoh, A. Kohyama, S. Nishio, B. Riccardi and M.S. Tillack, "Design and Material Issues for High Performance SiC$_f$/SiC-Based Fusion Power Cores," *Fus. Eng. Des.*, **55**, 55-95 (2001).

[2] L.L. Snead, Y. Katoh, A. Kohyama, J.L. Bailey, N.L. Vaughn and R.A. Lowden, "Evaluation of Neutron Irradiated Near-Stoichiometric Silicon Carbide Fiber Composites," *J. Nucl. Mater.*, **283-287**, 551-555 (2000).

[3] T. Hinoki, L.L. Snead, Y. Katoh, A. Hasegawa, T. Nozawa and A. Kohyama, "The Effect of High Dose/High Temperature Irradiation on High Purity Fibers and Their Silicon Carbide Composites," *J. Nucl. Mater.*, **307-311**, 1157-62 (2002).

[4] T. Nozawa, L.L. Snead, Y. Katoh and A. Kohyama, "Neutron Irradiation Effects on High-Crystallinity and Near-Stoichiometry SiC Fibers and Their Composites," *J. Nucl. Mater.*, **329-333**, 544-548 (2004).

[5] Y. Katoh, A. Kohyama, T. Hinoki and L.L. Snead, "Progress in SiC-Based Ceramic Composites for Fusion Applications," *Fus. Sci. Technol.*, **44**, 155-62 (2003).

[6] L.L. Snead, M. Balden, R.A. Causey and H. Atsumi, "High Thermal Conductivity of Graphite Fiber Silicon Carbide Composites for Fusion Reactor Application," *J. Nucl. Mater.*, **307-311**, 1200-04 (2002).

[7] Y. Lee, T. Nozawa, T. Hinoki and A. Kohyama, "Thermal and Thermo-Mechanical Properties of SiC/SiC," submitted to *Proceedings of the Sixth IEA Workshop on SiC/SiC Composites for Fusion Energy Application, International Energy Agency*.

[8] T. Nozawa, Y. Katoh, A. Kohyama and E. Lara-Curzio, "Specimen Size Effects in Tensile Properties of SiC/SiC and Recommendation for Irradiation Studies," *Proceedings of the Fifth IEA Workshop on SiC/SiC Composites for Fusion Energy Application, International Energy Agency*, 74-86 (2002).

[9] T. Ishikawa, K. Bansaku, N. Watanabe, Y. Nomura, M. Shibuya and T. Hirokawa, "Experimental Stress/Strain Behavior of SiC-Matrix Composites Reinforced with Si-Ti-C-O Fibers and Estimation of Matrix Elastic Modulus," *Compos. Sci. Technol.*, **58**, 51-63 (1998).

[10] W.A. Curtin, "Theory of Mechanical Properties of Ceramic-Matrix Composites," *J. Am. Ceram. Soc.*, **74**, 2837-45 (1991).

[11] E. Vagaggini, J-M. Domergue and A.G. Evans, "Relationship between Hysteresis Measurements and the Constituent Properties of Ceramic Matrix Composites: I, Theory," *J. Am. Ceram. Soc.*, **78**, 2709-20 (1995).

[12] J-M. Domergue, E. Vagaggini and A.G. Evans, "Relationship between Hysteresis Measurements and the Constituent Properties of Ceramic Matrix Composites: II, Experimental Studies on Unidirectional Materials," *J. Am. Ceram. Soc.*, **78**, 2721-31 (1995).

PREPARATION AND PROPERTIES OF DENSE SiC/SiC COMPOSITES

Michiyuki Suzuki , Mitsuhiko Sato and Norifumi Miyamoto
Ube Industries, Ltd.
1978-5 Kogushi
Ube, Yamaguchi, 755-8633, Japan

Akira Kohyama
Institute of Advanced Energy, Kyoto University
Gokasho
Uji, Kyoto 611-0011, Japan

ABSTRACT

SiC/SiC composites fabricated by a new process, named Nano-powder Infiltration and Transient Eutectoid (NITE) process has been developed and exhibited attractive thermo-mechanical properties and gas tightness. For the NITE-SiC/SiC composites, Tyranno Fiber SA and nano-sized SiC powder with certain amounts of sintering aids were employed as the reinforcement and the starting material of matrix, respectively. The NITE-SiC/SiC composites were densified by hot pressing with 20 MPa at 1800°C. In this study, effects of variation of the nano-sized SiC powders on mechanical properties of the NITE-SiC/SiC composites were evaluated. It was found that high specific surface area and carbon rich atomic ratio of the nano-sized SiC powder were necessary for the NITE-SiC/SiC composites with high strength and non-catastorophic fracture behavior. Another objective of the study was to develop near-net shape process and large-scale production process called pseudo-HIP process, which was a new type HIP using a carbon powder as the pressure transmitter and carbon mold with near-net shape cavity. Small and large diameter cylindrical shape ($\phi12/\phi10\times50$ mm, $\phi200/\phi194\times80$ mm) NITE-SiC/SiC composites can be fabricated by the pseudo-HIP process. Hence, the Pseudo-HIP process was found to be a useful process for near-net shape and large-scale NITE-SiC/SiC products.

INTODUCTION

Ceramic matrix composites (CMCs) have been investigated over the last two decades because of their attractive properties: high specific strength, stability at high temperatures and increased damage tolerance. In particular, SiC-fiber-reinforced SiC matrix composites (SiC/SiC composites) have been the most widely researched [1,2]. SiC/SiC composites are considered to be leading candidates for high-temperature, light-weight, structural applications to advanced combustion, nuclear energy and propulsion systems. For fabrication process for the SiC/SiC composites, chemical vapor infiltration (CVI), polymer impregnation and pyrolysis (PIP), melt infiltration (MI) and reaction sintering (RS) have been developed and researched. Recently the liquid phase sintering method densified by hot pressing at elevated temperatures in an inert gas, using nano-sized SiC powder with certain amounts of sintering aids as the starting material of matrix has been reported as a new type fabrication process of SiC/SiC composites using Tyranno SA fiber as the reinforcement. This process was called NITE (Nano-powder Infiltration and Transient Eutectoid) process originally developed by Kohyama et al [3-5]. The NITE-SiC/SiC composites were dense with excellent hermeticity and possessed relatively high thermal conductivity and excellent thermal stress figure of merit compared with other SiC/SiC composites fabricated by the processes mentioned above [5,6].

For the NITE process, heat resistance of the fiber is essentially important because the fiber must be stable during hot pressing under high-pressure at high temperatures. Tyranno SA fiber has been selected as the reinforcement because Tyranno SA fiber has high strength and modulus and shows no degradation at 1900°C in an inert gas [7]. Recently, present authors have found that the received nano-sized SiC powders have variation of properties such as specific surface area, which indicates particle size, C/Si atomic ratio, and those variations of the nano-sized SiC powders have an influence on microstructures and mechanical properties of the NITE-SiC/SiC composites [8]. The influence must be evaluated to stabilize the mechanical properties of the NITE-SiC/SiC composites for application. Another aspect of application of the NITE-SiC/SiC composites, development of near-net shape process and large-scale production process are inevitably necessary because of limitation of shape flexibility for products by hot pressing.

In this study, first objective is to evaluate the effects of variation of nano-sized SiC powders on mechanical properties of the NITE-SiC/SiC composites by fabricating the NITE-SiC/SiC composites using three kinds of nano-sized SiC powders. Second objective is to develop the near-net shape process and large-scale production process by called pseudo-HIP process, which was a new type HIP using a carbon powder as the pressure transmitter and carbon mold with near-net shape cavity.

EXPERIMENTAL PROCEDURE
Effects of variation of nano-SiC powders

Tyranno SA (Ube Industries, Ltd.; diameter 7.5 μm 1600 filaments per tow) fiber was used as a reinforcement. Three kinds of nano-sized SiC powders were evaluated as the matrix raw material, and certain amount of Al_2O_3, Y_2O_3 and SiO_2 mixed oxides were used as sintering additives. Table 1 shows the properties of three kinds of nano-sized SiC powders. All nano-sized SiC powders were β-phase, but specific surface area which indicated the particle size of SiC powders, C/Si atomic ratio were different among the nano-sized SiC powders.

Hereinafter, the three kinds of nano-sized SiC powders were donated as No.1 nano-sized SiC powder, No.2 nano-sized SiC powder and No.3 nano-sized SiC powder, respectively.

Table 1 Properties of evaluated nano-sized SiC powders

Lot No.	Specific surface area (m^2/g)	C (wt%)	O (wt%)	C/Si (at.)	Color
No.1	114	34	2.2	1.25	Black
No.2	93	31	2.7	1.09	Dark gray
No.3	37	28	2.6	0.96	Light gray

Fabrication process of the NITE-SiC/SiC composites is schematically shown in Fig. 1. First, carbon coating (0.5 μm) by CVD process on SA fiber was carried out. The coated SA fiber was continuously dipped in the matrix slurry, consisting nano-sized SiC powder and certain amount of Al_2O_3, Y_2O_3 and SiO_2 mixed oxides, and wound on the drum to produce uni-directional (UD) prepreg sheets. The UD prepreg sheets were cut and stacked so that the fibers were aligned uni-directionally, and then hot pressed with 20 MPa at 1800°C. The fiber volume fraction of the composite was 45 vol%. The sample was cut and ground into 4×3×36 mm for three-point flexural test with a support span of 30 mm. Fracture surfaces after the three-point flexural test were observed by scanning electron microscopy (SEM).

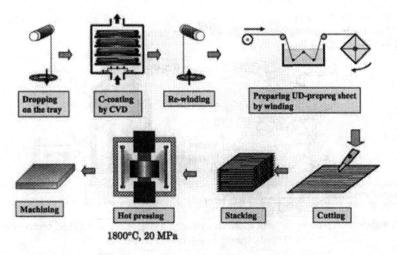

Fig. 1 Fabrication process of the NITE-SiC/SiC composites.

Develop the near-net shape process and large-scale production process

Fabrication process of small diameter cylindrical shaped NITE-SiC/SiC composites is schematically shown in Fig. 2. The cut UD prepreg sheets mentioned above (No.1 nano-sized SiC powder in Table 1 was used.) were stacked on the graphite die with fiber angles to the axial direction of ±15° and 30° and the preform was obtained. Then the preform was densified by pseudo-HIP process, which was a new type HIP using a carbon powder as the pressure transmitter and carbon mold with near-net shape cavity. In this process, pressure was applied to the upper graphite die same as hot pressing, however, the pressure was transmitted through the carbon powder filled in the mold. For the small diameter cylindrical shape, pressure was designed to be transmitted from outside to inside as shown in Fig. 2. The pseudo-HIP was conducted at 1800°C with pressures of 30 and 50 MPa. The fiber volume fraction of the composites was 45 vol%. The pseudo-HIPed sample was machined to $\phi12/\phi10\times50$ mm tube shape. Density and diametral ring compression strength of the $\phi12/\phi10\times50$ mm tube were measured. Specimen size for the diametral ring compression test was $\phi12/\phi10\times10$mm. To evaluate the diametral ring compression strength, plate shape (50×50×5 mm) composite with ±30° fiber arrangement was also fabricated by hot pressing with 20 MPa at 1800°C, and three-point flexural strength was measured by the same procedure described in previous section. The diametral ring compression strength was the strength against tensile stress generated at the top and bottom insides of the ring during the test, hence, the diametral ring compression strength and the three-point flexural strength could be compared.

Fabrication process of large diameter cylindrical shaped NITE-SiC/SiC composites is schematically shown in Fig. 3. The cut UD prepreg sheets mentioned above (No.1 nano-sized SiC powder in Table 1 was used.) were stacked inside of the graphite die with fiber angle to the axial direction of ±45°. Then the preform was densified by the pseudo-HIP process. For the large diameter cylindrical shape, pressure was designed to be transmitted from inside to outside as shown in Fig. 3. The pseudo-HIP was conducted at 1800°C with a pressure of 20 MPa. The fiber volume fraction of the composite was 35 vol%. The pseudo-HIPed sample was machined to φ200/φ194×80 mm shape.

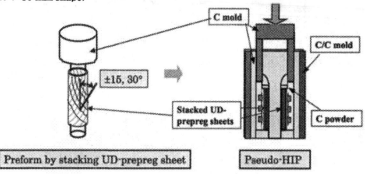

Fig. 2 Fabrication process of small diameter cylindrical shaped NITE-SiC/SiC composites.

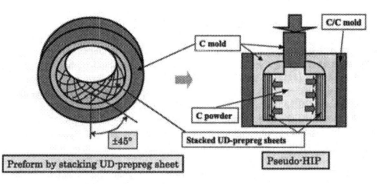

Fig. 3 Fabrication process of large diameter cylindrical shaped NITE-SiC/SiC composites.

RESULTS AND DISCUSSION

Effects of variation of nanao-sized SiC powders

Figure 4 shows the flexural strengths of the NITE-SiC/SiC composites using three kinds of nano-sized SiC powders. The flexural strengths of the NITE-SiC/SiC composites depended strongly on the type of nano-sized SiC powders. The highest strength was obtained by the NITE-SiC/SiC composite using No.1 nano-sized SiC powder, and the intermediate strength was obtained by the composite using No.2 nano-sized SiC powder. The lowest strength was obtained by the composite using No.3 nano-sized SiC powder, which was only 43 % of the strength of the composite using No.1 nano-sized SiC powder. Figure 5 shows the load-displacement curves from the three-point flexural test. From Fig. 5, the NITE-SiC/SiC composite using No.1 nano-sized SiC powder, which showed the highest strength, exhibited non-catastorophic fracture behavior, on the other hand, the composite using No.3 nano-sized SiC powder, which showed the lowest strength, exhibited brittle fracture behavior. Fracture surfaces of the NITE-SiC/SiC composites after the flexural test were shown in Fig. 6. As for No.1 nano-sized SiC powder, slight fiber deformation was detected which was believed to be occurred during the hot pressing, and some fiber pull-out observed on the fracture surface, which was considered to correspond with the highest strength and non-catastorophic fracture behavior of the composite. In case of No.2 and No.3 nano-sized SiC powders, flat fracture surface, fiber deformation and thinning and disappearing of carbon interphase were observed, in particular for No.3. Those features must be correlated with lower strength and brittle fracture of the composites.

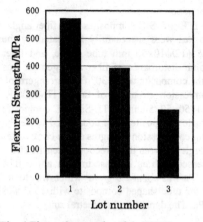

Fig. 4 Flexural strengths of the NITE
- SiC/SiC composites.

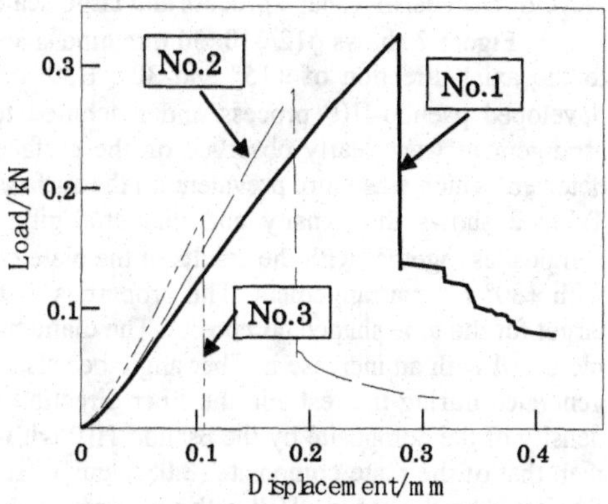

Fig. 5 Load-displacement curves from the
three-point flexural test.

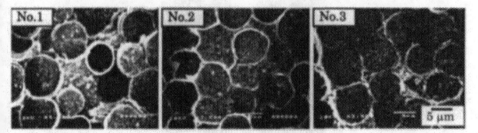

Fig. 6 Fracture surfaces of the NITE-SiC/SiC composites after the flexural test.

From Fig. 6, thinning and disappearing of carbon interphase are observed on the fracture surfaces of the composites using lower C/Si atomic ratio nano-sized SiC powders (No2. and No.3). Those phenomena indicated that certain amount of carbon was consumed during sintering process by the hot pressing, thus, the excess carbon within the raw matrix materials to avoid the consumption of carbon interphase was considered to be needed. Detailed mechanism of the carbon consumption during the sintering process is now being investigated. As for the deformation of fiber which was prevalent in the composites using No.2 and No.3 nano-sized SiC powders, smaller specific surface area of the nano-sized SiC powders, that is, larger particle size (The particle sizes of the SiC powders observed by SEM were 10-20 nm for No.1, 20-30 nm for

No.2 and 40-80 nm for No.3, respectively.), might be responsible because nano-sized SiC powder with small specific surface area could not be well infiltrated into fiber bundles during the sintering process so that the fiber might deform to fill in the space between the fibers. From the discussion above, high specific surface area, that is, small particle size, and carbon rich atomic ratio of nano-sized SiC powder are necessary for the NITE-SiC/SiC composites with high strength and non-catastorophic fracture behavior.

Develop the near-net shape process and large-scale production process

　　　Figure 7 shows $\phi12/\phi10\times50$ mm tube shaped NITE-SiC/SiC composites with fiber angles to the axial direction of ±15° and 30°. Both composites were successfully fabricated by the developed pseudo-HIP process and machined to the $\phi12/\phi10\times50$ mm tube shape, and fiber arrangement was clearly observed on the surface. Some disturbance of fiber arrangement was detected, which was more prevalent on the surface of the composite with ±30° fiber arrangement. Table 2 shows the density and diametral ring compression strengths of the NITE-SiC/SiC composites together with the results of the plate shaped ($50\times50\times5$ mm) NITE-SiC/SiC composite with ±30° fiber arrangement. The properties of the plate composite were considered to be the target for the tube shaped composite. The diametral ring compression strengths of the composites increased with an increase in fiber angle because the angle between the direction of tensile stress generated during the test and the fiber direction decreased with an increase in fiber angle. The density of the composite by the pseudo-HIP with 30 MPa was about 2.9 g/cm^3 which was lower than that of the plate composite (3.08 g/cm^3). Hence, the tube shaped composite with ±30° was fabricated by the pseudo-HIP with a pressure of 50 MPa. The density and diametral ring

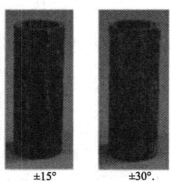

±15°　　　　　±30°.

Fig. 7 $\phi12/\phi10\times50$ mm tube shaped NITE-SiC/SiC composites.

Table 2 Density and diametral ring compression strengths of the NITE-SiC/SiC composites

Type of the composite	Density (g/cm^3)	Diametral ring compression strength (MPa)
Fiber angle ±15°, HIP:30 MPa	2.92	46.5
Fiber angle ±30°, HIP:30 MPa	2.86	83.5
Fiber angle ±30°, HIP:50 MPa	3.02	108
Plate : Fiber angle ±30°, Hot press:20 MPa	3.08	135 (3-point bending)

compression strength of the composite were improved by an increase in pressure of the pseudo-HIP. The density was very close to that of the plate composite and the diametral ring compression strength was about 80 % of the three-point flexural strength of the plate composite. From the results, higher pressure is needed for the pseudo-HIP than that of the hot pressing for fabricating the composites with same level of density and mechanical properties. It was considered that pressure loss occurred within the carbon powder as the pressure transmitter and/or interface between the mold and carbon powder or the preform and carbon powder. The carbon powder was found to become rigid and loose the fluidity needed as the pressure transmitter during the pseudo-HIP so that the carbon powder could not well follow the shrinkage of the preform during densification, which could cause the pressure loss. This issue remains as a part of the future work.

Figure 8 shows $\phi200/\phi196\times80$ mm cylindrical shaped NITE-SiC/SiC composite with fiber angle to the axial direction of $\pm45°$. The large composite was successfully fabricated by the developed pseudo-HIP process and machined to the $\phi200/\phi196\times80$ mm cylindrical shape. The disturbance of fiber arrangement was less compared with the small tube shape shown in Fig. 7 in spite of higher fiber angle, indicating the design for pressure transmitted direction was an important factor to avoid the disturbance of fiber arrangement. Because of the first trial for large size composite, the pressure of the pseudo-HIP was 20 MPa for safe fabrication, which was not enough to fabricate dense cylindrical shaped NITE-SiC/SiC composites from the previous results described above. The density of the composite was 2.7 g/cm^3, hence the fabrication with higher pressure accompanied with modified mold design for increased pressure is the next work.

Fig. 8 $\phi200/\phi196\times80$ mm cylindrical shaped NITE-SiC/SiC composite.

CONCLISION

Dense SiC/SiC composites were fabricated by a new process, named Nano-powder Infiltration and Transient Eutectoid (NITE) process. For application of the NITE-SiC/SiC composites, evaluation of nano-sized SiC powders was conducted to stabilize the microstructures and mechanical properties of the composites. Properties of the nano-sized SiC powders have large influence on mechanical properties of the NITE-SiC/SiC composites. For the NITE-SiC/SiC composites, fine grain size (10-20 nm) and about 25 % carbon rich atomic ratio were found to be required properties of nano-sized SiC powder for high performance of the NITE-SiC/SiC composites.

For another aspect of application of the NITE-SiC/SiC composites, near-net shape process and large-scale production process was developed, which was called pseudo-HIP process. Small and large diameter cylindrical shaped NITE-SiC/SiC composites were fairly well fabricated by the pseudo-HIP process. Although some issues remains as future work, the pseudo-HIP process can be a useful process for near-net shape and large-scale parts production for the NITE-SiC/SiC composites.

ACKNOWLEDGENTS

This work was supported by Advanced Ceramics Composite Integration Research Activity for Gas-Cooled Fast Reactor (ACCIRA-GFR), Ministry of Education, Culture, Sports, Science and Technology.

REFERENCES

[1] J. J. Brennan, "Interfacial Characterization of a Slurry-Cast Melt-Infiltrated SiC/SiC Ceramic-Matrix Composite," *Acta mater.*, **48**, 4619-28 (2000).

[2] G. N. Morscher, "Tensile Stress Rupture of SiC_f/SiC_m Minicomposites with Carbon and Boron Nitride Interphases at Elevated Temperatures in Air," *J. Am. Ceram. Soc.*, **80**, 2029-42 (1997).

[3] S. Dong, Y. Katoh, and A. Kohyama, "Preparation of SiC/SiC Composites by Hot Pressing, Using Tyranno-SA Fiber as Reinforcement," *J. Am. Ceram. Soc.*, **86**, 26-32 (2003).

[4] Y. Katoh, A. Kohyama, S. Dong, T. Hinoki, and J-J. Kai, "Microstucture and Properties of Liquid Phase Sintered SiC/SiC Composites," *Ceram. Eng. Sci. Proc.*, **23[3]**, 363-370 (2002).

[5] A. Kohyama, S. Dong, and Y. Katoh, "Development of SiC/SiC Composites by Nano-Infiltration and Transient Eutectoid (NITE) Process," *Ceram. Eng. Sci. Proc.*, **23[3]**, 311-318 (2002).

[6] Y. Hirohata, T. Jinushi, Y. Yamaguchi, M. Hashida, T. Hino, Y. Katoh, and A. Kohyama, "Gas Permeability of SiC/SiC Composite as Fusion Reactor Materials," *submitted to Fusion Eng. & Design*.

[7] T. Ishikawa, Y. Kohtoku, K. Kumagawa, T. Yamamura, and T. Nagasawa, "High-Strength Alkali-Resistant Sintered SiC Fibre Stable to 2200°C," *Nature*, **391**, 773-775 (1998).

[8] M. Sato, M. Suzuki, and N. Miyamato, *unpublished data*.

FABRICATION OF SIC/SIC WITH DISPERSED CARBON NANO-FIBERS COMPOSITE FOR EXCELLENT THERMAL PROPERTIES

Tomitsugu Taguchi, Naoki Igawa, Shiro Jitsukawa, Shin-ichi Shamoto, Yoshinobu Ishii
Japan Atomic Energy Research Institute
2-4, shirakata-shirane
Tokai, Ibaraki, 319-1195, Japan

ABSTRACT

The SiC/SiC with dispersed carbon nano-fibers (CNFs) composites were fabricated in order to improve the thermal conductivity. The SiC/SiC with and without dispersed CNFs composites were fabricated by reaction bonding (RB) process. The effect of dispersed CNFs on the thermal properties was investigated.

The volume fraction of CNFs was approximately 4 %. The CNFs were successfully dispersed in the SiC/SiC composites without reducing the volume fraction of SiC fibers. It is, therefore, considered that the strength of SiC/SiC with dispersed CNFs composites is not reduced. The thermal conductivities of the composites were measured in the temperature range from room temperature to 1000 °C. The thermal conductivities of SiC/SiC with dispersed CNFs composites were twice as high as those without CNFs; the value was approximately 90 W/mK at room temperature. The thermal conductivity of SiC/SiC with dispersed CNFs composites at 800 °C, which is the operation temperature range of fusion reactors, was more than 40 W/mK. The thermal conductivities of SiC/SiC with dispersed CNFs composites in this study satisfy completely the assumed design criterion for fusion reactors, which is 15-20 W/mK at the operation temperature of fusion reactors.

INTRODUCTION

Ceramic matrix composites show excellent mechanical properties at high temperature and non-catastrophic failure behavior. These materials, therefore, are expected to be used for structural applications at high temperature. In particular, the continuous silicon carbide fiber reinforced silicon carbide matrix (SiC/SiC) composites have been developed as structural materials for fusion power reactors in part due to their high temperature strength and low residual radioactivity to neutron irradiation[1-3]. Furthermore, it is reported that the SiC/SiC composites using highly crystalline and near-stoichiometric SiC fiber fabrics (Hi-Nicalon Type S[4] or Tyranno SA[5]) as reinforcement have excellent stability about microstructure and mechanical properties against neutron irradiation up to 10 dpa[6,7]. High thermal conductivities in SiC/SiC composites are also required for high efficiency of heat exchange and reducing thermal stress as structural materials for fusion power reactors[1-3]. However, the thermal conductivities of SiC/SiC composites fabricated by a conventional process such as chemical vapor infiltration (CVI) or polymer impregnation and pyrolysis (PIP) process are lower than those of the assumed design criteria for fusion reactors (15-20 W/mK at 1000°C[2,3]). We reported that the thermal conductivities of SiC/SiC composites fabricated by reaction bonding (RB) process were approximately twice as high as those by CVI and PIP processes[8]. Yamada et al. reported that the thermal conductivities of

SiC/SiC composites were increased by using three dimensional (3D) SiC fiber fabrics[9]. Unfortunately, the volume fraction of fibers parallel to external stress (x and y directions) decreased because of weaving fiber bundles of z direction. Therefore, the strength of SiC/SiC composites with 3D SiC fiber fabrics might be decreased.

Recently, carbon nano-fibers (CNFs), which have excellent thermal and mechanical properties, have been developed. The average diameter of CNFs is very small compared to that of SiC fiber. It is, therefore, possible to disperse the CNFs in the matrix of SiC/SiC composites without decreasing the volume fraction of fiber parallel to external stress.

In this study, the SiC/SiC with or without dispersed CNFs composites were fabricated by RB process and the effect of dispersed CNFs on the thermal conductivities of SiC/SiC composites was investigated.

EXPERIMENTAL PROCEDURE
Materials and fabrication process

The CNFs were obtained from Showa Denko K. K. (Vapor Grown Carbon Fiber, VGCF™). Two different highly crystalline and near-stoichiometric SiC fiber fabrics were examined in this study; 2D-plane weave Tyranno SA (Ube Industries, Ltd., Japan) and 2D-8 harness satin Hi-Nicalon Type S (Nippon Carbon Co., Ltd., Japan). Figure 1 shows the flow chart of process of fabricating the SiC/SiC with or without dispersed CNFs. composites

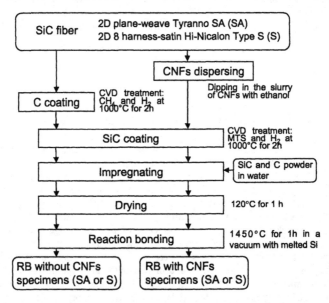

Fig. 1. The flow chart of fabricating process of the SiC/SiC with or without CNFs composites

The CNFs were dispersed in the both SiC fiber fabrics by dipping fiber fabrics in the slurry of CNFs with ethanol. The SiC fiber fabrics dispersed with CNFs were coated with only β-SiC by chemical vapor deposition (CVD) treatment. On the other hand, the SiC fiber fabrics without CNFs were coated with carbon in addition to β-SiC on the fibers. The precursors were CH_4 for carbon and methyltrichlorosilane (CH_3SiCl_3, MTS) for SiC deposition. The carbon and β-SiC interphase layer was deposited on to the fibers. The flow rate for carbon deposition were 0.5 l/min CH_4 and 0.5 l/min H_2 for the carrier gas at 1050 °C and 0.05 MPa. The SiC layer was formed with decomposition of MTS with the flow rate of about 0.2 g/min MTS and 0.7 l/min H_2 for carrier gas at 1050 °C and 0.05 MPa.

The slurry for RB process was an aqueous solution containing α-SiC and C powder. The green composites were obtained by impregnating this slurry in a vacuum to the both coated SiC fiber fabrics with and without CNFs by CVD treatment. The SiC/SiC composites were obtained by heating at 1450 °C in a vacuum for 1 hour being contacted with excess molten silicon.

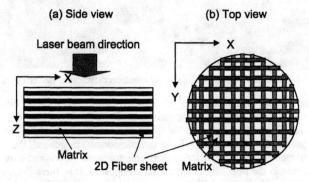

Fig. 2. Schematics of fiber configurations and the laser beam direction.

Evaluation of thermal properties

The thermal diffusivities of specimens (nominally 10 mm in diameter and 2 mm in thickness) were measured in the temperature range from room temperature to 1000 °C in a vacuum using a laser flash thermal diffusivity analyzer (TC-7000, Ulvac Sinku-Riko, Inc., Japan). The through-the-thickness thermal diffusivity, α, was obtained by $t_{1/2}$ method using the following equation:

$$\alpha = \frac{1.338 \cdot L^2}{\pi^2 \cdot t_{1/2}}$$

(1)

where $t_{1/2}$ is the time required to reach half of the total temperature rise on the rear surface of the specimen, and L is the specimen thickness. The C_p and ρ are the specific heat and the bulk density of the specimen. The specific heat, C_p, of monolithic SiC was used for SiC/SiC composites fabricated in this study[10]. Figure 2 shows the schematics of fiber configurations and the laser beam direction. In the thermal diffusivity measurement, the laser beam is incident on specimens along the Z-direction.

RESULTS AND DISCUSSION

Figure 3 shows the TEM microphotograph of CNFs and the selected electron diffraction pattern used in this study, indicating their high crystallinity. The length of CNFs was more than 10 μm. The average diameter of CNFs was approximately 150 nm. The intertube distance of CNFs was 0.34 nm, corresponding to the lattice constant of graphite, c=0.34 nm.

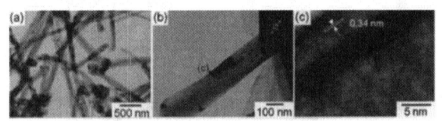

Fig. 3. TEM microphotographs and selected electron diffraction pattern of CNFs

Table I shows the characterization of SiC/SiC with and without dispersed CNFs composites fabricated by RB process. The densities of SiC/SiC composites decreased by the presence of dispersed CNFs. On the other hand, the volume fractions of SiC fibers were not influenced by the presence of dispersed CNFs in the composites. In general, the load is mainly maintained by unfractured fibers and friction between fractured fibers and matrix above the proportional limit stress in the fiber reinforced matrix composites. It is, therefore, considered that the strength of SiC/SiC with dispersed CNFs composites is not reduced. The volume fraction of CNFs was approximately 4 % in both SiC/SiC composites.

Table I. Characterization of SiC/SiC with and without CNFs composites by RB process

Specimen ID	Fiber volume fraction (%)	Density (g/cm³)
RB without CNFs (S)	29.1±0.5	2.78±0.02
RB without CNFs (SA)	35.9±1.1	2.65±0.03
RB with CNFs (S)	33.2±1.8	2.38±0.06
RB with CNFs (SA)	28.4±1.2	2.56±0.06

Figure 4 shows the cross-sectional SEM microphotographs of the SiC/SiC with and without dispersed CNFs composites. Although a few pores existed around fibers in both the specimens, SiC interphase layers were formed fully within the fiber bundle by CVD treatment. The SEM observation revealed that the interphase produced by CVD treatment consisted of carbon and β-SiC layers in the composites without dispersed CNFs while only β-SiC layer was formed in the composites with dispersed CNFs. Almost all the fibers were distinguished from the matrix in all the specimens. Although the melted Si in the composites had very high reactiveness during the RB process, the β-SiC produced by CVD treatment prevented SiC fibers from contacting the molten Si and adhering to the matrix. This result indicates that the CVD treatment is effective to reduce effectively adherence of the fibers to the matrix.

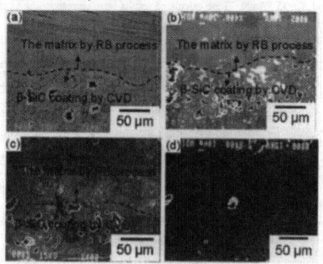

Fig.4. Cross-sect ional SEM microphotographs of the SiC/SiC with and without dispersed CNFs composites; (a) RB without CNFs (S), (b) RB without CNFs (SA), (c) RB with CNFs (S) and (d) RB with CNFs (SA).

The thermal diffusivities and thermal conductivities of SiC/SiC with and without dispersed CNFs composites were given in Fig. 5. The thermal diffusivities and thermal conductivities of SiC/SiC composites fabricated by conventional processes were also given in Fig. 5[11]. The thermal diffusivities and thermal conductivities of specimens in this study were much higher than those of the SiC/SiC composites fabricated by conventional process. This is because the pore volume fractions in the composites fabricated by RB process (less than 10%) were much smaller than those of composites by conventional processes (~20%). Independent of presence of the dispersed CNFs, the thermal diffusivities and thermal conductivities of the composites with Tyranno SA SiC fibers were slightly higher than those of the composites with Hi-Nicalon Type S SiC fibers. The reason is that the thermal conductivity of Tyranno SA SiC fiber (60 W/mK at room temperature) was higher than that of Hi-Nicalon Type S SiC fiber (24 W/mK at room temperature).

For the simplest geometry, a series of layers (like a Fig. 2 (a)), the total thermal conductivity of composite, K_{cl} is given by

$$K_{cl}=K_1K_2/(V_2K_1+V_1K_2) \qquad (2)$$

Where V_1, V_2, K_1 and K_2 are the volume fraction and thermal conductivity of phase 1 and 2. In this study, the phases 1 and 2 are matrix and fibers. The V_f was about 0.3. In the case of K_f=24 (Hi-Nicalon Type S) and K_m=35 W/mK, the value of K_{cl} at room temperature is 30 W/mK by equation (2). And the value of K_{cl} at room temperature is 40 W/mK in the case of K_f=60 (Tyranno SA) and K_m=35 W/mK. These calculated results were corresponding to experimental results of RB without CNFs speicmens.

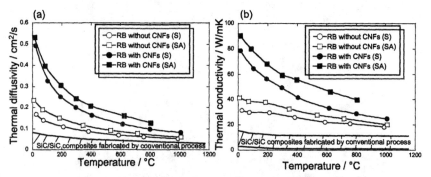

Fig. 5. (a) thermal diffusivity and (b) thermal conductivity of SiC/SiC with and without dispersed CNFs fabricated by RB process.

Although the volume fraction of CNFs, V_{CNF}, was only 4 % and the density of RB with CNFs specimen was smaller than that of RB without CNFs specimen, the thermal conductivities of RB with CNFs specimens were twice as high as those of RB without CNFs specimens. The values were more than 80 W/mK at room temperature. The K_{cl} of RB with CNFs specimen calculated by equation (2) is hardly increased in the case of K_{CNF}=more than 10000 W/mK, K_m=40 W/mK (RB without CNFs (SA) specimen) and V_{CNF}=0.04. For the heat flow parallel to the plane of the layers in the simplest geometry, the total thermal conductivity, K_{c2}, is given by

$$K_{c2}=V_1K_2+V_1K_2 \qquad (3)$$

In the case of K_{CNF}=1300 W/mK, K_m=40 W/mK (RB without CNFs (SA) specimen) and V_{CNF}=0.04, the K_{c2} of RB with CNFs (SA) specimen calculated by equation (3) is 90 W/mK at room temperature. And the K_{c2} of RB with CNFs (S) is 80 W/mK at room temperature in the case of K_{CNF}=1300 W/mK, K_m=30 W/mK (RB without CNFs (S) specimen) and V_{CNF}=0.04. These calculated results by equation (3) were corresponding to the experimental results in this study. Based on these results, the thermal conductivity of CNFs was approximately estimated to be 1300 W/mK. This value is equivalent to that of the high crystalline graphite fibers[12]. The experimental result in the composites with dispersed CNFs can be simulated by equation (3), not by equation (2). It is, therefore, important that paths of high heat flow through the thickness in the specimen exist in order to increase the thermal conductivity. The photographs of two types of SiC fiber fabrics and the schematic of cross-section in the composite were given in Fig. 6. There are the void (white region in Figs. 6 (a) and (b)) among the SiC fiber bundles in both SiC fiber fabrics. Since the CNFs are existing in the void and connecting to other dispersed CNFs layers, they may function as good paths of heat flow through the thickness (Z direction) in the specimen.

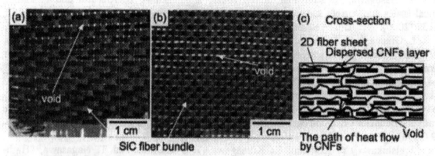

Fig. 6. Photographs of two types of SiC fiber fabrics;(a) 8HS Hi-Nicalon Type S and (b) PW Tyranno SA, and schematic of cross-section in the composite.

The thermal conductivity of RB with CNFs (SA) at 800 °C, which is the operation temperature range of fusion reactors, was more than 40 W/mK. The thermal conductivities of SiC/SiC with dispersed CNFs composites in this study satisfy completely the assumed design criterion for fusion reactors, which is 15-20 W/mK at the operation temperature of fusion reactors. However, the thermal conductivities in SiC/SiC composites and SiC are reduced to approximately 50 % of their unirradiated values by neutron irradiation at 800-1000 °C. The thermal conductivity of twice as high as the assumed design criterion (30-40 W/mK) is, therefore, required for fusion structural materials[12,13]. The thermal conductivities of SiC/SiC with dispersed CNFs composites are found to satisfy this high requirement.

CONCLUSION

The SiC/SiC with and without dispersed CNFs composites were fabricated by RB process. The effect of dispersed CNFs on the thermal properties in the SiC/SiC composites was investigated.

(1) The volume fraction of CNFs was approximately 4 %. The CNFs were able to be dispersed in the SiC/SiC composites without reducing the volume fraction of SiC fibers. It is, therefore, considered that the strength of SiC/SiC with dispersed CNFs composites is not reduced.
(2) The CVD treatment is effective to reduce effectively adherence of the fibers to the matrix.
(3) The thermal conductivities of SiC/SiC with dispersed CNFs composites were twice as high as those without CNFs; the value was approximately 90 W/mK at room temperature.

(4)   The thermal conductivity of SiC/SiC with dispersed CNFs composites at 800 °C, which is the operation temperature range of fusion reactors, was more than 40 W/mK. The thermal conductivities of SiC/SiC with dispersed CNFs composites in this study satisfy completely the assumed design criterion for fusion reactors.

REFERENCES

[1]A. Hasegawa, A. Kohyama, R. H. Jones, L. L. Snead, B. Riccardi and P. Fenici, "Critical issues and current status of SiC/SiC composites for fusion," *J. Nucl. Mater.*, **283-287**, 128-137 (2000).

[2]L. Giancarli, G. Aiello, A. Caso, A. Gasse, G. Le Marois, Y. Poitevin, J. F. Salavy and J. Szczepanski, "R&D issues for SiC$_f$/SiC composites structural material in fusion power reactor blankets," *Fus. Eng. Des.*, **48**, 509-520 (2000).

[3]L. Giancarli, H. Golfier, S. Nishio, R. Raffray, C. Wong and R. Yamada, "Progress in blanket designs using SiC$_f$/SiC composites," *Fus. Eng. Des.*, **61-62**, 307-318 (2000).

[4]M. Takeda, A. Urano, J. Sakamoto and Y. Imai, "Microstructure and oxidative degradation behavior of silicon carbide fiber Hi-Nicalon type S," *J. Nucl. Mater.*, **258-263**, 1594-1599 (1998).

[5]T. Ishikawa, Y. Kohtoku, K. Kumagawa, T. Yamamura and T. Nagasawa, "High-strength alkali-resistant sintered SiC fibre stable to 2200°C," *Nature*, **391**, 773-775 (1998)

[6]T. Taguchi, N. Igawa, S. Miwa, E. Wakai, S. Jitsukawa, L.L. Snead and A. Hasegawa, "Synergistic effects of implanted helium and hydrogen and the effect of irradiation temperature on the microstructure of SiC/SiC composites," *J. Nucl. Mater.*, **335**, 508-514 (2004).

[7]R. H. Jones, L. Giancarli, A. Hasegawa, Y. Katoh, A. Kohyama, B. Riccardi, L. L. Snead and W. J. Weber, "Promise and challenges of SiC$_f$/SiC composites for fusion energy applications," *J. Nucl. Mater.*, **307-311**, 1057-1072 (2002).

[8]T. Taguchi, N. Igawa, R. Yamada, M. Futakawa and S. Jitsukawa, "Mechanical and thermal properties of dense SiC/SiC composite fabricated by reaction bonding process," *Ceram. Eng. Sci. Proc.*, **22[3]**, 533-538 (2002).

[9]R. Yamada, N. Igawa and T. Taguchi, "Thermal diffusivity/conductivity of Tyranno SA fiber- and Hi-Nicalon Type S fiber-reinforced 3-D SiC/SiC composites," *J. Nucl. Mater.*, **329-333**, 497-501 (2004).

[10]Specific heat, nonmetallic solids, in: Y.S. Touloukian, E.H. Buyco (Eds.), Thermophysical Properties of Matter, The TPRC Data Series, vol. 5, IFI/Plenum, New York (1970) 448.

[11]R. Yamada, T. Taguchi, N. Igawa, "Mechanical and thermal properties of 2D and 3D SiC/SiC com posites," *J. Nucl. Mater.*, **283-287**, 574-578 (2000).

[12]L. L. Snead, M. Balden, R. A. Causey and H. Atsumi, "High thermal conductivity of graphite fiber silicon carbide composites for fusion reactor application," *J. Nucl. Mater.*, **307-311**, 1200-1204 (2002).

[13]G. E. Youngblood, D. J. Sensor and R. H. Jones, "Effects of irradiation and post-irradiation annealing on the thermal conductivity/diffusivity of monolithic SiC and f-SiC/SiC composites," *J. Nucl. Mater.*, **329-333**, 507-512 (2004).

TRANSPIRATION COOLING STRUCTURE EFFECTS ON THE STRENGTH OF 3D-WOVEN SiC/SiC COMPOSITES UNDER THERMAL CYCLING

Toshimitsu Hayashi
3rd Research Center, Technical Research&Development Institute, Japan Defense Agency
1-2-10 Sakae, Tachikawa, Tokyo, 190-8533, JAPAN

ABSTRACT
Evaluation of strength degradation under thermal cycling conditions is an important issue for applying the CMC to hot section parts of gas turbine engines. In this study, the strength degradation of the 3D-woven CMC (SiC/SiC) with and without the transpiration cooling structure was investigated in a burner rig test. Each specimen was subjected to the test for 200 thermal cycles with the same heating (from the front side of the specimen) and cooling (from the back side of the specimen) conditions. Four-point bending strength before and after the burner rig test results indicated that the transpiration cooling structure was effective in reducing the strength degradation of the CMC under thermal cycling conditions.

INTRODUCTION
Aircraft jet engines need high heat resistant materials for increasing turbine inlet temperature or decreasing the quantity of cooling air in order to improve performance. The ceramics matrix composite (CMC) has been anticipated for application to the jet engine hot section parts as one high heat resistant material. To prevent heat damage to cold section parts surrounding the hot section parts, the outer side (opposite side of combustion side) of hot section parts should be cooled down by air. Even if the Ni-based metal parts are replaced with the CMC, the cooling air will still be necessary but should be reduced to a minimum by applying the cooling structure in the CMC parts as in the metal parts. The conventional machined cooling hole which has often been applied to Ni-based metal parts is, however, unsuitable for the CMC. The machined holes cut fibers and decrease the total strength of the CMC, and the oxidation along the cooling holes may result in strength degradation in the CMC. In this sense, the cooling structure for the CMC has not yet been well established.

In this study, the transpiration cooling structure that was the most effective among the cooling structures [1] was created in the CMC, and the effects of the cooling structure on the durability under thermal cycling were investigated. Two types of 3D-woven fabrics, "3D-orthogonal" and "one-sided chain stitching", were used to fabricate the CMC. The 3D-woven fabrics are favorable for making the structure, because they have many cavities between fiber bundles. This cooling structure was created by the cavities without the machining of holes. The air passages were observed by micro-focused X-ray CT scan. Specimens were tested under thermal cycling using a burner rig and were compared to the results of a specimen without the cooling structure.

MATERIALS AND EXPERIMENTAL PROCEDURES
Materials and Specimens for the Burner Rig Test

The materials used in this study are SiC_f/SiC_m composites with two kinds of 3D weaving structures, "3D-orthogonal(3D)" and "one-sided chain stitching (CS)". The structures are schematically shown in Fig.1. The SiC fibers are Tyranno™ ZMI except for the CS where Z-fiber is the Tyranno™ LOX-M. Fiber volume fraction of X,Y and Z is 1:1:0.1. The X and Y fibers are stacked in the direction [90/0/90/0/90/0/90] and woven with the Z fiber. The woven performs are coated with a carbon interface by chemical vapor infiltration (CVI) process. The performs are densified with pure SiC matrix by CVI process and much SiC matrix by polymer infiltration process (PIP)[2]. The number of PIP cycles can control the rate of cavity. The air passages can be formed with a connection of the cavities in PIP matrix. The air passages were verified by the air-leakage test mentioned later. The 3D and CS fabrics with 4 cycles of PIP were called the specimen 3D-L and CS-L, respectively, and the 3D with 8 cycles of PIP was the specimen 3D-H. The fiber architectures of the specimens are listed in Table 1. After the PIP, the specimens were cut into 50 x 50mm and coated by SiC with CVI process to protect the surface. The thickness of the specimen 3D and CS were approximately 2 and 3mm, respectively.

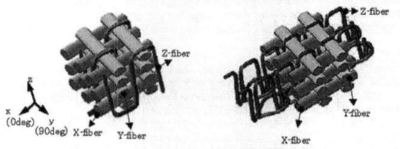

3D-orthogonal(3D)   One-sided chain stitching(CS)
Fig.1 Schematic view of weaving structure

Table 1   Fiber architecture of specimens

Type	PIP Cycles	Pitch X,Y	Filaments/yarn X,Y	Filaments/yarn Z	V_f %	Structure
3D-L	4	2	1600	1600	35-40	3D-orthogonal
3D-H	8	2	1600	1600	35-40	3D-orthogonal
CS-L	4	4	3200	1200	30-35	One-sided Chain Stitching

Transpiration Cooling Structure of Specimens

To observe the transpiration cooling structure of the specimen 3D-L and CS-L, a micro-focused X-ray CT scan (Toshiba IT control,TOSCANNER-32250μhd) was used [3]. The CT-scan images were taken with 10 x 10 x 2mm size of specimens.

336

Air flow test

Air flow tests were carried out to verify the air passages in the specimens. Samples were installed as shown in Fig.2. Nitrogen gas was used to make differential pressure of 5kPa at maximum.

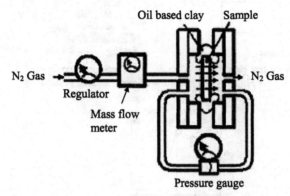

Fig.2 Schematic view of the air flow test apparatus

Burner Rig Test

A burner rig test bench is shown in Fig.3. Each specimen was mounted on holder and softly supported at 4 points using ceramics pins. The front surface of the specimen was heated by LNG furnace. The flow rates of LNG and oxygen were 45 and 110 l/min, respectively. The furnace temperature was approximately 1700℃. The back surface of the specimen was continuously cooled by air of 40l/min. The front surface of the specimen was cooled quickly by air only in the cooling process. Front and back surface temperatures were measured by pyrometer and thermocouples, respectively. The temperature history at front surface is shown in Fig.4. The same testing conditions were applied to each specimen. The temperature T_{max} in Fig.4 is dependent on a thermal balance in each specimen and the temperature T_{min} is room temperature. The thermal cycles were repeated for 200 cycles.

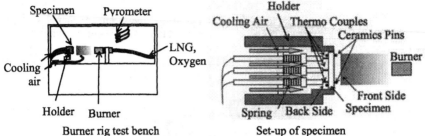

Burner rig test bench   Set-up of specimen

Fig.3 Schematic view of experimental system

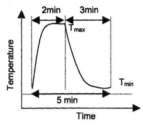

Fig.4 Temperature history of the burner rig test

4-Point Bending Test

    Four-point bending test with load span 10mm and support span 20mm were carried out with a testing machine (Instron, Model5505) at room temperature in air under a constant displacement speed of 0.5 mm/min. The specimens were prepared by cutting out from samples tested in the burner rig and untreated samples. The dimension of the specimen was 50mm length x 4mm width x 2(3D) or 3(CS) mm thickness.

RESULTS AND DISCUSSION

Transpiration Cooling Structure of Specimens

    Two CT slice images, shown in Fig.5, were obtained from the specimen 3D-L at heights of 1.06mm and 1.26mm from the bottom in the XY plane. The same pattern of images was observed along the specimen thickness or Z direction repeatedly. Many large cavities are observed among fiber yarns in the figure. The transpiration cooling structure consists of these cavities, for example, a cavity A, a cavity B and a passage A-B that connects those cavities. The passage A-B appears to be one of the smallest areas along air-passages and is dependent on the quantity of Z-yarn's filaments and matrix. Accordingly, Z-yarn's filaments and matrix could adjust the area and thus control the air flow rate. A similar structure was also observed in the specimen CS-L.

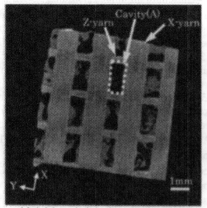 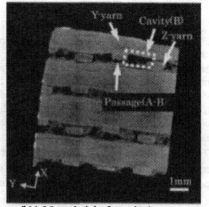

(a) 1.06mm height from the bottom   (b)1.26mm height from the bottom

Fig.5 Micro-focused X-ray CT scan images of the specimen 3D-L

Air Flow Test

The air flow rate is plotted vs. the differential pressure ΔP in Fig.6. Both of the specimens 3D-L and CS-L showed increased air flow rate with increasing ΔP. However, no air flow was observed in the specimen 3D-H even at $\Delta P=5kPa$. Therefore it can be concluded that 4 cycles of PIP could leave the air passages open in the CMC and that 8 cycles of PIP could not. Since the mechanical strength of the CMC increases with the number of PIP cycles, it is important to take into account the balance between the quantity of the air passage and the mechanical strength when the number of PIP cycles is determined.

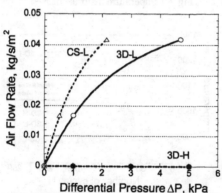

Fig.6 Air flow rate vs. differential pressure

Temperature History in the Burner Rig Test

The temperature histories of N=1 and the trends of the maximum front/back temperatures, T_F and T_B, in the burner rig test are shown in Figs.7-8. The maximum front surface temperatures, T_F, of each specimen are almost exactly the same. The maximum back surface temperature, T_B, of the specimen CS-L is the lowest, followed by 3D-L and 3D-H(highest T_B). The reason for the same value of T_F of the specimens can be explained as follows. When the 3D-L and 3D-H are compared, the thermal conductivity of 3D-L is lower than that of 3D-H due to 3D-L's large cavity ratio. If the difference between 3D-L and 3D-H is only the thermal conductivity, the front surface temperatures T_F of 3D-L will increase like that of 3D-L' as illustrated in Fig.9. Then the cooling effect which may occur in the 3D-L decreases the inside temperature of the CMC and thus resulted in the decrease of the value of T_F. As shown in Fig.8, the temperature range ΔT (T_F-T_B) of CS-L and 3D-L is larger than that of 3D-H, which also can be explained by the low thermal conductivity of 3D-L and CS-L due to the large cavity ratio. The largest ΔT of the CS-L appears to be caused by its greater thickness compared to 3D type specimens.

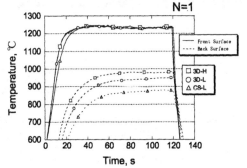

Fig.7 Temperature histories of N=1

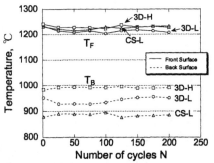

Fig.8 Maximum front/back surface temperature
during thermal cycling

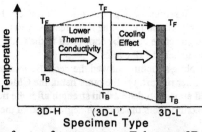

Fig.9 Comparison of the front surface temperature T_F between 3D-H and 3D-L(Schematic)

Bending Strength after cyclic thermal loading in the Burner Rig Test

The 4-point bending strength measured after 200 thermal cycles of the burner rig test is shown in Fig.10 in conjunction with the initial strength obtained by virgin specimens. The initial strength of 3D-H is larger than that of 3D-L. This indicates that the incremental PIP cycles increased the bending strength. The specimen without the cooling structure 3D-H showed a 25% decrease in strength after the burner rig test. However, the specimens with the transpiration cooling structure showed almost no decrease in strength. Therefore it is concluded that the transpiration cooling structure is effective in reducing strength degradation under thermal cycling conditions. Since the maximum front surface temperatures, T_{max}, of 3D-L and 3D-H are almost the same and the difference in ΔT is not large, these temperature trends would not affect the strength of the CMC. The thermal stress caused by the internal temperature distribution or the existence of coolant air above the front surface has the possibility of affecting the strength of the CMC. Further measurement of surface temperatures and gas temperature above the front surface will be needed to verify those possibilities. The verification will also clarify the transpiration cooling efficiency. The relationships between the cooling efficiency, surface temperatures and their effect on the strength are one of the next objectives in this study.

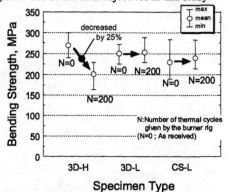

Fig.10 Bending strength before and after 200 thermal cycling
given by the burner rig test

CONCLUSION

The effects of the transpiration cooling structure on the strength of the CMC under thermal cycling conditions were investigated. The main results are as follows:

(1) The transpiration cooling structure could be controlled in the CMC with 3D and CS woven fabrics by adjusting the number of PIP cycles.

(2) The CMC without the cooling structure showed the decrease in strength by 25 % after 200 thermal cycling in the burner rig test. On the other hand, the CMC with the transpiration cooling structure showed almost no decrease in strength after the thermal cycling. Thus the transpiration cooling structure alters the strength degradation of the CMC under thermal cycling.

ACKNOWLEDGEMENTS

The author would like to acknowledge the efforts of Sunao Sugimoto of Japan Aerospace Exploration Agency(JAXA) for his aid in making the observation with the Micro-focused X-ray CT scan.

REFERENCES

[1] A.H.Lefebvre, Gas Turbine Combustion, Taylor & Francis Ltd.

[2] T. Araki, K. Watanabe, T Yoshida, S. Nishide, and S. Masaki, "High Temperature Properties of SiC Fiber Reinforced SiC Matrix Composites for Turbine Rotor Application", Ceramic Eng. Sci. Proc., 23, 581-588(2002)

[3] S. Sugimoto, T. Aoki, Y. Iwahori, and T. Ishikawa, " Nondestructive Evaluation of Composites Using Micro-Focused X-Ray CT Scanner", Review of Progress in Quantitative Nondestructive Evaluation, 24, 2005 in printing

CONSTITUENT PROPERTIES DETERMINATION AND MODEL VERIFICATION FOR CERAMIC MATRIX COMPOSITE SYSTEMS

G. Ojard, K. Rugg, L. and M. Colby
Pratt & Whitney
400 Main Street
East Hartford, CT 06108

L. Riester
Oak Ridge National Labs
Bethel Valley Road
Oak Ridge, TN 37831

Y. Gowayed
Auburn University
Auburn, AL 36849

ABSTRACT

The increased interest in ceramic matrix composites requires the knowledge and models to manufacture them for their intended use. Prediction of the elastic moduli of composites requires estimates of the *in situ* properties of each of the phases of the material: fiber, matrix, and interface. Fiber tow testing was done to determine the elastic modulus of fibers. This was followed by nano-indentation of the fibers in the actual composite systems. Nano-indentation was also done on the matrix to determine its elastic properties. The influence of the interface was modeled through its effect on the fiber transverse and shear moduli. All of these properties were entered into the PcGina model system to predict the elastic properties of the CMC systems. The result of the model was compared to a series of mechanical tests to determine the key elastic properties of the composite. This work was performed on the MI SiC/SiC system, the SiC/SiNC system and on an Oxide/Oxide system.

INTRODUCTION

Ceramic matrix composites are being developed, designed and considered for extended high temperature uses such as combustor liners and turbine vanes[1,2]. These applications take benefit of the high temperature capability of the material with the added benefit of weight reduction (density) as well as durability improvement[3]. This can be done without the need for cooling air that is needed for the nickel base superalloys that are being considered for replacement.

All of the above takes into account certain assumptions about the base properties of the composite that need to be confirmed. A specific property of interest that will be explored in this paper is the elastic properties. These properties are based on interactions between the fiber, fiber interface coating and the matrix. As a starting point, a rule of mixtures approach is possible. Even with this simple approach, assumptions are needed as to what the individual properties are of the phases. For the most part, the fiber properties are some of the most understood but not thoroughly.

In this paper, testing has been performed by various methods to determine the elastic properties of the fiber and matrix in the effort to better understand the resulting bulk elastic properties (E_{11}, G_{12} and E_{33} for this effort.). The individual elastic properties are a crucial part of the understanding of the overall system. This is needed for the understanding and insertion of components fabricated out of ceramic matrix composites.

PROCEDURE

Material Description

In this paper, the following composite systems were interrogated at various levels: Melt Infiltrated In-Situ BN SiC/SiC composite (MI SiC/SiC), SiC/SiNC composite and an Alumina/Alumina Silicate (Oxide/Oxide) composite. The MI SiC/SiC system has a stochiometric SiC (Sylramic™) fiber in a multiphase matrix of SiC deposited by chemical vapor deposition followed by slurry casting of SiC particulates with a final melt infiltration of Si. The specific MI SiC/SiC tested for this effort had 36% volume fraction fibers using a 5 HS weave at 20 EPI. The SiC/SiNC system uses a non-stochiometric SiC (CG Nicalon™) fiber in a matrix of Si, N and C that is arrived at by multiple iterations of a polymer pyrolisis process. The specific SiC/SiNC tested for this effort had 42% volume fraction fibers using an 8 HS weave at 24 EPI. The Oxide/Oxide system is an Al_2O_3 fiber in a sol gel matrix of Alumina Silicate. This last system if the only one that does not have a weak interface coating (like BN or C) and instead relies on the high percent of porosity (25%) in the matrix for crack deflection. The specific Oxide/Oxide tested for this effort had 46% volume fraction fibers using an 8 HS weave at 27 EPI. All samples for this effort used a cross ply lay-up. A cross section of each material is shown in Figure 1.

a) MI SiC/SiC   b) SiC/SiNC   c) Oxide/Oxide

Figure 1. Cross sections of the composite systems being characterized in this effort

Fiber Tow and Fiber

Elastic property determination of the fiber was done by two different methods. For the two SiC fiber systems being investigated, fiber tows were available and a series of different gage lengths were tested to determine the system compliance as well as the modulus of the fiber in question (ASTM C1557). This effort had to assume a fiber diameter to determine the cross section area (10 μm diameter and 800 fibers per tow for Sylramic™ fibers, 14 μm diameter and 500 fibers per tow for CG-Nicalon™ fibers). Also, nano-indentation work was done to determine the fiber modulus of the fiber in mounted polished cross sections of the ceramic matrix composite (all systems). This was done by looking at the un-load portion of the load-displacement output from a nano-indenter[4]. The slope of the un-load portion of the curve is proportional to the modulus[4]. Since micro-structural cross sections were made of the cross ply samples, it was possible to determine the elastic properties along and perpendicular direction to the axis of the fiber.

Matrix

The matrix elastic properties were only determined by nano-indentation techniques consistent with the effort described above for the fiber. Multiple regions were interrogated since the fabrication processes for the materials being tested are not uniform. Due to this fact, more measurements (nano-indentation) were taken of the matrix than the fiber.

Composite

The bulk elastic properties were determined by tensile and compressive testing of the material. E_{11} and G_{12} were determined by tensile tests according to ASTM procedures (C1359-96 for E_{11} and D3518-94 for G_{12}). E_{33} was determined by machining multiple disks from plates of material and stacking them followed by compression of the full stack. By use of an extensometer on the full stack (stack height > 25 mm), the stress-strain could be determined allowing a fit of the linear curve to determine the through thickness modulus. This through thickness modulus tests was developed by NASA –GRC[5].

Modeling

Modeling was done using the Graphical Integrated Numerical Analysis (pcGINA) that has been developed to model the mechanical and thermal behavior of textile composites. This is a two-part model. First, a geometrical model is used to construct the textile preform and characterize the relative volume fractions and spatial orientation of each yarn in the composite space. Data acquired from the geometrical analysis is used by a hybrid finite element approach to model the composite mechanical and thermal behavior.

The geometrical model used in pcGINA starts by modeling the preform forming process – weaving or braiding. An ideal fabric geometrical representation is constructed by calculating the location of a set of spatial points "knots" that can identify the yarn centerline path within the preform space. A B-spline function is utilized to approximate a smooth yarn centerline path relative to the identified knots due to its ability to minimize the radius of curvature along its path and its C^2 continuity. The final step in this model is carried out by constructing a 3-D object (i.e. yarn) by sweeping a cross section along the smooth centerline forming the yarn surface.

A repeat unit cell of the modeled preform is identified from the geometric model and used to represent a complete yarn or tow pattern. A hybrid finite element approach is used to divide the unit cell into smaller subcells. Each subcell is a hexahedral brick element with fibers and matrix around each integration point. A virtual work technique is applied in the FE solution to calculate the properties of the repeat unit cell. The unit cell properties are considered to be representative of the composite properties.

RESULTS

Tow Testing

The tow testing compliance check was done per the ASTM procedure as mentioned above. The resulting data for this effort is shown in Figure 2. As can be seen, the stochiometric Sylramic™ fiber has the higher modulus than the CG-Nicalon fiber. By this method, the Sylramic™ fiber was determined to have a modulus of 394 GPa. The CG-Nicalon™ fiber was determined to have a modulus of 190 GPA. These values are consistent with reported values for these fibers[6].

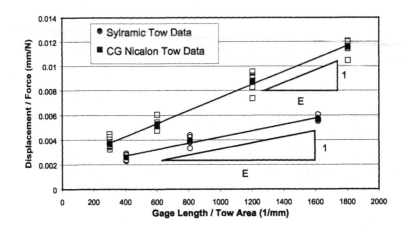

Figure 2. Tow testing data showing modulus difference between Sylramic™ and CG-Nicalon tows™

Nano-indentation of fibers

Indentation work was done on three different fibers in their respective composite. Typical load-displacement curves are shown in Figure 3. As can be seen, the stoichiometric SiC fiber (Sylramic™) has the steepest curve indicating the high modulus of the material. The CG-Nicalon™ and the Alumina fiber appear to be very similar. The modulus data from these curves and multiple other measurements are shown in Table I. For the SiC fibers, these numbers are in good agreement with the tow testing. All the data in Table I was for the fiber in circular cross section. In addition, the fibers were also tested perpendicular to this direction. This data did not show any significant difference in the resulting value. All of the modulus data reported here was based on determination at displacements of 100 nm.

Table I. Elastic Properties of Fibers by Nano-indentation

Fiber	Elastic Modulus (GPa)		# of Data Points
	Average	St. Dev.	
Sylramic™ (SiC)	395.2	27.6	9
CG Nicalon™ (SiC)	204.7	2.1	8
Alumina	257.2	7.4	5

Nano-indentation of matrix (SiC/SiNC and Alumina/Alumina Silicate)

Indentation was also done on the matrix of both the SiC/SiNC material as well as the Alumina/Alumina Silicate system. Both of these systems have a relatively single-phase material matrix but the Alumina/Alumina Silicate system has a significant percentage of porosity present. Typical load-displacement curves for this effort are shown in Figure 4. Figure 4 clearly shows the low modulus nature of the Alumina Silicate matrix (due to the presence of porosity). The modulus data from these curves and multiple other measurements are shown in Table II.

Nano-indentation of matrix (MI SiC/SiC)

As pointed out in Table I, the MI SiC/SiC system has multiple matrix phases present, chemical vapor infiltrated SiC around the fiber tows, SiC particulates and melt infiltrated Silicon. Nano-indentation was done on all of these three distinct phases. Typical curves for these specific phases are shown in Figure 5. The results of the testing are in Table III.

346

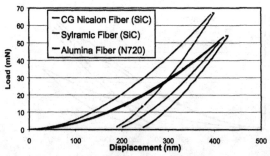

Figure 3. Load displacement curves for indentation on different fibers

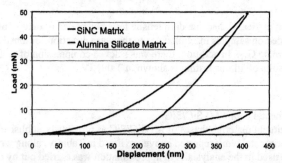

Figure 4. Load displacement curves for indentation on different matrices

Table II. Elastic Properties of Matrix by Nano-indentation

Matrix	Elastic Modulus (GPa)		# of Data Points
	Average	St. Dev.	
SiNC	139.7	18.3	16
Alumina Silicate	56.9	8.7	9

Table III. Elastic Properties of MI SiC/SiC Matrix Phases by Nano-indentation

Matrix Phase	Elastic Modulus (GPa)		# of Data Points
	Average	St. Dev.	
CVI Matrix	439.1	25.8	6
SiC Particulate	405.9	16.6	5
Silicon	165.0	19.7	6

Table IV. Tensile and Compressive Test Results (Various Moduli)

Property	MI SiC/SiC	SiC/SiNC	Oxide/Oxide
E_{11}	272.6	107.0	70.4
G_{12}	93.8	48.3	13.1
E_{33}	72.5	55.9	28.3
All modulus values are in units of GPa			

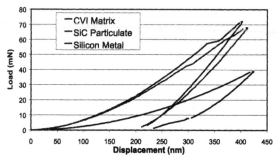

Figure 5. Load displacement curves for indentation of the matrix phases of MI SiC/SiC

Composite Testing

The 3 composites listed earlier, all had standard tensile tests performed to determine E_{11}. Tensile tests were also performed on samples machined 45° to the warp direction for tensile testing of these coupons to determine G_{12}. In addition, there was stacked compression of multiple disks to determine E_{33}. The results of all this testing is shown in Table IV.

DISCUSSION

Comparison to experimental data – MI SiC/SiC

Properties of constituent material phases of MI SiC/SiC as previously published and confirmed by the experimental work reported above are listed in Table V, along with relative volume fractions were used in the analysis[7]. An approximation was carried out by dividing the composite into two parts – coated fiber comprised of iBN-Sylramic™ fibers coated with Si-doped BN and a matrix formed from CVI-SiC, SC-SiC and Si. Properties of the coated fiber were calculated using the rule of mixtures utilizing data from Table V. Micrographic imaging of the SiC/SiC composite revealed a shiny material (Si) mixed with another grayish color material (SC-SiC) in the place between the yarns with some dark areas that are most probably voids. Based on micrographs an in-series model was used to calculate the combined properties of slurry cast SiC and Si (iso-stress model) and an in-parallel model (iso-strain model) was used to combine these properties with the properties of the CVI-SiC. Utilizing this approach the matrix properties were calculated as E_m= 330.6 GPa, G_m = 139.4 GPa, and v_m=0.182.

Calculation of room temperature elastic properties using pcGINA with 225 eight-noded hexahedra brick elements was compared to experimental data as shown in Table VI. This range of results in Table VI for the effect of voids is caused by the lack of information on the exact location of voids. Typically, the upper bound values are for voids that exist as spherical voids in the matrix away from the fibers, while lower bound values are for voids that are aligned within yarns and at yarn cross-over points. It can be seen from the Table that estimates for in-plane tensile moduli (E_{11} and E_{22}), shear modulus (G_{12}) and Poisson's ratio (v_{12}) are very close to the experimental data. The calculation for the through thickness modulus (E_{33}) is over 3 times higher than the experimental data. Nevertheless, it should be noted that the through thickness modulus experimental value is less than half the value of the lowest constituent material modulus (Si) as listed in Table V.

348

Table V: Properties of constituent material at room temperature (24°C)

Property	Sylramic® fiber (β-SiC)	BN Coating (Si-doped BN)	SiC-CVI (β-SiC)	SiC-MI		Porosity
				α-SiC	Si	
E (GPa)	380.3	27.6	425.3	460.3	166.3	
G (GPa)	162.5	11.3	181.7	196.7	68.2	
ν	0.17	0.22	0.17	0.17	0.22	
V_f	0.36	0.072	0.23	0.177	0.135	0.026

Table VI: Experimental and analytical modeling of elastic properties at 24°C (GPa)

Property	Without voids	With 2.7% voids		Experiment
		Lower bound	Upper bound	
E_{11}, E_{22}	269.2	227.0	268.9	272.6
E_{33}	214.0	204.1	214.7	72.5
G_{12}	87.4	83.7	87.4	93.8

Comparison to experimental data – SiC/SiNC and Oxide/Oxide

The effective elastic constants of the ceramic matrices were determined from the composite and fiber moduli by way of a back calculation in pcGINA. For the oxide-oxide composite, the calculation is straightforward since there is no interface phase. The Nextel 720 fiber was assumed to have a modulus of 260 GPa (and confirmed by testing to be 257.6 GPa). The tow size and spacing and the layer thickness were measured from optical micrographs. The pcGINA 8-harness satin weave model was created and the fiber volume fraction of 46.5% was confirmed. The elastic modulus and Poisson's ratio of the matrix were then guessed and the composite moduli calculated. The routine was iterated until a good match was found to the experimental data. The calculated values are shown in Table VII for a matrix Young's modulus of 7 GPa and a Poisson's ratio of 0.1, reasonable values for a high porosity oxide. The value shown in Table II was not used as the effect of porosity was not clearly seen in the measurement curve.

For the non-oxide composites, the interface phase(s) need to be considered. There is no way to account for the high shear modulus and low through thickness modulus otherwise. The simplest way of considering the interface is through its contribution to coated fiber properties. If the interface is cracked, porous, or otherwise compliant, the coated fiber would have considerably lower transverse and shear moduli than the uncoated fiber.

The SiC-SiNC composite was analyzed using this approach. An optimization algorithm was constructed to run pcGINA while allowing the fiber transverse and shear moduli and the matrix elastic modulus to vary. Optimization was carried out for E_{11} = 103.4 GPa, E_{33} = 34.5 GPa, and E_{45} = 103.4 GPa. It was found that these settings resulted in a higher than expected composite shear modulus, so G_{12} = 45.5 GPa was added to the optimization rules. These numbers were based on an earlier analysis of the experimental data. The matrix Poisson's ratio was set to 0.2 and the matrix shear modulus was calculated from G = E/2(1+ν). The fiber axial modulus was set to 210 GPa and Poisson's ratio to 0.2. The 8-HS weave geometry was set with tow spacing of 1.07 mm in both directions. From optical microscopy, the elliptical tows have axes of 0.79 mm and 0.16 mm. The layer thickness was set to 0.0125". This analysis did not factor in the up to 5% pore volume fraction in SiC/SiNC.

The optimal values of the input variables are Em = 159 GPa, Ef(trans) = 17.2 GPa, and Gf = 32.1 GPa. The calculated matrix modulus is close to the 127 ± 20 GPa measured by nano-indentation. The resulting composite properties are shown in Table V. These are in generally good agreement with experimentally determined values. The in plane Poisson's ratio of 0.07 is lower than the experimental values of 0.214 and the out of plane shear modulus of 29.3 GPa is higher than the accepted value of 7.6 GPa.

Table VII. Experimental and Calculated Elastic Properties for Oxide/Oxide and SiC/SiNC

Property	Oxide/Oxide		SiC/SiNC	
	Experiment	Calculated	Experiment	Calculated
E_{11}	70.4	73.4	107.0	107.6
G_{12}	13.1	10.2	48.3	44.8
E_{33}	28.3	25.7	55.9	46.4
All modulus values are in units of GPa				

CONCLUSIONS

The overall elastic modeling and experimental effort indicated that E_{33} for composites with a fiber interface coating pose the biggest challenge at present. Review of the experimental work has not raised testing issues and other experimentalists have confirmed these numbers by through transmitted ultrasonic means[8]. Additional work is planned to further refine the measurements taken. Specifically, pure matrix samples will be fabricated to further refine the matrix modulus.

ACKNOWLEDGMENTS

The Materials & Manufacturing Directorate, Air Force Research Laboratory under contract F33615-03-2-5200 and contract F33615-01-C-5234 sponsored portions of this work. Research work at ORNL was sponsored by the Assistant Secretary for Energy Efficiency and Renewable Energy, Office of FreedomCAR and Vehicle Technology Program, as part of the High Temperature Materials Laboratory User Program, Oak Ridge National Laboratory managed by UT-Battelle, LLC for the U.S. Department of Energy under contract number DE-AC05-00OR22725

REFERENCES
[1]Brewer, D., Ojard, G. and Gibler, M., "Ceramic Matrix Composite Combustor Liner Rig Test", ASME Turbo Expo 2000, Munich, Germany, May 8-11, 2000, ASME Paper 2000-GT-0670.
[2]Calomino, A., and Verrilli, M., "Ceramic Matrix Composite Vane Sub-element Fabrication", ASME Turbo Expo 2004, Vienna, Austria, June 14-17, 2004, ASME Paper 2004-53974.
[3]Bouillon, E.P., Ojard, G.C., Habarou, G., Spriet, P.C., Arnold T., Feindel, D.T., Logan, C., Rogers, K., and Stetson, D.P., "Engine Test and Post Engine Test Characterization Of Self Sealing Ceramic Matrix Composites For Nozzle Applications in Gas Turbine Engines", ASME Turbo Expo 2004, Vienna, Austria, June 14-17, 2004, ASME Paper 2004-53976.
[4]W.C. Oliver, G.M. Pharr J. Mater. Res., Vol. 7, No.6, June 1992
[5]Anthony Calomino, NASA-Glenn Research Center, Cleveland, OH, Personal Communication.
[6]Jurf, B., COI Ceramics, San Diego, CA, Personal Comminication.
[7]Murthy, P., Mital, S. and DiCarlo, J., "Characterizing the Properties of a Woven SiC/SiC Composite Using W-CEMCAN Computer Code", NASA/TM—1999-209173.
[8]John, R., Buchanan, D., Knapeke, D., and Rudell, M., Unpublished Research.

SHORT-FIBER REINFORCED CMCS: POTENTIALS AND PROBLEMS

Roland Weiss
Schunk Kohlenstofftechnik GmbH
Rodheimer Strasse 59
35452 Germany

Martin Henrich
Schunk Kohlenstofftechnik GmbH
Rodheimer Strasse 59
35452 Germany

ABSTRACT

Short fibre reinforced CMCs have a tremendous market potential in future applications which will be demonstrated as an example for brake discs in automotive industry. Market penetration in serial production is strongly depending on the final costs. Therefore, the usable technologies are cost driven and determine finally the properties of the products. CMC brake discs are lifetime parts which have to fulfill the safety requirements.

As known from literature (1), the CMCs with long and short fibre reinforcement show a high residual strength after 10^6 cycles on a level of 60 % to 80 % of the initial value. Therefore, mechanical quality assurance can be performed at ambient temperature.

One of the main problems of these materials are the inhomogeneities caused by the fibre distribution, the different fibre lengths and the fibre sticks, which result from the compounding technique itself. These sticks and their orientation have a tremendous effect on the mechanical properties and the failure behaviour of the short fibre reinforced CMCs. Furthermore, a precalculation of the mechanical behaviour via FEM is more or less impossible. This is mainly caused by the lack of testing standards, because standards available for endless fibre reinforced CMCs cannot be directly applied on the short fibre reinforced ones. The problems of mechanical testing will be discussed in detail for flexural tests as well as tensile tests of short fibre reinforced CMCs.

One of the main problems in testing short fibre reinforced CMCs is the selection of a representative sample cross-section in order to get mechanical properties independent of the size and geometry of the testing samples. The influence of the cross-sections of the testing samples on the mechanical properties will be shown and discussed in more detail.

Furthermore, the mechanical properties of samples with identical cross-sections will be discussed from the perspective of the applied standard testing method.

INTRODUCTION

The presentation will be limited on C/C-SiC materials as CMCs. The mechanical properties of C/C-SiC materials can be varied by the fibre reinforcement, the manufacturing routes as well as the silicon treatment technology for conversion of C/C to C/C-SiC. The market penetration of innovative materials and products depend on their cost effectiveness and/or their unique selling proposition. In particular, cost effectiveness is the limiting factor for applications in automotive industry.

The market demand is the driving force for the manufacturing technology.

POTENTIALS FOR SHORT CARBON FIBRE REINFORCED C/C-SIC

Typical markets of C/C-SiC materials are ballistic systems, industrial furnaces and brake materials for automotives, transportation, and industrial applications or trains. The most promising demand is an application for brake discs in high cost and luxury cars.

SGL have published a marketing study (2) on car production in Western Europe with a total of 15 Mio. of passenger cars. The market segment of luxury cars is 0.62 million/year, that means approximately 4 % of the total volume.

Although no exact numbers of manufactured CMC discs are published, the existing market is below 30 000 discs/year. This is approximately 1.3 % of the potential market.

The application of CMC discs by Porsche, Ferrari and Mercedes have already demonstrated and proven the superior properties of CMC discs in comparison to the metallic ones (see Figure 1).

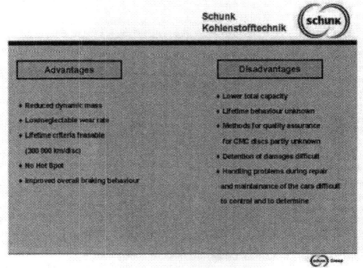

Figure 1: Advantages of CMC discs in comparison to cast iron ones

The automotive industry estimates the market penetration in the next 5-10 years between 5 % and 50 % of the luxury segment (4), resulting in a market demand of 120 000 to 1 200 000 discs per year.

Assuming an average weight of 5 kg/disc, the CMC volume is between 600 and 6000 to/year. The limitations for market penetration are the high costs of the CMC discs of 1000 to 1500 Euro and the uncertainty of acceptance by the final customer.

Based on the existing knowledge, the prices for CMC discs for luxury passenger vehicles will be in the area of 500 to 1000 Euro/disc. Nevertheless, due to the unique selling propositions of the CMC discs a remarkable increase in sales is estimated within the coming years. The development of mass production technologies opens further markets on low cost levels due to different requirements compared to CMC discs.

MANUFACTURING ROUTES
Figure 2 shows a manufacturing scheme of carbon/carbon composites.

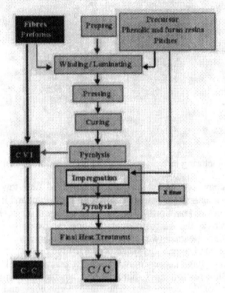

Fig. 2: Manufacturing scheme of carbon/carbon composites

Post treatments with silicon result in C/C-SiC composites. The degree of conversion and the amount of remaining free silicon can be tailored by the method and the parameters of the silicon post treatment. The post treatment is required in order to obtain the best tribological properties combined with a minimum or negligable wear rate of the discs.

The costs of CMCs with carbon fibres are controlled from the material side by the costs of fibre preforms, the forming process and the required densification steps.

Best mechanical properties are achieved with endless C-fibre placement. The prepreg route, however, is also combined with the highest costs.

Low cost qualities as requierd for mass series applications can only be realized by using short fibre reinforcements and manufacturing technologies as known from carbon and glass fibre reinforced polymers (see Figure 3).

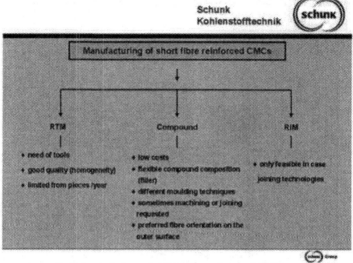

Fig. 3: Low cost manufacturing techniques for short fibre reinforced CMCs

RTM as well as the compound technology use the so-called fibre tow sticks as reinforcement. The sticks can vary in length (from milled materials up to 25 mm) and in the number of filaments (3K, 6K, 12 K, ... to heavy tow fibres). The size and shape of the sticks and their composition influence the morphology, the microstructure, the mechanical properties and the degree and homogeneity of the silicon treatment and thereby the final material and tribological properties.

In case of the compound technology (state of the art) the procedure has to preserve the shape of sticks during compounding in order to avoid a strong silicon attack on the individual fibres. A direct conversion of the reinforcing C-fibres by SiC formation is always combined with an extreme embrittlement of the composites. The final morphology and microstructure of CMC brake discs vary among the manufacturers. SGL is milling the compound before moulding and modifies the composition by mixing different stick lengths after separation. The CMC discs possess a finer microstructure than in case of discs manufactured by Schunk, which are made from compounds with longer sticks. These sticks and their orientation have a tremendous effect on the mechanical properties and the failure behaviour of the short fibre reinforced CMCs. Furthermore, a precalculation of the mechanical behaviour via FEM is more or less impossible. This is mainly caused by the lack of testing standards, because standards available for endless firbre reinforced CMCs cannot be directly applied on the short fibre reinforced ones.

One of the main problems in testing short fibre reinforced CMCs is the selection of a representative sample cross-section in order to get mechanical properties independent from the size and geometry of the testing samples.

TESTING PROCEDURE
The short fibre reinforced CMC was tested under flexural and tensile load according to EN 658 for long-fibre reinforced composites.

The flexural tests were performed as 3 point bending tests with a span to depth ratio of 20:1. The representative volume elements varied from 1800 mm³ up to 60 000mm³ with cross sections from 30 mm² up to 300 mm².

The cross section of the tensile samples was from 30 mm² (10 mm x 3 mm) up to 200 mm² (20 mm x 10 mm).

The representative volume elements were in between 3000 and 20 000 mm³.

RESULTS
Table I shows the mechanical properties of the flexural tests as a function of the increasing cross section and representative volume element.

Table I: Flexural bending properties of short fibre reinforced CMCs as a function of the representative volume element

Cross section mm²	30	60	60	90	100	120	180	200	300
	(10x3)	(20x3)	(10x6)	(30x3)	(10x10)	(20x6)	(30x6)	(20x10)	(30x10)
Volume Element (mm³)	1800	3600	7200	5400	20000	14400	21600	40000	60000
Flexural Strength (MPa)	51	51	46	49	44	53	45	47	56
Structural Deviation	15	13	11	10	8	10	5	4	6
Strain [%]	0.30	0.35	0.22	0.33	0.17	0.22	0.25	0.17	0.20
Flexural Modulus [GPa]	22	20	28	17	32	30	24	34	34
Standard Deviation	5	3	5	2	2	5	4	2	2

The flexural strength did not show a clear tendency as a function of the selected volume element of cross section, respectively. However, the higher the representative volume element, the lower the resulting standard deviation.

However, an influence of the sample size could be observed for the strain volumes and the Young's Modulus as a function of the sample thickness as demonstrated in Figures 4 and 5.

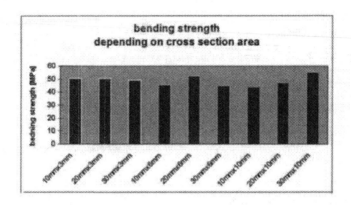

Figure 4: Bending strain data in dependence of the sample thickness

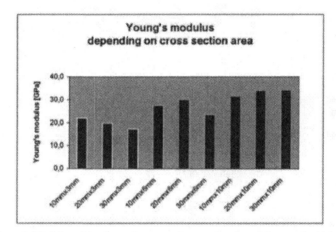

Figure 5: Young´s Modulus of the flexural test samples as a function of the sample thickness

Increasing sample thickness is combined with a decrease in strain and increase in Young's modulus.

The microstructure of the short fibre reinforced CMC materials is responsible as can be seen from Figure 6.

Fig. 6: Crack initiation by perpendicular orientation of fibre sticks in respect to the load direction

Such crack initiation shrinks can be observed in SEM due to the random orientation of the sticks. Their influence should increase with decreasing cross section.

Whereas in case of flexural tests no clear influence of the cross section area (representative volume element) could be found, the tensile properties decrease with decreasing representative volume elements. Crack initiation by perpendicular orientated fibre sticks becomes more dominant (compare Figures 7 and 8).

CONCLUSION

It has to be concluded, that in case of compound based materials with a microstructure dominated by sticks with total lengths of some millimeters, the flexural as well as the tensile properties are influenced by the selected volume element (sample size). The values are depending on failure initiation mechanisms caused by perpendicular orientations of the fibre sticks. This effect dominates the tensile behaviour and is also responsible for increased standard deviations of flexural properties in test samples with lower volume elements.

Standard test methods as defined for endless reinforced CMC materials cannot be directly applied to obtain realistic size independent material properties. Additional effects are required to develop suitable test methods.

357

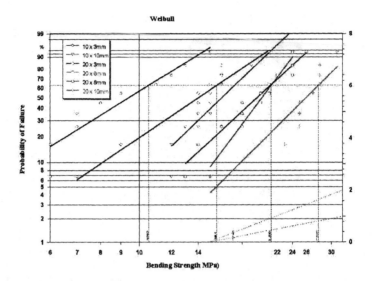

Figure 7: Tensile behaviour and failure probability of short fibre reinforced CMC samples with different sample sizes.

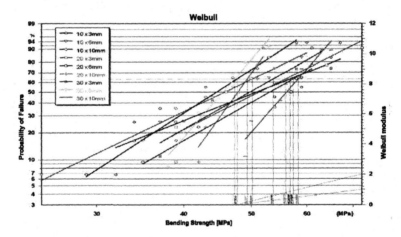

Figure 8: Flexural behaviour and failure probability of short fibre reinforced CMC as a function of the sample size

REFERENCES

[1]A. Neubrand, B. Thielicke, and A. Kienzle, "Four Point Bending Tests of a C/SiC Material for Industrial Applications under Quasistactic and Cyclic Loading," *Proceedings of the 5th International Conference on High Temperature Ceramic Matrix Composites,* 205-210 (2004).

[2]SGL Brakes GmbH, Presentation at the University of Darmstadt, "Carbon-Keramik-Bremsscheibe,"presented on 15 January 2003.

[3]Private information.

Laminated Ceramics and
Particulate-Reinforced Ceramics

CERAMIC LAMINATES WITH HIGH MECHANICAL RELIABILITY BY DESIGN

Vincenzo M. Sglavo
DIMTI, University of Trento
Via Mesiano, 77
38050 Trento (Italy)

Massimo Bertoldi
Eurocoating Spa
Via Al Dos de la Roda, 60
38057 Pergine Valsugana (Italy)

ABSTRACT

A procedure for designing innovative ceramic laminates with high mechanical performances is proposed in this work. A fracture mechanics approach has been considered to define the stacking sequence, thickness and composition of the different laminae on the basis of the requested strength and of the defect size distribution. Once the different laminae are stacked together a residual stress profile is generated upon cooling after sintering because of the differential thermal expansion coefficient. Such residual stress profile is conceived in order to allow stable growth of surface defects upon bending and guarantee limited strength scatter. As an example, the proposed approach is then used to design and produce a ceramic laminate in the alumina-mullite system.

INTRODUCTION

Many efforts have been made in the past to increase the mechanical reliability of glasses and ceramics. Higher fracture toughnesses have been achieved through the exploitation of the reinforcing action of grain anisotropy or second phases, or the promotion of crack shielding effects associated to phase-transformation or micro-cracking.[1] As an alternative, fracture behavior of ceramics has been improved by introducing low-energy paths for growing crack in laminated structures,[2-6] or by introducing compressive residual stresses.[7,8] Laminates presenting threshold strength have been also successfully produced by alternating thin compressive layers and thicker tensile layers.[9] Unfortunately, the most important limitations of such laminates is that they can be used only with specific orientations with respect to the applied load and, for example, they are not easily suitable to produce plates, shells or tubes as usually required in typical applications.

The idea that surface stresses can hinder the growth of surface cracks has been extensively exploited in the past especially on glasses.[10-11] Sglavo and Green have recently proposed that the creation of a residual stress profile in glass with a maximum compression at a certain depth from the surface can arrest surface cracks and result in higher failure stress and limited strength variability.[12-14]

Residual stresses in ceramic materials can arise either from differences in the thermal expansion coefficient of the constituting grains or phases, uneven sintering rates or martensitic phase transformations. As described below, if the development of the residual stresses in ceramic multilayer is opportunely controlled, materials characterized by high fracture resistance and limited strength scatter can be designed and produced.

THEORY

The aim of the present work is to set up a design procedure useful to produce ceramic components with high mechanical reliability, *i.e.* characterized by limited strength scatter and, possibly, high fracture resistance. In order to reach such a target the idea is advanced to promote the stable growth of defects before final failure. In this way, regardless the initial flaw size, an invariant final strength could be obtained. For example, stable crack propagation is possible when fracture toughness, T, is a growing function of crack length, c, steeper than the applied stress intensity factor generally defined as $K_{ext} = \psi \sigma c^{0.5}$, where ψ is the shape factor and σ the applied stress. It has been demonstrated elsewhere that the stability range, if any, is finite.[12] This is shown schematically in Fig.1 where the interval $[c_A, c_B]$ represents the range where cracks can grow in a stable fashion under the effect of an external load. As a direct consequence, all the defects included in such an interval propagate upon loading to the same maximum value before final failure, thus leading to a unique strength value.

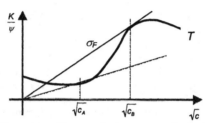

Figure 1. *T*-curve that allows the stable growth phenomenon in the interval (c_A, c_B). Straight lines are used to evaluate the stable growth interval and final strength, σ_F.

The presence of residual stresses within the material can be responsible for a *T*-curve like in Fig. 1. If the simple model shown in Fig. 2 is considered, corresponding to a surface crack in a laminate ceramic, residual stresses are associated to the stress intensity factor:[15]

$$K_{res} = 2\left(\frac{c}{\pi}\right)^{0.5} \int_0^c Y(x/c)\sigma_{res}(x)dx. \tag{1}$$

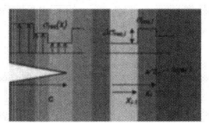

Figure 2. Crack model considered in the present work.

In the present analysis, discontinuous stepwise stress profiles are taken into account according to the multilayered laminate structures subjected to bending loads of relevance here (Fig. 2). Perfect adhesion between different laminae is also assumed.

Under the influence of the external load (K_{ext}) crack propagation occurs when the sum ($K_{res} + K_{ext}$) equals the fracture toughness, K_C, of the material. If the residual stresses are virtually considered as a material property, the "apparent" fracture toughness can be defined as:

$$T = K_C - K_{res}.$$ (2)

It is clear from Eqs (1) and (2) that for compressive residual stresses (negative) there is a beneficial effect on T. In addition, given a proper residual stress profile, it should be possible to obtain T being a steep growing function of c. Moreover, as surface flaws have been considered, the T-curve is characteristic of the laminate structure and independent of any specific defect and therefore can be considered as fixed with respect to the surface position. Therefore, crack length (c) and depth from the surface (x) will be regarded as identical quantities in the following analysis.

In order to understand the effect of residual stress intensity and location on T it is useful to analyze some special cases. For the simple situation corresponding to a simple step profile (Fig. 3(a)), T can be analytically calculated. As shown in Fig. 3(b) a stability range exists between x_1 and the tangent point. One can observe that for larger x_1 the strength decreases and the stability interval width increases. Since both high strength and large stable growth interval are desired, an intermediate value of x_1 has to be considered in the perspective laminate design. On the other side, an increase of σ_R is useful to increase both the stable growth range and the maximum stress. In addition, if K_C increases, the maximum stress is higher but the stability range decreases though one must consider that K_C is a parameter that depends on the material selection and it is not usually modified in the design approach.

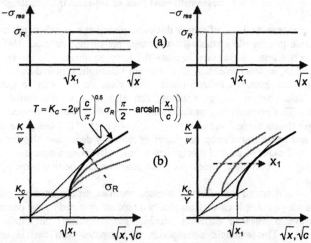

Figure 3. Step residual stress profile (a) and corresponding apparent fracture toughness (b). The effects of intensity (left) and location (right) of the residual stress are shown.

The natural development of the step profile is the square-wave profile (Fig. 4). In this case the T-curve can be calculated both analytically and by using the principle of superposition.[1,15] The square-wave profile can be considered in fact as the sum of two simple step profiles with stresses of identical amplitude but opposite sign placed at different depths (x_1 and x_2). This special case is useful to discuss an important point. Depending on the width (x_2-x_1) of the compressive layer, the tangent point can fall beyond the position x_2. In this case the stable growth range is automatically defined by the interval [x_1, x_2] and the maximum stress is lower than the tangent stress. Strength and instability point become therefore mutually independent within a certain degree.

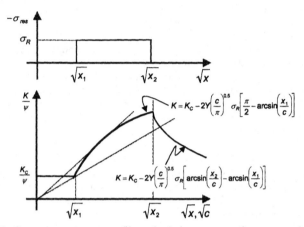

Figure 4. Square-wave stress profile and relative apparent fracture toughness.

A single compressive layer of proper thickness placed at a certain depth from the surface is therefore suitable to generate a stable growth range for surface defects. Unfortunately, this simple solution is not actually practical because the force equilibrium in the component is not satisfied. In addition, in order to obtain high strength values the required compressive stress is usually very high and intense localized interlaminar shear stresses can be generated and be responsible for delamination. Edge cracking can also arise at the boundaries of highly compressed layer. This phenomenon was analyzed in a previous work[16] to occur when the layer thickness is larger than a critical value, $t_c = K_C^2 / [0.34\ (1+\nu)\ \sigma^2]$, ν being the Poisson's ratio. Layer thickness and compressive stress are therefore mutually dependent also for this reason and it is not possible to design the desired mechanical behavior by using a square-wave stress (single layer) profile, only. Opportunely, almost all these problems can be overcome by considering a multilayered structure.

In order to understand the effect of a more complex profile on the T-curve trend it is useful to analyze another simple case. Consider two profiles obtained by the combination of two simple square-wave profiles of different (double, for simplicity) amplitude and identical extension (Fig. 5). This example corresponds to laminates with two layers of different composition and same thickness. The actual order of the two layers is the only difference between the two examined profiles.

It is clear from Fig. 5 that the order of the compressive layers is important for both the final strength and the stability interval. Such a consideration is general and a conclusion can be drawn that one way to obtain a properly designed T-curve is to have continuously growing compressive intensity in successive layers.

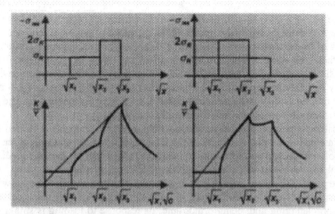

Figure 5. Residual stress and corresponding T-curve for two simple square-wave profiles placed in different order.

At this point, the principle of superposition can be used to calculate the T-curve for a general multi-step profile. Such approach can be extended in fact to n layers provided that n step profiles with amplitude $\Delta\sigma_j$ (Fig. 2) equal to the stress increase of layer j with respect to the previous one are considered. A general equation, which defines the apparent fracture toughness for layer i in the interval $[x_{i-1}, x_i]$ (Fig. 2), is obtained:

$$T = K_c{}^i - \sum_{j=1}^{i} \left[2Y \left(\frac{c}{\pi}\right)^{0.5} \Delta\sigma_j \left[\frac{\pi}{2} - \arcsin\left(\frac{x_{j-1}}{c}\right)\right]\right] \quad x_{i-1} < x < x_i \tag{3}$$

where i indicates the layer rank and x_j is the starting depth of layer j. Equation (3) represents a short notation of n different equations, the sum being calculated for different number of terms for each i. This represents a mathematical translation of the "memory" effect of stress history that deeper layers maintain with respect to the layer previously encountered by the propagating crack. Regardless of the layer order, since $2n$ parameters (x_i, $\Delta\sigma_i$) are now available and two conditions have to be satisfied (forces equilibrium and equivalence between the sum of single layer thickness and the total laminate thickness), $2n-2$ are the remaining degrees of freedom suitable to define the desired T-curve.

It is important to point out that in the calculations carried out to obtain Eq. (3) the approximation is made that the elastic modulus of the different layers is constant. Nevertheless, it has been demonstrated elsewhere that the approximation in T estimate does not exceeds 10% if the Young's modulus variation is less than 33%.[17,18]

LAMINATE DESIGN

Equation (3) suggests some considerations about the conditions that a proper stress profiles should possess to promote stable growth. The stable propagation of surface defects is possible only when the T-curve is a monotonic increasing function of c and this requires a continuous increase of compressive stresses from the surface to internal layers. A stress-free or slightly tensile stressed layer is also preferred on the surface since this allows to decrease the lower boundary of the stable growth interval. It is important to point out that, according to Eq. (3), the effect of the surface layer is transferred to all the internal layers. The surface tensile layer has in fact a reducing effect on the T-curve for any crack length and for this reason its depth and tensile stress intensity must be limited, the maximum stress being, otherwise, too low. In addition, by using multi-step profiles it is possible to reduce the thickness of the most stressed layer with the introduction of intermediate stressed layers before and beyond it. The risk of edge cracking and delamination phenomena are reduced accordingly.

The residual stress profile that develops within a ceramic laminate is related either to the composition/microstructure and thickness of the laminae and to their stacking order. According to the theory of composite plies,[19] in order to maintain flatness during in-plane loading, as in the case of biaxial residual stresses developed upon processing, a laminate structure has to satisfy some symmetry conditions. If each layer is isotropic, like in ceramic laminae with fine and randomly oriented crystalline microstructure, and the stacking order is symmetrical, the laminate remains flat upon sintering and, being orthotropic, its response to loading is similar to that of a homogeneous plate.[19]

Regardless of the physical source of residual stresses, their presence in a co-sintered multilayer is related to constraining effect. Under the condition that the different layers perfectly adhere each other, every lamina must deform similarly and at the same rate of the others. The difference between free deformation or free deformation rate of the single lamina with respect to the average value of the whole laminate accounts for the creation of residual stresses. Such stresses can be either viscous or elastic in nature and can be relaxed or maintained within the material depending on temperature, cooling rate and material properties. With the exception of the edges, if thickness is much smaller than the other dimensions, each layer can be considered to be in a biaxial stress state.

The fundamental task to properly design a symmetric multilayer is the estimate of the biaxial residual stresses. In the common case where stresses are developed from differences in thermal expansion coefficients only, the following conditions must be fulfilled:

- stress equilibrium: $\quad \sum_{i=1}^{n} \sigma_i t_i = 0$ (4a)

- compatibility: $\quad \varepsilon_i = e_i + \alpha_i \Delta T = \varepsilon$ (4b)

- constitutive model: $\quad \sigma_i = E_i^* e_i$ (4c)

where α_i is the thermal expansion coefficient, $E_i^* = E_i /(1 - v_i)$ (v_i = Poisson's ratio, E_i = Young modulus), e_i the elastic strain, ε_i the deformation. Equations (4) represent a set of $3n+1$ equations and $3n+1$ unknowns (σ_i, ε_i, e_i, ε). The solution of such linear system allows to calculate the residual stress in the generic layer i (among n layers) as:

$$\sigma_i = E_i^* \left(\overline{\alpha} - \alpha_i\right) \Delta T \qquad (5)$$

where $\Delta T = T_{SF} - T_{RT}$ (T_{SF} = stress free temperature, T_{RT} = room temperature) and $\overline{\alpha}$ is the average thermal expansion coefficient of the whole laminate, defined as:

$$\overline{\alpha} = \sum_1^n E_i^* t_i \alpha_i \Big/ \sum_1^n E_i^* t_i \qquad (6)$$

t_i being the layer thickness. In this specific case the residual stresses are therefore generated upon cooling after sintering. It has been shown in previous works that T_{SF} represents the temperature below which the material can be considered to behave as a perfectly elastic body and visco-elastic relaxation phenomena do not occur.[20]

APPLICATION OF THE DESIGN PROCEDURE

Equation (3) represents a fundamental tool for the design of a ceramic laminate with pre-defined mechanical properties. Different ceramic layers can be stacked together in order to develop after sintering a specific residual stress profile that can be evaluated by Eq. (5) if the elastic constants, the thermal expansion coefficient and the thickness of each layer are known. Since the stress level in Eq. (5) does not depend on stacking order, the sequence of laminae can be still changed provided the symmetry condition is maintained to tailor the T-curve and promote stable growth of defects. Once the stress profile is defined, the apparent fracture toughness can be estimated by Eq. (3) and strength and fracture behavior are directly defined. By changing the stacking order and composition of the layers, it is therefore possible to produce a material with unique and predefined failure stress.

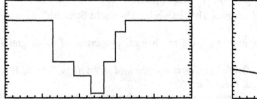 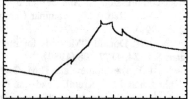

Figure 6. Designed residual stress profile (a) and corresponding T-curve (b) of the engineered laminate.

As an example, a ceramic laminate composed of different layers belonging to the alumina/mullite system has been designed and produced. Starting from the surface the layer sequence (composition/thickness) in the laminate is: AM0/41, AM20/44, AM30/48, AM40/43, AM20/44, AM10/42, AM0/540; the sequence is then repeated in the reverse order. In the used notation, "AMz/y", z corresponds to the volume percent content of mullite and y the thickness in microns. The composition and thickness of the layers and their stacking order was selected to produce a ceramic laminate with a "constant" strength of $\approx$ 400 MPa as shown in Fig. 6(b). The apparent fracture toughness curve and corresponding residual stress profile were correspondingly tailored in order to promote the stable growth of surface defects as deep as $\approx$ 180 μm. The stress distribution was calculated by Eq. (5) from geometrical and thermo-elastic properties of the materials[21] and the corresponding T-curve estimated according to Eq. (3). Since T-curve was calculated step by step, the corresponding diagram is discontinuous at the boundary between layers, this reflecting the discontinuities in the σ_{res} diagram.

The material was produced by stacking and sintering together laminae obtained by tape casting.[21] The engineered composite was characterized by a strength of 456 ± 32 MPa, even when large indentation flaws were produced on its surface[21]. This illustrates the power of this approach and the potential of ceramic laminates with high inherent mechanical reliability by design.

REFERENCES

[1] B.R. Lawn, Fracture of brittle solids, Cambridge Univ., Cambridge, UK (1993).

[2] J.B. Davis et al., "Fabrication and crack deflection in ceramic laminates with porous interlayer", *J. Am. Ceram. Soc.*, **83** [10] 2369-74 (2000).

[3] W.J. Clegg et al., "A simple way to make tough ceramics", *Nature*, **347**, 455-57 (1990).

[4] W.M. Kriven et al., "High-strength, flaw-tolerant, oxide ceramic composite", U.S. Pat. No. 5,948,516, Sept. 7, 1999.

[5] R.E. Mistler et al., "Strengthening alumina substrates by incorporating grain growth inhibitor in surface and promoter in interior", U.S. Pat. No. 3,652,378, March 28, 1972.

[6] M.P. Harmer et al., "Unique opportunities for microstructural engineering with duplex and laminar ceramic composites", J. Am. Ceram. Soc.. **75** [7] 1715-28 (1992).

[7] C.J. Russo et al., "Design of laminated ceramic composite for improved strength and toughness", *J. Am. Ceram. Soc.*, **75** [12] 3396-400 (1992).

[8] R. Latkshminarayanan et al., "Toughening of layered ceramic composites with residual surface compression", *J. Am. Ceram. Soc.*, **79** [1] 79-87 (1996).

[9] M.P. Rao et al., "Laminar Ceramics That Exhibit a Threshold Strength", *Science*, **286**, 102-5 (1999).

[10] I.W. Donald, "Methods for improving the mechanical properties of oxide glasses", *J. Mater. Sci.*, **24**, 4177-208 (1989).

[11] R.F. Bartolomew and H.M. Garfinkel, *Glass science and technology*, Vol. 5, Ed. by D. R. Uhlmann and N.J. Kreidl, Academic Press, New York (1980).

[12] V.M. Sglavo et al., "Flaw-insensitive ion-exchanged glass: I, theoretical aspects", *J. Am. Ceram. Soc.*, **84** [8] 1827-31 (2001).

[13] V.M. Sglavo and D.J. Green, "Flaw-insensitive ion-exchanged glass: II, production and mechanical performance", *J. Am. Ceram. Soc.*, **84** [8] 1832-38 (2001).

[14] D.J. Green et al., "Crack arrest and multiple cracking in glass using designed residual stress profiles", *Science*, **283**, 1295-97 (1999).

[15] D.J. Green, *An introduction to the mechanical properties of ceramics*, Cambridge Univ. Press, Cambridge, UK (1998).

[16] S. Ho et al., "Surface cracking in layers under biaxial, residual compressive stress", *J. Am. Ceram. Soc.*, **78** [9] 2353-59 (1995).

[17] R.J. Moon et al., "R-curve behaviour in alumina/zirconia composites with repeating graded layers", *Eng. Fract. Mech.*, **69**, 1647-65 (2002).

[18] T.J. Chung et al., "Fracture toughness and R-curve behaviour of Al_2O_3/Al FGMs", *Ceramic Transaction*, Vol. 114, 789-96 (2001).

[19] J.C. Halpin, *Primer on composite materials analysis*, Technomic Publ., USA (1992).

[20] M.P. Rao et al., "Residual stress induced R-curves in laminar ceramics that exhibit a threshold strength", *J. Am. Ceram. Soc.*, **84** [11] 2722-24 (2001).

[21] V.M. Sglavo et al., "Tailored residual stresses in high reliability alumina-mullite ceramic laminates", *J. Am. Ceram. Soc.*, to be published.

METHODS OF RESIDUAL STRESS ANALYSIS IN LAYER COMPOSITES AND THEIR APPLICATION

Martin Wenzelburger, Maria José Riegert-Escribano and Rainer Gadow

Institute for Manufacturing Technologies of Ceramic Components and Composites
University of Stuttgart
Allmandring 7 b
Stuttgart (Germany), D-70569

ABSTRACT

Residual stresses occur in most coating deposition and layer composite manufacturing processes. Main reasons are temperature gradients during the manufacturing and in the cooling period as well as different thermophysical properties of the component materials, but also solidification and phase transformation processes. In case of metal-ceramic or glass-ceramic composites, the mismatch in the thermophysical properties is of high importance. Residual stresses even can lead to crack formation and failure of the materials and components. Therefore, residual stress analysis is an important tool and essential for the optimization of layer composites and ceramic manufacturing processes in order to ensure stability and reliability of technical systems.

This paper gives an overview and introduction to the most customary residual stress measurement techniques. The application ranges of the different methods are given with respect to the underlying measurement principle and the physical boundaries. Their advantages and disadvantages are compared with respect to the application on industrial machine parts. The incremental-step hole drilling and milling method is presented in more detail. Thereby, the measurement principle, analysis of the measured data and different possibilities for the determination of calibration curves are explained. Additionally, an application example for the hole drilling method is given from the field of process optimization in ceramic coating development with the atmospheric plasma spraying (APS) technique.

INTRODUCTION

Residual stresses arise in every material and machine part that is exposed to variable temperatures, dynamic heat transfer, phase transformation or mechanical loading in the manufacturing process. Additionally, during the whole operation life-time mechanical loading, recrystallization, creep, damage, etc. affect the state of residual stresses in metallic, ceramic, cermet and composite components. Residual stresses have a significant influence on material properties and component behavior. Particularly the interaction of residual stresses and operational load stresses can lead to a reduction of strength and to increased fatigue. On the other hand, it also can lead to load reduction and an increase in lifetime. Therefore, the knowledge about residual stresses in components and their formation during manufacturing is of great importance for planning and optimization of component design and manufacturing processes.

There are various experimental methods for residual stress analysis in materials which are based on different physical principles (mechanical, diffraction, acoustic methods and others), and therefore, their measurement ranges reach from microscopic stresses on the atomic scale to macroscopic stresses which cover a multiple of the grain size or even the whole component (global stresses). They are either destructive, quasi-destructive or non-destructive and of different quantitative, partly quantitative or qualitative character [1]. The characteristical properties of residual stress measurement systems have to be considered both for scientific use and also for their practical implementation in industrial process development or in quality control systems. Several methods are already used in industrial applications, while others only exist on experimental scale up to date. Therefore, a short summary of the most common methods, giving their advantages and disadvantages with respect to the usability for engineering process development is given below, highlighting the hole drilling and milling method. For a more detailed description, please refer to Scholtes [1], Macherauch and Hauk [2], Peiter [3] or others.

EXPERIMENTAL RESIDUAL STRESS ANALYSIS

The mechanical methods for residual stress measurement are based on determination of macroscopic deformations as a result of destructive interference of the stress equilibrium in a material by material removal. The deformations are generally measured on the surface by the use of strain gages, and the measured strains are used for stress calculation. The mechanical methods include dissection, layer removal, and slitting for components of simple geometry as well as cutting, ringcore drilling and hole drilling for more complex geometries. All methods are destructive with different degrees of destruction. The hole drilling technique is partly destructive or quasi non-destructive if the bore hole can be closed after measurement. Strip curvature is a real non-destructive method that uses the macroscopic deformation (bending) of geometrically simple specimens. Therefore, it can be used for process development and parameter optimization, but generally not for stress analysis on components. The mechanical methods are suitable for the determination of stress depth profiles in the range of some mm from the surface, see Fig. 1. But, there also can be problems with strain gage adhesion and limited measurement accuracy in case of plastic material deformation.

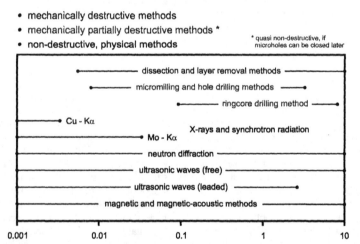

Fig. 1: Depth range of different residual stress analysis techniques [4]

The diffraction methods, X-ray, synchrotron and neutron diffraction, are based on the measurement of crystal lattice deformations of the material, which are caused by microscopic and macroscopic stresses. These deformations lead to a change of the material's Bragg angle, which can be measured directly by diffraction experiments. The lattice deformations can be used in order to quantitatively calculate stresses in the material. X-ray diffraction is the most commonly used technique, but it is limited to a depth range of approx. 20-30 μm, while neutron diffraction has a higher penetration depth, comp. Fig. 1, but a lower spatial resolution. By using pulsed, linearly polarized synchrotron X-rays with a high intensity and low divergency, the spatial resolution and penetration depth of X-ray diffraction can be increased. The diffraction methods are non-destructive, but, in order to obtain stress depth information, they have to be combined with layer removal methods. The sample size is also limited, and therefore, it is difficult to analyze residual stresses in real machine parts. Only crystalline materials with well-known elastic constants can be analyzed, and only grains of the same orientation are included in the measurement result. Microscopic and macroscopic stresses have different influences on the result, depending on the interpretation method, but with some effort, they can be separated, which leads to a high accuracy of results.

Acoustic residual stress measurement by the use of ultrasonic waves is a non-destructive method and is also based on the detection of crystal lattice deformations, which affect the elastic properties and the acoustic velocity of materials. By the use of multiple wave sources that are applied in different angles, residual stresses at the point of intersection can be analyzed with good spatial resolution, while the minimum size of the analyzed volume as well as the maximum depth penetration are limited, see Fig. 1. Furthermore, it is difficult to separate the influences of microscopic stresses and macroscopic stresses on the acoustic velocity, so that only an overall value is detected. Texturization and phase gradients in the component as well as interfaces also have an influence on the measurement results. Therefore, the ultrasonic wave method is only applicable to homogeneous materials to date, but nevertheless, this technique is of increasing importance in technical use. It offers the advantage of easy, time efficient handling, and it can be applied on every component geometry, also in the form of field tests with small equipment [5].

A group of non-destructive, magnetic and magneto-acoustic techniques operates with the measurement of stress effects on the magnetic hysteresis in ferromagnetic materials. The Barkhausen noise method is the most popular of them and is already in industrial use. Stress effects on the magnetic properties of materials are very complex and also difficult to separate from other effects. Therefore, calibration measurements are necessary in order to get utilizable, qualitative results, while whole series of calibration tests have to be performed to achieve quantitative measurement accuracy. Barkhausen noise is a time efficient method with a great potential for comparative measurements in serial production quality control systems.

Optical stress measurement with birefringent coatings is of great importance in industry for the analysis of operational stresses. However, it is not a practicable method in the case of residual stress analysis. Other optical methods which are based on the observation of fine, geometrically defined surface patterns are also mainly used for operational stress analysis.

THE INCREMENTAL-STEP HOLE DRILLING AND MILLING METHOD

The mechanical methods for residual stress analysis offer some advantages, like good practicability even for big components, good depth resolution, good quantitative accuracy and the analysis of macroscopic (I. order) stresses, while most other techniques yield integral results regarding the spatial resolution as well as the micro-/macroscopic stress character.

The incremental-step high-speed hole drilling and circular milling technique has further advantages that make it a suitable method for the analysis of residual stress depth profiles with a high accuracy in depth resolution and measured stress values, and therefore, for residual stress measurement during process development and optimization, including coating composite development. The measuring setup is comparably small, and it can be applied on a variety of complex machine parts for on-site services as well as for sampling tests. The measurement of surface strains by the use of strain gages is a well-known technique with a high accuracy of results. Therefore, even without knowing the elastic properties of materials, the comparison of measured surface strains can yield valuable information about the residual stresses in components. The stepwise material removal in increments of 5-20 μm leads to a good depth resolution. The combination of high speed drilling and circular milling, see Fig. 2, reduces the mechanical loading of the material at the bottom of the bore hole as well as the thermal input into the material. Hence, the influence of stress manipulation during measurement is reduced. Additionally, this material removal technique leads to a well-defined geometrical shape of the bore hole, which results in a higher accuracy of the calculated stresses. The difficulties in the experimental implementation of the hole drilling method are mainly due to preparation of the surface by grinding without influencing the state of residual stresses and the application (adhesive bonding) of the strain gages to the surface.

There already exists a standardized procedure ASTM E837 for the implementation of hole drilling measurements, which, however, only describes the standard center hole drilling technique without additional milling step. In order to achieve an optimized reliability of the hole drilling technique, Grant, Lord et al. [6] currently evaluate this method in comparison with other mechanical and non-mechanical techniques.

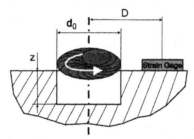

Fig. 2: Schematic illustration of the incremental hole drilling and milling method [7]

Interpretation of measurement data (residual stress calculation)

Results of hole drilling measurements consist of a table of surface micro-strain values for the three strain gages of a strain gage rosette (0°, 45°, 90° strain gages: $\varepsilon_{0/45/90}$) and for the single depth increments. These surface strains have to be correlated with the corresponding residual stress release on the bottom of the hole in order to receive residual stress depth profiles. Therefore, as a first step, apparent strains on the bottom of the hole (which are called nominal strains ε_n) are determined for every drilling step, depending on the measured surface strains and transfer coefficients (polynomial functions that are also called calibration curves $\varepsilon_{cal.}$, or K). The calibration curves have to be determined separately, as described below. With these nominal strains and according to Hooke's law, residual stresses in the directions of the three strain gages can be calculated for every drilling step. And finally, by means of the relations of Mohr's circle, the main residual stresses $\sigma_{1,2}$ as well as the angle of the main stress axes α can be determined. The main stresses are used for calculation of the residual stresses σ_x, σ_y and τ_{xy} referring to the specimen axes.

There is a number of analysis methods for residual stress calculation from the measurement results of hole drilling tests. One of the most commonly used analysis techniques, which is also applied here, was described by Schwarz [8] and is based on the incremental strain method that was introduced by Soete and Vancombrugge [9] and further developed by Kelsey [10] and König [11]. The incremental strain method applies the surface strain gradients of two consecutive measurement steps for the calculation of the respective nominal strains, without considering the absolute strain values. This procedure can lead to an increase of measurement tolerances during residual stress calculation. However, a smoothening of the measured strain depth curves (e. g. by means of cubic spline functions) can reduce this inaccuracy. On the other hand, the incremental strain method is an easy to operate technique which has been optimized in order to compensate plastic deformation effects and anisotropy of the material properties in the residual stress calculation procedure [11]. Finally, Schwarz applied two different calibration curves, taking into account the material deformation in longitudinal (K_1) and transverse (K_2) stress direction.

Determination of calibration curves

Calibration curves are determined either experimentally by hole drilling measurements during tensile or bending tests, or by finite element calculations. They depend on the drilling depth z, the diameter of the bore hole d_0, and the distance of the bore hole edge to the strain gages (or rather, to the diameter of the strain gage rosette D). In order to reduce parameters, calibration curves are normally described as a function of the normalized drilling depth $\xi = z/d_0$. This means, that a set of calibration curves for different ratios D/d_0 can be applied for every measurement, hole diameter and strain gage, provided that the material properties are homogeneous. For layer composites, calibration curves additionally depend on the ratio of the Young's moduli of coating and substrate material E_C/E_S, and on the thickness of the coating. This would lead to an enormous extent of measurements in order to achieve a representing set of calibration curves experimentally, and therefore, numerical calculation of calibration curves by finite element simulation of hole drilling experiments has to be preferred [12].

Besides the easy and time efficient variation of geometrical boundary conditions and material properties, modeling and simulation of hole drilling measurements offers further advantages. Calibration curves should be determined under ideally homogeneous stress conditions and independently of component geometry. It is difficult to generate a homogeneous stress distribution in the sample material in experimental tests, while it is simple to define homogenous stress fields and quasi-infinite extension of the sample in a finite element model. The information about the strain field in the whole component during hole drilling and material removal can also yield additional knowledge about the begin and the regions of plastic deformation in metals. Furthermore, the possibility of varying material properties and customizing the model to a specific component or composite design means an increase in accuracy of calibration curves for specific applications.

EXPERIMENTAL PROCEDURE

In order to demonstrate the suitability of the hole drilling and milling method for the determination of residual stress depth profiles in ceramic layer composites, an optimization problem was chosen which occurred during the development of thermal barrier coatings for metal forming tools. The coating material was YSZ (ZrO_2-Y_2O_3 93/7 wt.%), which was applied on rectangular plates of stainless austenitic steel ($100 \times 100 \times 4$ mm^3) by atmospheric plasma spraying (APS), see Fig. 3. The technical requirements of the coating were density (no corrosion channels to the substrate surface), mechanical stability, good bonding strength, and reliability of the coating process. Residual stress formation exerts an influence on all of these properties.

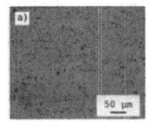 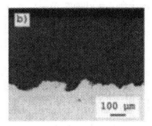

Fig. 3: Micrographs of a thermally sprayed YSZ coating on stainless austenitic steel substrate, a) top view, b) polished cross section

In order to analyze the influence of the temperature gradients during coating deposition on the residual stress formation, the substrate was heated to different temperatures prior to the coating process (preheating). Additionally, the influence of cooling of the surface by compressed air (air jet cooling) during coating deposition was investigated. The parameter variations are summarized in Tab. I. The distance of the APS torch to the substrate surface was 125 mm for all samples. The coating thicknesses were in the region 400-500 μm.

Tab. I: Thermal spray parameters for APS-coating of stainless austenitic steel substrates with YSZ (ZrO_2-Y_2O_3 93/7 wt.%) coating material

Specimen	Cooling (Air Jet)	Preheating Temp. (°C)
No. 1	Yes	180
No. 2	Yes	160
No. 3	Yes	150
No. 4	Yes	no preheating
No. 5	No	no preheating

The incremental hole drilling and milling technique as described above was used in order to measure residual surface strain data for the five different specimens. The drilling depth for all measurements was 0.8 mm with depth increments of 20 μm. The drilling equipment was manufactured by MTU Aero Engines (Munich, FRG). The drilling/milling tools were made by Gebr. Brasseler (Lemgo, FRG). Manufacturer of the strain gages was Vishay Measurements Group (Malvern/PA, USA). Residual stress depth profiles were calculated for the directions of the specimen axes using numerically determined calibration curves, and finally, residual equivalent stresses according to von Mises were calculated.

RESULTS AND DISCUSSION

The measured micro-strain (surface) vs. drilling depth profiles are displayed in Fig. 4. These curves give a first overview about the states of residual stress in the specimens. While compressive residual stresses will result in positive strain values, tensile stresses result in negative strain values. The finally calculated stresses depend on the gradient of the strain curve. Therefore, specimen no. 5 will show compressive stresses in the coating, while specimens no. 1-4 will show low tensile residual stresses. The reason for this difference lies in the lack of cooling and preheating for specimen no. 5, which (in combination with the strong heat flow to the substrate during coating deposition) leads to high temperature gradients between coating and substrate, but also in the coating material itself. High temperature gradients lead to high tensile stress formation in the coating during solidification and shrinkage of the coating material with simultaneous expansion of the substrate. Tensile stresses that exceed the strength of the coating material can yield micro-crack formation and stress relaxation. As a result, the component shrinkage during cooling, which in any case leads to a compensation and decline of tensile thermal stresses, can finally yield compressive stress values. For specimens no. 1-4, the reduction of heat flux into the substrate by means of simultaneous surface cooling as well as the reduction of temperature gradients by means of component preheating prevents coating damage and, therefore, the formation of compressive residual stresses. The calculated residual (equivalent) stress vs. drilling depth profiles, which are displayed in Fig. 5, confirm these interpretation of the measured micro-strains.

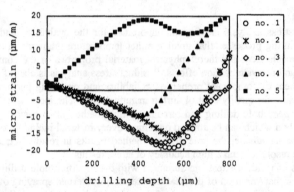

Fig. 4: Measured micro-strain (0° strain gage) vs. drilling depth curves, YSZ (ZrO_2-Y_2O_3 93/7 wt.%) coatings on stainless austenitic steel substrates, variation of spray parameters

The residual stress analysis proves a small influence of the variation of preheating temperature in the range of 150-180 °C on the final residual stress values in coating and substrate. However, samples without preheating show a change in the residual stresses. For specimens no. 4 and no. 5, the depth of the change of sign of stress values is directly at the interface, while the application of preheating moves this point into the substrate material, see Fig. 5. Thus, preheating can have an additional influence on coating adhesion and life-time of the component. Additionally, the residual stress measurements prove the necessity of surface cooling during the coating process. Specimen no. 5, for which cooling was omitted, shows a significantly different character of the state of residual stresses (compressive instead of tensile stresses in the coating), which must be attributed to coating damage during the process due to high temporary thermal stresses.

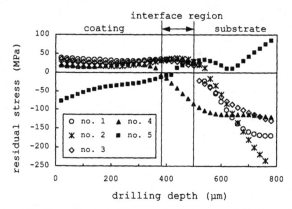

Fig. 5: Residual stress depth profiles (von Mises equivalent stresses) for thermally sprayed YSZ coatings (ZrO_2/Y_2O_3 93/7 wt.%) on steel substrates, variation of pre-heating and surface cooling

CONCLUSIONS

Residual stress analysis is of great importance for the evaluation and optimization of ceramic manufacturing processes that involve either temperature gradients in combination with gradients in the mechanical and thermophysical material properties of the composite materials, or solidification and recrystallization effects. Residual stress analysis is also of great importance for the improvement of component properties, reliability and operating performance as well as process stability, and for quality control during manufacturing. The incremental step, quasi non-destructive high-speed hole drilling and circular milling method is an easy-to-handle, time and cost efficient method which can be applied on almost every material including coatings and layer composites. As an advantage, the hole drilling method results in residual stress depth profiles with a high depth resolution and a high accuracy of stress results.

By means of residual stress measurement with the modified hole drilling method, it has been demonstrated that variation of the heat transfer during thermal spraying of ceramic coating materials as well as thermal pre-treatment of the substrate have a huge influence on the final residual stresses in the component. High temperature gradients lead to high thermal stress formation, while the application of preheating and/or simultaneous cooling of the component during the process can reduce temperature gradients and residual stress formation, and can even lead to a change of sign of residual stress values.

REFERENCES

[1]B. Scholtes, "States of residual stresses in mechanically deformed edge layer materials (Eigenspannungen in mechanisch randschichtverformten Werkstoffzuständen)", State doctorate, University of Karlsruhe, Germany, 1990.

[2]E. Macherauch and V. Hauk (eds.), "Residual stresses in science and technology", Papers presented at Int. Conf. on Residual Stresses, 1986, Garmisch-Partenkirchen (FRG), Oberursel : DGM Informationsges., 1987.

[3]A. Peiter, "Practical handbook on stress measurement (Handbuch Spannungsmesspraxis)", Braunschweig : Vieweg, 1992.

[4]H. Kockelmann, "Mechanical methods of determining residual stresses", Conf. on Residual Stresses Proceedings, Darmstadt, 1990, pp. 37-52.

[5]E. Schneider, "Nondestructive evaluation of stress states in components using ultrasonic and electromagnetic techniques", 11th Int. Symp. on Nondestructive Characterization of Materials, June 24-28, 2002, Berlin.

[6]P.V. Grant, J.D. Lord, and P.S. Whitehead, "The measurement of residual stresses by the incremental hole drilling technique", NPL Good Practice Guide No. 53, National Physical Laboratory (NPL), Materials Centre, Great Britain, August 2002.

[7]M. Buchmann, M.J. Escribano, and R. Gadow, "Residual stress analysis in thermally sprayed layer composites, using the hole milling and drilling method", *J. Therm. Spray Technol.*, **14** [1], 100-108 (2005).

[8]T. Schwarz, "Residual stress analysis on isotropic, anisotropic as well as inhomogeneous, layered composite materials by means of hole drilling and ringcore drilling method (Beitrag zur Eigenspannungsermittlung an isotropen, anisotropen sowie inhomogenen, schichtweise aufgebauten Werkstoffen mittels Bohrlochmethode und Ringkernverfahren)", Dissertation, University of Stuttgart, 1996.

[9]W. Soete and R. Vancombrugge, "An industrial method for the determination of residual stresses", *Proceedings SESA*, **8** [1], 17-28 (1950).

[10]R.A. Kelsey, "Measuring non-uniform residual stresses by the hole drilling method", *Proceedings SESA*, **14** [1], 181-194 (1955).

[11]G. König, "A contribution to further development of partially destructive residual stress measurement techniques (Ein Beitrag zur Weiterentwicklung teilzerstörender Eigenspannungsmessverfahren)", Dissertation, University of Stuttgart, 1991.

[12]M.J. Escribano and R. Gadow, "Residual stress measurement and modeling for ceramic layer composites", *Ceram. Eng. Sci. Proc.*, **24** [3], 615-622 (2003).

POTENTIAL USE OF MULLITE-SiC WHISKERS-SiC PARTICLES MULTI-COMPOSITE AS HIGH TEMPERATURE SPRINGS

Wataru Nakao, Koji Takahashi and Kotoji Ando
Yokohama National University
79-5, Tokiwadai, Hodogaya-ku
Yokohama, Kanagawa, 240-8501

Masahiro Yokouchi
Kanagawa Industrial Technology Research Institute
705-1 Imaizumi,
Ebina, Kanagawa, 243-0435

ABSTRACT

Mullite has high fracture strength at high temperature, excellent resistance to heat, corrosion and oxidation as well as low Young's modulus. Thus, mullite is well suited for cushions and springs used at high temperatures. However, the low fracture toughness and reliability of mullite prevents use in these applications. If the reliability of mullite is improved through self-crack-healing ability, the use of mullite would be possible in these applications. In this study, mullite/SiC particle/SiC whisker multi-composites with improved fracture toughness and excellent self-crack-healing ability where the crack-healed part is stronger than the base materials were developed. Fracture strength and Young's modulus of mullite/15vol% SiC whiskers/10 vol% SiC particles multi-composite were found to be 750 MPa and 272 GPa, respectively. The maximum strain of the composite was evaluated to be 0.79 %. The value is almost 2 times larger than the monolithic mullite. The mullite multi-composite has 2.2 times larger fracture toughness of monolithic mullite. Moreover, mullite/15vol% SiC whiskers/10 vol% SiC particles multi-composite can completely heal a pre-crack of 100 μm in surface length by exposure for 2 h in air at 1573 K. The crack-healed part was also found to be stronger than the base materials up to 1573 K.

INTRODUCTION

Ceramics springs made of mullite based composites have high deformability, because mullite has the lowest Young's modulus among structural ceramics. Moreover, mullite has excellent heat resistance. Thus, mullite springs could be used for various high-temperature applications. However, mullite is a low reliability material, because surface cracks caused by machining significantly deteriorate the fracture strength due to low fracture toughness. In this work, the authors propose self-crack-healing ability using oxidation of SiC to enhance the reliability of these composite materials. When mullite[1-4], alumina[5-7] and silicon nitride[8-10] combined with SiC are exposed to air at high temperatures, cracks are completely healed by the reaction products and the heat of the reaction which is the oxidation of SiC located on the crack surface. As a result, fracture strength reduced by cracking is completely recovered. Moreover, it is necessary to prevent crack formation by an increase in fracture toughness for using mullite in spring type of applications.

In a previous study[1], mullite composites reinforced by SiC whiskers were sintered for increasing fracture toughness and good crack-healing ability. Fracture toughness was found to increase linearly as a function of the reinforcing SiC whiskers content. It was observed that a mullite composite reinforced with 25 vol% SiC whiskers had 2.5 times higher fracture toughness of monolithic mullite. Adding 20 vol% SiC whiskers could provide the self-crack-healing ability where fracture strength is completely recovered. However, it has been also found that the crack-healed parts formed from SiC whiskers alone became weaker than the base materials at high temperatures. Therefore, it is necessary to improve the reliability at high temperatures in addition to higher self-crack-healing ability.

This study aims at developing new mullite composites having high fracture toughness and excellent crack-healing ability. For this purpose, mullite/SiC particles/SiC whiskers multi-composite were sintered. Crack-healing abilities and mechanical properties of these sintered composites were investigated. From the obtained results, the usefulness of the mullite composite as spring was discussed.

EXPERIMENTAL
Mullite/15 vol% SiC whiskers/ 5 vol% SiC particles and Mullite/ 15 vol% SiC whiskers/ 10 vol% SiC particles multi-composites, abbreviated MS15W5P and MS15W10P, respectively, were prepared in this study. The mullite powder (KM 101, Kioritzz Co. Ltd., Japan) used in this study has an average particle size of 0.2 μm and Al_2O_3 content of 71.8 %. The used SiC whisker (SCW, Tateho Chemical Industry Co. Ltd., Japan) have a diameter of 0.8 μm to 1.0 μm and a length of 30 μm to 100 μm. The SiC powder (Ultrafine grade, Ibiden Co. Ltd., Japan) has a mean particle size of 0.27 μm. To prepare the raw powder for MS15W5P and MS15W10P, SiC powder was added to mullite powder and the mixture was blended well in alcohol for 12 h. Subsequently, SiC whiskers were added to the mixture and the mixtures of mullite/ SiC whisker/ SiC particle blended for 12 h. Rectangular plates of 50 mm ×50 mm × 9 mm were hot pressed in Ar at 1973 K and 40 MPa pressure for 1 h. The sintered plates were cut into 3 mm × 4 mm × 22 mm rectangular bar specimens. The specimens were polished to a mirror finish on one face and

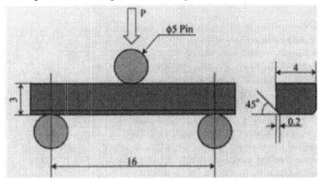

Figure 1 Dimensions of the specimen and the three-point loading system used for this investigation

the edges of specimens were beveled 45°, as shown in Fig. 1, to reduce the likelihood of edge initiated failures.

A semi-elliptical surface crack of 100 μm in surface length was made at the center of the tensile surface of specimens with a Vickers indenter, using a load of 19.6 N. The ratio of depth (a) to half the surface length (c) of the crack (aspect ratio a/c) was 0.9. The cracked specimens were subjected to a crack-healing treatment at 1573 K for 2 h in air, where the crack-healing condition is described in the previous study[4]. The crack-healing treatments were also carried out from 1073 K to 1473 K for 10 h so as to investigate the crack-healing rate as a function of temperature.

All fracture tests of the crack-healed specimens were performed on a three-point loading system with a span of 16 mm, as shown in Fig. 1, at room temperature and temperatures from 873 K to 1573 K. The cross-head speed was 0.5 mm/min. As-received heat-treated specimens exposed to the crack-healing treatment in order to heal the crack introduced during machining were also tested as above. Young's modulus was evaluated from the relation between the applied stress and the strain measured by strain gauge on the crack-healed part. Fracture toughness was obtained by the indentation fracture method expressed by the following equation,

$$K_{IC} = 0.026 \cdot E^{1/2} \cdot P^{1/2} \frac{a_v}{c^{3/2}} \tag{1},$$

where E is the Young's modulus of MS15W10P.

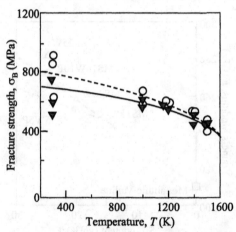

Figure 2 Temperature dependence of the bending strength of the crack-healed and heat-treated MS15W10P

RESULTS

Figure 2 shows the temperature dependence of the bending strength of the crack-healed and heat-treated MS15W10P. The shaded triangle and circle indicate the bending strength of the crack-healed and heat-treated specimens, respectively. The bending strengths of the crack-healed and heat-treated specimens decrease gradually up to 1473 K, above which these decrease rapidly. The bending strength of crack-healed specimen was equal to that of heat-treated specimen over

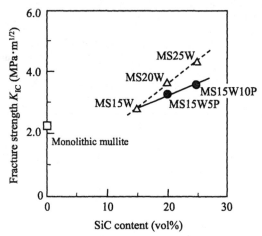

Figure 3 Fracture toughness as a function of SiC contents

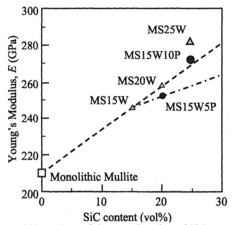

Figure 4 Young's modulus as a function of SiC contents

the measured temperature range. Moreover, the crack-healed MS15W10P was never fractured from the pre-crack which had been healed. If the pre-crack healed does not have enough stuffiness, fractures propagate at the pre-crack healed. This suggests that the self-crack-healing ability of mullite is enhanced at elevated temperature by the addition to SiC particles and SiC whiskers, compared to monolithic mullite.

Figure 3 shows the fracture toughness of mullite as a function of the total SiC contents. The shaded circles and the open triangles indicate the fracture toughness of the crack-healed mullite/SiC multi-composites (MS15WxP) and the crack-healed mullite/SiC whisker composite (MSxW), respectively. Fracture toughness increases gradually with increasing SiC whisker content due to the combined phenomena of whisker bridging and pulling-out. Also, fracture toughness is increased by adding SiC particles to a mullite-SiC whiskers composite. It can be seen that the fracture toughness of mullite is improved by a combination of SiC particles and SiC whiskers.

Figure 4 shows the improvement in the Young's modulus of the monolithic mullite as a function of SiC content. The values of MS25W and MS15W10P were obtained by the experimental results. The Young's moduli of the other composites were evaluated to prorate the values[11] of mullite, SiC whisker and SiC particle. The measured values were slightly higher than that of evaluation values.

DISCUSSION

The application of mullite/SiC particle/SiC whisker multi-composite as a ceramics spring are discussed below. Initially, the shear deformation abilities of the composites were estimated. The maximum shear strains, γ_{max}, are evaluated from Eq. (2),

$$\gamma_{max} = \tau / G \qquad (2),$$

where τ is share strength corresponding to the bending strength, σ_B, and G is shear modulus. The

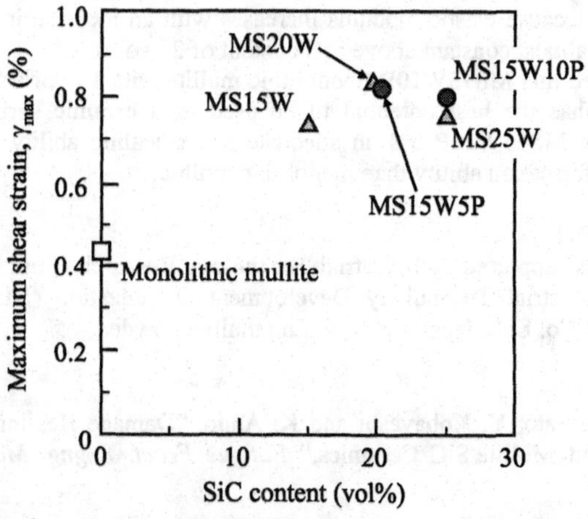

Figure 5 Maximum shear strains of the mullite/ SiC composites as a function of SiC content

385

values of G was evaluated from Eq. (3),

$$G = E / 2 (1 + \nu) \tag{3},$$

where ν is Poisson's ratio. A value of 0.3 was used as an estimate. Figure 5 shows the maximum shear strains of the mullite/SiC composites as a function of SiC content. The value of γ_{max} shows a maximum at SiC content of 20 vol%, above which it slightly decreases, because Young's modulus increases with an increase in SiC content but the fracture strengths are almost constant above a SiC content of 20 vol%. However, the crack-healing abilities of MS20W and MS15W5P were not adequate. In contrast, MS15W10P has an adequate crack-healing ability as well as almost 2 times higher shear deformation ability than monolithic mullite. This suggests that MS15W10P has the best potential for use as a ceramic spring used at elevated temperatures.

CONCLUSION

For improvement in fracture toughness as well as endowment with excellent self-crack-healing ability, mullite was hot pressed with SiC particles and SiC whiskers. Crack-healing abilities and mechanical properties of these hot pressed composites were investigated. From the results obtained, formulations of mullite/SiC whiskers/SiC particles with properties suited for composite springs were developed.

MS15W10P can completely heal the pre-crack at lower temperature than MS25W in spite of the same SiC content. Over the entire temperature range measured, the bending strength of crack-healed specimen was equal to that of heat-treated specimen. The addition of 15 vol% SiC whiskers and 10 vol% SiC particles to monolithic mullite could provide enough self-crack-healing ability to form a stronger and tougher crack-healed part than monolithic mullite at elevated temperatures.

Fracture toughness was observed to increase with increasing SiC contents over the whole SiC content range prepared in the present study. The value of γ_{max} shows a maximum at SiC content of 20 vol%, because elastic modulus increases with an increase in SiC content but the fracture strengths are almost constant above SiC content of 20 vol%.

It was observed that MS15W10P (monolithic mullite with 15 vol% SiC whiskers and 10 vol% SiC particles) has the best potential to be used as a ceramic spring used at elevated temperatures, because MS15W10P has an adequate crack-healing ability as well as almost 2 times higher shear deformation ability than monolithic mullite.

ACKNOWLEDGMENT

This study was supported by Industrial Technology Research Grant Program in '04 from New Energy and Industrial Technology Development Organization (NEDO) of Japan. We acknowledge Kioritzz Co. Ltd., Japan for providing mullite powder.

REFERENCE

[1]M.C. Chu, S. Sato, Y. Kobayashi and K. Ando, "Damage Healing and Strengthealing Behavior in Intelligent Mullite/SiC Ceramics," *Fatigue Fract. Engng. Mater. Struct.*, **18** [9], 1019-1029, (1995).

[2]K. Ando, M.C. Chu, K. Tuji, T. Hirasawa, Y. Kobayashi and S. Sato, "Crack Healing Behaviour and High-Temperature Strength of Mullite/SiC Composite Ceramics," *J. Eur. Ceram. Soc.*, **22**, 1313-19, (2002).

[3]K. Ando, K. Furusawa, M.C. Chu, T. Hanagata, K. Tuji, and S. Sato, "Crack Healing Behavior Under Stress of Mullite/Silicon Carbide Ceramics and the Resultant Fatigue Strength," *J. Am. Ceram. Soc.*, **84** [9], 2073-78, (2001).

[4]M. Ono, W. Ishida, W. Nakao, K. Ando, S. Mori and M. Yokouchi, "Crack-Healing Behavior, High Temperature Strength and Fracture Toughness of Mullite/SiC Whisker Composite Ceramic," *J. Soc. Mater. Sci. Jpn.*, (accepted).

[5]B.S. Kim, K. Ando, M.C. Chu and S. Saito, "Crack-Healing Behavior of Monolithic Alumina and Strength of Crack-Healed Member," *J. Soc. Mater. Sci. Jpn.*, **52** [6], 667-673, (2003).

[6]K. Ando, B.S. Kim, S. Kodama, S.H. Ryu, K. Takahashi and S. Saito, "Fatigue Strength of An Al$_2$O$_3$/ SiC Composite and A Monolithic Al$_2$O$_3$ Subjected to Crack-Healing Treatment," *J. Soc. Mater. Sci. Jpn.*, **52** [11], 1464-1470, (2003).

[7]K. Takahashi, M. Yokouchi, S.K. Lee and K. Ando, "Crack-Healing Behavior of Al$_2$O$_3$ Toughened by SiC Whiskers," *J. Am. Ceram. Soc.*, **86** [12], 2143-47, (2003).

[8]K. Ando, M.C. Chu, Y. Kobayashi, F. Yao and S. Sato, "The Study on Crack Healing Behavior of Silicon Nitride Ceramics," *Jpn. Soc. Mech. Eng. Intl. J.*, **64** A, 1936-1942, (1998).

[9]F. Yao, K. Ando, M.C. Chu and S. Sato, "Static and Cyclic Fatigue Behaviour of Crack-Healed Si$_3$N$_4$/SiC Composite Ceramics," *J. Eur. Ceram. Soc.*, **21**, 991-997, (2001).

[10]K. Ando, K. Takahashi, S. Nakayama and S. Sato, "Crack-Healing Behavior of Si$_3$N$_4$/SiC Ceramics Under Cyclic Stress and Resultant Strength at The Crack-Healing Temperature," *J. Am. Ceram. Soc.*, **85** [9], 2268-72, (2002).

[11]H. Okuda, T. Hirai and O. Kamigaito, "*Engineering Ceramics*", pp. 11, Fine Ceramics Technology Series 6, Japan, (1987).

Joining

BRAZING AND CHARACTERISATION OF ADVANCED CERAMIC JOINTS

Jolanta Janczak-Rusch
EMPA
Ueberlandstrasse 129
Duebendorf, CH-8600

ABSTRACT

The possibilities of application of the brazing technology for the manufacturing of ceramic joints are shown on examples of TiN particle reinforced Si_3N_4 and SiC fibre reinforced borosilicate glass composite joints. These composite materials were brazed to other ceramics or metals in a one-step process using active filler metals. In case of ceramic-metal joints the main attention was paid to the control of the residual stresses. Novel filler metals were developed to minimize the residual stresses and to improve the joint. Numerical and experimental methods were applied to evaluate the residual stress state in the joints. The joints were characterised in their temperature service range in bending- and tensile- tests. The characterisation of the thermo-mechanical behaviour was accompanied by fractography work. The influence of the filler metal, of the specimen preparation (especially chamfering), processing parameters and joint design on the joint strength were studied. To clarify these effects, the microstructure of the joints, especially of the brazing zone was characterised.

INTRODUCTION

Suitable joining technologies are required to assemble complex components and to expand the application possibilities of advanced ceramics, glass-ceramics and their composite materials. Standardized testing methods are further needed for safe operation of joined structures. However, only a few of the procedures developed for joining traditional ceramics or glasses are directly transferable to their composites. The situation becomes even more complicated when joining ceramic matrix composites with metals. Besides the typical problem due to differences in physical and thermal properties of metals and ceramics, the different wetting behaviour of the composite constituents (e.g. fibre and matrix) as well as the compatibility of the joining process with the composite fabrication technique have to be taken into account when joining composite materials. Among different joining processes as mechanical interlocking, diffusion bonding and joining by preceramic polymers, the brazing technology, widely used in the joining of monolithic ceramics to other monolithic ceramics or metals has been proposed to be suitable also for a wide range of ceramic-matrix composites (CMCs) [1,2]. As shown by the example of a ceramic and a glass-ceramic composite material, high strength, reliable ceramic-ceramic and ceramic-metal joints can be produced by active brazing when a suitable filler metal is found, the joint design and processing parameters are optimised. The materials used in the evaluation were

a) a fine-grained Si_3N_4-TiN ceramic composite with 30 wt% TiN content hot pressed with yttria and alumina additives [3]

b) an unidirectional reinforced borosilicate (Duran®) glass matrix composite with 40 vol.% Nicalon® NL 202 SiC fibres (average diameter: 0.015 mm) manufactured by the sol-gel slurry and hot-pressing method [4] (Schott Glaswerke, Mainz, Germany).

For ceramic-metal joints a 14NiCrNi steel was used as a joining partner for the Si_3N_4-TiN ceramic and molybdenum for the borosilicate glass matrix composite. The selection of the metals was done keeping in mind the middle temperature range applications of the investigated ceramic materials like cutting tools (Si_3N_4) or tools for the handling of hot glassware and non-ferrous metals (SiC/Duran).

Vacuum brazing technology was used to produce the joints, the process pressure was always lower than 10^{-5}mbar.

BRAZING AND CHARACTERISATION OF Si_3N_4-TiN CERAMIC COMPOSITES

Selection of the brazing filler

Several commercially available as well as developmental active filler metals were evaluated. The chemical compositions of the brazing filler metals are given in Table 1. The content of the active element (Ti) varied in these fillers between 0.6 wt% for the AgInTi (CB6) and 10 wt% for the CuSnTiZr filler metal. The CB6 filler is recommended by the manufacturer for the joining of Si_3N_4 ceramics.

Table 1. Composition of the brazing filler metals

Brazing alloy	Ag [wt-%]	Cu [wt-%]	In [wt-%]	Sn [wt-%]	Ti [wt-%]	Zr [wt-%]	form	brazing temp.[°C]
CuSnTiZr	-	74.5	-	14.0	10.0	1.5	paste	930
AgCuInTi (Incusil ABA®, Wesgo)	59.0	27.3	12.5	-	1.25	-	foil	740
AgInTi (CB6®, Brazetec)	98.4	-	1.0	-	0.6	-	foil	1010

The resultant 4-point room temperature bending strength of the ceramic-ceramic and ceramic-metal joints are depicted in Fig. 1 (average, minimum and maximum strength values). The bending tests were carried out based on the standard EN 843-1 for ceramic flexural testing, Fig 1 left, with specimen geometry of 3x4x50mm,. According to this standard the edges of the specimens were chamfered (size 0.12±0.03mm). Additionally, for one batch of specimens (used for the filler evaluation, Fig.1), the edges of the intersection were chamfered (angle: 45°).

Whereas more possibilities exist when choosing the brazing filler for the ceramic-ceramic joints, the selection must be done very carefully when brazing ceramic with metals. In particular, an optimum content of the active element has to be found to ensure wetting of the ceramic and to avoid extensive reactions between the brazing filler and the metal joining partner. Furthermore, optimised brazing filler should have physical properties matching well the properties of the joining partners and show good plastic deformation capacity to absorb the residual stresses.

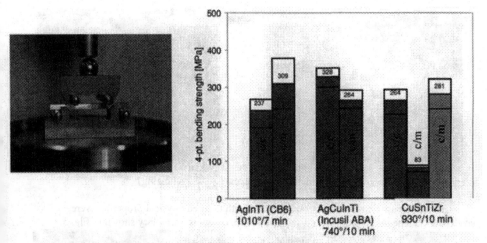

Fig. 1: Four point bending strength of ceramic-ceramic (c/c: blue) and ceramic-metal (c/m pink) joints for different brazing filler metals used. Cu-interlayer was additionally used when brazing ceramic to metal with CuSnTiZr (pink striped).

In our study all used brazing fillers allowed to produce ceramic-ceramic joints of a moderate strength in the range between an average of 237 MPa for the recommended CB6 brazing filler with the lowest Ti-content (0.6%) and 328 MPa for the Incusil ABA with the slowly higher Ti-content (1.25%). The ceramic-ceramic joints with Incusil ABA showed not only the highest strength but also the lowest deviation of the measured values. Besides the CB6 joints all ceramic-ceramic joints failed in the ceramic. The strength values for ceramic-metal joints differ between 83 MPa for the CuSnTiZr brazing filler and 309 MPa for the CB6 alloy. Brittle Laves-phases (Fe_2Ti) formed in the brazing zone of metal-ceramic joints when brazed with CuSnTiZr, Fig 2a, hardening the Cu-based brazing alloy and the interface zone to steel. With a soft interlayer working as an diffusion barrier and at the same time allowing the relaxation of the residual stresses, ceramic-metal joints of strength comparable to the one of the ceramic-ceramic joints may be produced. Particularly, by use of a 0.5 mm thick Cu-interlayer the strength of ceramic-metal joints brazed with the CuSnTiZr brazing filler was improved by a factor 3.5 (from 83 MPa without Cu-interlayer to 281 MPa with Cu). However an additional processing step (brazing of the interlayer) was needed to realize this design.

Development of composite filler metals

The use composite brazing fillers was shown to be an effective method for the optimisation of residual stresses and the control of the braze zone microstructure in ceramic-metal joints[5].

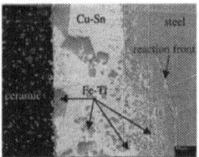

Fig.2: Microstructure of the braze zone of the Si₃N₄-TiN/14NiCr joint brazed with CuSnTiZr filler. a) without an interlayer b) with a Cu-interlayer (two-steps brazing)

Ceramic particle reinforced brazing alloys with tailored physical properties were developed[6,7,8]. The microstructure of such composite solders is exemplary shown in Fig. 3.

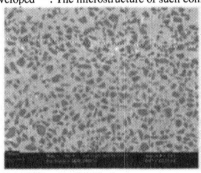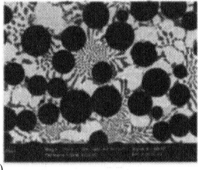

a)   b)

Fig.3: SEM-micrographs of composite brazing fillers. a) SiC particle reinforced CuSnTiZr filler, b) Glass reinforced AgCuInTi filler

The use of composite fillers for brazing Si₃N₄-TiN with steel leads to a significant improvement of the joints properties when compared to joints produced using non-composite fillers. Specifically, with an AgCuInTi/30%SiC$_p$ composite filler in a sandwich structure, the room temperature bending strength was improved by 20%, the maximum service temperature increased by 50°C and the reliability (measured by the strength scatter) by 35%. Even with a small volume content of particles, it was possible to improve the strength values above the room temperature and to increase joints temperature service range. The effect of the modification of the AgCuInTi brazing filler by adding 10 vol.% of SiC particles on the joint strength is shown as a function of the testing

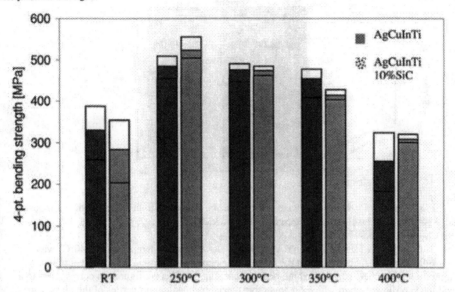

Fig. 4: Bending strength of Si_3N_4-TiN/14NiCr14 joints brazed with composite filler metal ($AgCuInTi/10\%SiC_p$) and non-composite filler metal (AgCuInTi) in comparison for testing temperatures: RT-400°C; non-chamfered specimens.

However, depending on the joint requirements, the optimum particle volume content as well as the joint design has to be evaluated. The calculated (FE method) residual stress distribution in the braze zone of the Si_3N_4-TiN/steel joint brazed with a $AgCuInTi/30\%SiC_p$ composite filler in a sandwich structure with is shown in Fig.5.

BRAZING AND CHARACTERISATION OF GLASS CERAMICS

The only brazing filler materials ensuring wetting of the glass matrix composite below the glass transformation temperature of the matrix material (550°C for the borosilicate glass) are glass brazes. However, as own investigations have shown, joints brazed with glass solder did not allow to utilise the full potential of the glass matrix composite[9]. With optimised processing parameters it is possible to avoid the degradation of the properties of the glass matrix composite even when brazing at temperatures much higher then the glass transition point of the joining partner.

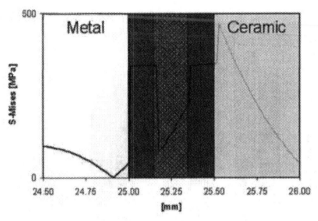

Fig.5: Residual stresses in a braze zone of a Si_3N_4-TiN/steel joint brazed with a composite filler (AgCuInTi-AgCuInTi/30%SiC$_p$-AgCuInTi sandwich).

The AgCuInTi filler (Incusil ABA) with a moderate brazing temperature (740°C) was successfully used to produce joints of the borosilicate glass composite with molybdenum in a configuration with fibers parallel to the braze surface, Fig 6a. This configuration corresponds to a typical load configuration in which the favorable thermomechanical properties of the glass matrix composite can be exploited, for example in components for the handling of hot glassware [10]. However, for this configuration standardized mechanical tests to assess the joint strength do not exist.

Fig. 6: Brazed joints of SiC fibre reinforced borosilicate glass a) and their qualitative characterization b).

A qualitative test, Fig 6b, was performed to investigate joints behavior. The joints showed a high loading capacity, Fig.6. The study results indicated that the joint's bond strength is higher than the interlaminar strength of the glass matrix composite. Failure always occurred through the glass matrix composite by significant delamination, as the fracture surface in Fig. 7 indicates. The brazing zone as well as the Mo substrate remained intact. Moreover, no degradation of the glass matrix composite material due to thermal effects during the brazing process was observed.

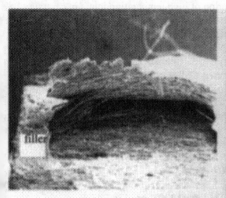 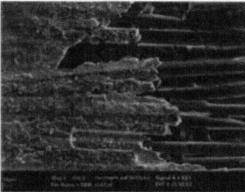

Fig. 7: Failure behaviour of brazed SiC$_f$/borosilicate glass-Mo joints.

However, such behaviour can be observed only when brazing in the fibre direction. In case the fibers are perpendicular to the joined surface, the joint failure behaviour is completely different [9]. The joints fail at the brazing filler-glass matrix composite interface, glass matrix composite delamination is not observed, fracture occurs by brittle failure at the interface.

CONCLUSIONS

High strength joints of Si_3N_4-TiN ceramics were produced by vacuum brazing with Cu- and Ag-based active fillers. When brazing ceramics with metals the Ti-content of the filler metal was shown to play a critical role. Composite brazing fillers were developed and successfully applied to braze Si_3N_4-TiN ceramics with metals. Reliable joints with improved thermomechanical properties were produced in one-step brazing.

Active brazing was shown as an effective method to produce joints of a SiC fibre reinforced glass composite with molybdenum in configuration adapted to the relevant loading conditions. In this configuration the strength of the brazed glass composite/Mo joints is higher than the delamination stress of the composite and the properties of the glass composite can be fully utilised. Although the working temperature of Incusil ABA is higher than the glass transformation temperature, it is a suitable filler metal for joining borosilicate glass matrix composites with metals.

ACKNOWLEDGEMENTS

The author gratefully acknowledge the Gebert Rüf Foundation, Switzerland for financial support of the "Verbundlote" project, Project No. P-001/01. V.Bissig, H.-R.Elsener, D.Piazza are thanked for their contribution to this work. The author would also like to acknowledge the cooperation with Dr.A.Boccaccini (Imperial College, UK), his help and critical comments.

REFERENCES

[1] P J. D. Cawley, "Joining Ceramic Matrix Composites", *Ceram. Bull.* **68** 1619-1623 (1989).

[2] M. Singh, "Design, fabrication and characterization of high temperature joints in ceramic composites", *Key Eng. Mater.* **164-1**, 415-419 (1999).

[3] G. Blugan, M. Hadad, J.Janczak-Rusch, J. Kübler, T. Graule, "Microstructure, Mechanical Properties and Fractography of Commercial Silicon Nitride-Titanium Nitride Composites for Wear Applications", *J.Am.Cer.Soc.*, in print..

[4] W. Pannhorst, M. Spallek, R. Brückner, H. Hegeler, C. Reich, G. Grathwohl, B. Meier, D. Spelmann, "Fibre-Reinforced Glasses and Glass-Ceramics Fabricated by a Novel Process", *Ceram. Eng. Sci. Proc.* **11** [7-8] 947-963(1990).

[5] J.Janczak-Rusch, "Filler metals, solders and joints of next generation", *Empa Activity reports*, in print.

[6] J.Janczak-Rusch, B.Zigerlig, U.Klotz, et.al., "Joining of Si_3N_4/TiN Ceramics with Steel Components Using Particle Reinforced Active Brazing Alloys", *Proc. of 6th Int. Conf. on Joining Ceramics, Glass and Metal*, Munich, 30 Sept.-1 Oct. 2002, pp. 41-48.

[7] H.R. Elsener, J. Janczak-Rusch, V. Bissig, U.E. Klotz, B. Zigerlig, *„Partikelverstärkte Aktivlote: Grundlagen", in Verbundwerkstoffe"*, Hrsg. Degischer, 2003, pp. 738-743.

[8] G.Blugan, J.Janczak-Rusch, J.Kübler, "Properties and fractography of Si_3N_4/TiN ceramic joined to steel with active single-layer and double-layer braze filler alloys", *Acta Materialia*, 52/15 4579-4588 (2004).

[9] J.Janczak-Rusch, D.Piazza, A.R. Boccaccini, "Brazing of SiC fibre reinforced borosilicate glass matrix composites with molybdenum", *Journal of Materials Science*, in print.

[10] W. Beier, J. Heinz, W. Pannhorst, W., „Langfaserverstärkte Gläser und Glaskeramiken: eine neue Klasse von Konstruktionswerkstoffen," *VDI Berichte* Nr. 1021 255-267 (1993).

9

DEVELOPMENT OF JOINING AND COATING TECHNIQUE FOR SiC AND SiC/SiC COMPOSITES UTILIZING NITE PROCESSING

T. Hinoki, N. Eiza, S. Son, K. Shimoda, J. Lee and A. Kohyama
Institute of Advanced Energy
Kyoto University, Gokasho, Uji, Kyoto 611-0011, Japan

ABSTRACT

For joining, the same SiC nano-powder and sintering additive with NITE processing to fabricate SiC were used. Silicon carbide or SiC/SiC plates were joined by hot-pressing at 1700-1900 °C for 1 hour under 20 MPa. The joined bars were evaluated by tensile test and asymmetric four point flexural test. For the coating, W is used considering the application. Starting materials were both fine powders and by lay-up two zones of powders (slurry) followed by hot pressing, two layers of W and SiC were produced with strong bonding.

The joint thickness was controlled to approximately 10 μm. The tensile strength for the joint of monolithic SiC was approximately 40 MPa with 40% porosity at the interface. The shear strength of the joint of monolithic SiC could not be obtained, since the crack at joint propagated into the substrate, but it was more than 90 MPa. After improving the technique, the dense joint was formed and then, the tensile strength increased to more than 250 MPa. For W coated SiC, SiC, W and their interface reaction layers can be controlled to make acceptable bonding by optimizing the materials used and process condition. Using the optimized condition, W coated SiC/SiC composites were developed, although the optimization of processing for the composites is still required.

INTRODUCTION

SiC/SiC composites are considered for use in extremely harsh environments at high temperature primarily due to their excellent thermal, mechanical and chemical stability, and the exceptionally low radioactivity following neutron irradiation [1,2]. In particular, recent improvement in the crystallinity and purity of SiC fibers, the developments and improved composite processing have improved physical and mechanical performance under harsh environments [3-5].

The novel processing called Nano-powder Infiltration and Transient Eutectic-phase (NITE) Processing has been developed based on the liquid phase sintering (LPS) process modification [1,6]. The NITE processing can achieve both the excellent material quality and the low processing cost. The important issues to use the NITE SiC/SiC composites for industry are development of joining and coating technique. Several kinds of joining techniques have been developed for joining of SiC and SiC/SiC composites using polymer [7], glass-ceramics [8,9] and reaction bonding [10]. One of the key for the development is the stability of the joining at application temperature. Using the SiC for joint of SiC or SiC/SiC composites has the advantage at the high temperature due to the no coefficient of thermal expansion (CTE) mismatch. In this work, monolithic SiC, which corresponds to the matrix SiC formed by NITE process, was used for the joint. Refractory armored materials (RAM's) have been considered for the plasma facing material in the fusion blanket system to reduce the introduction of power-sapping impurities into the plasma [11,12], and to reduce erosion. Considering the CTE mismatch and melting point, tungsten (W) was selected as the refractory armor material for SiC. The objective of this work is to develop joining and coating technique for SiC and SiC/SiC composites utilizing NITE processing.

EXPERIMENTAL

The substrate material for joining was Hexoloy® SA [5] SiC (sintered α-SiC) and SiC/SiC composites fabricated by NITE processing. The substrates with dimension 23 mm (long) × 2.7 mm (wide) × 3 mm (thick) were machined from plate. The substrate SiC bars were joined with the slurry including SiC nano-powder (<20nm) and the sintering additive of Al_2O_3, Y_2O_3, SiO_2. They were hot-pressed at 1800 °C with the pressure of 15-30 MPa in Ar environment. Butt joint was applied to the SiC bars and 46 mm (long) × 2.7 mm (wide) × 3 mm (thick) bars were formed. Mechanical properties of the joint were evaluated using the bars by tensile test according to ASTM C1275 and asymmetric four points bend according to ASTM C1469. For the tensile test, the gauge section was 20 mm-long in the middle of the specimen. The specimens were gripped using a pair of wedge-type grips and aluminum end tabs, which were adhesively bonded to the specimen to promote uniform stress in the gripping area. The magnitude of the clamping pressure was sufficiently large to prevent slippage between the grips and a specimen. The grips were connected to the load train using universal joints to promote self-alignment of the load train during the movement of crosshead and to reduce unwanted bending strains in the specimen. All tests were conducted at a cross-head speed of 0.3 m/min at ambient temperature. Asymmetric four point flexural test was conducted using the same specimen for the tensile test. Inner span and outer span of the asymmetric four points test were 8 mm and 44 mm, respectively. Shorter inner span of 4 mm was also applied to enhance the fracture at joint.

The tungsten coated SiC was fabricated by one processing. Both SiC powder and tungsten powder was hot-pressed with appropriate sintering additive to form SiC. Various processing temperature with the range of 1700-1875 °C and various processing time with the range of 10-120 min were applied to understand the effect of the processing condition on the microstructure of the interphase and the shear strength for the coating. The mechanical properties of the coating were evaluated by shear test and indentation test. The NITE-SiC/SiC composites were also coated by tungsten using the same technique. Mechanical properties of tungsten coated SiC were evaluated by shear tests. Shear tests were conducted under 0.5 mm/min of crosshead speed with a specimen size of 3 (long) × 3 (wide) × 2 (thick) mm.

The microstructure and fracture surfaces following mechanical test of the joint, coating and their interfaces were observed by optical microscopy (OM) and field emission scanning electro microscopy (FE-SEM), and analyzed by energy dispersive X-ray spectroscopy (EDS) and X-Ray Diffraction (XRD).

RESULT AND DISCUSSION

Silicon carbide substrates were succesfully joined with forming thin NITE-SiC of approximately 10 μm-thick at interface. Figure 1 shows the backscattering image of the joint. No concentration of sintering addtive was seen, while small white dots including the sintering additive was observed.

Tensile strength and shear strength were obtained using the batt joint spesimens. In tensile test, specimens failed at the joint. The tensile strength was approximately 40 MPa on average, while relatively large scatter of data was seen. Appearent shear strength was approximately 50 MPa on average with relatively small scatter of data when the inner span was 8 mm. Indeed the specimens did not fail at the joint. Actual shear strength was more than 50 MPa. The apparent shear strength increased to approximately 90 MPa, when the inner span was reduced to 4 mm. However it still did not failed at the joint, and it was found that the actual shear strength was more than 90 MPa. The other test method or the notched specimen at the joint for the assymetric four point flexural test is required to evaluated actual shear strength.

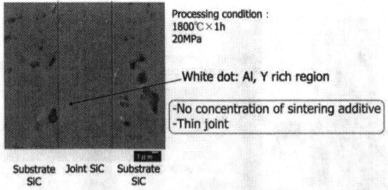

Processing condition :
1800°C × 1h
20MPa

White dot: Al, Y rich region

- No concentration of sintering additive
- Thin joint

Substrate   Joint SiC   Substrate
SiC   SiC

Figure 1: Interface of joint of monolithic SiC

Fracture surfaces following the tensile test were observed as shown in Figures 2. It was found the joint had approximately 40 % of porosity including the large pore, which might induced the stress concentration during the loading. The reason of the relatively small tensile strength of 40 MPa for SiC was due to the large non-bonded region, although even with the large porosity and the stress concentration the joint showed large strength compared with the other joint for SiC or SiC/SiC composites. Recently the porosity in the joint for monolithic SiC was significantly reduced and tensile strength increased to more than 250 MPa. Because of too large strength, specimens failed not at the joint but the substrate as shown in Figure 3. Well-machined subsrate is required to evaluate the actual tensile strength.

SiC/SiC composites were also joined using the same technique as shown in Figure 4. The

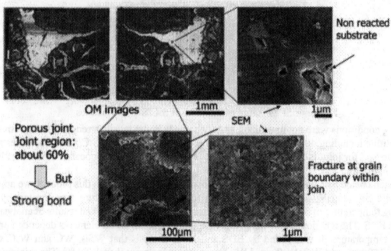

Non reacted substrate

OM images   1mm

SEM   1µm

Porous joint
Joint region:
about 60%

⬇ But

Strong bond

Fracture at grain boundary within join

100µm   1µm

Figures 2: OM images and SEM images of fracture surfaces of the joint
following tensile test

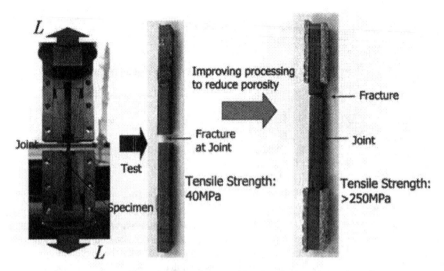

Figure 3: Effect of process improvement on fracture behavior of joined specimens following tensile test

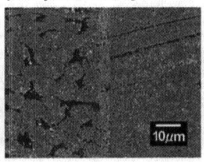

Figure 4: SEM image of the joint of SiC/SiC composites

processing conditions were not optimized, and it showed similar tensile strength with the strength for monolithic including 40 % porosity at interface. Major difference of SiC/SiC composites from monolithic SiC for joint is the surface roughness. Pores were likely to appear compared with the joint for monolithic SiC.

Tungsten coated SiC was formed using SiC powder and W powder. In this process, we applied to fabricate SiC and W at the same time. So the sintering temperature was limited by the processing temperature of SiC. Figures 5 show interphases of SiC and W and results of analysis by XRD and EDS. The reaction layers were formed and their thickness and elements depended on the sintering temperature. It was found by EDS and XRD analysis that WSi_2, WC and W_2C were formed at 1700 °C, and WC, W_5Si_3 and W_2C were formed at more than 1780 °C. The thickness of the reaction layers decreased with decreaing the sintering temperature. The thickness of the relaction layers also depended on the reaction time. The relation between the total thickness of

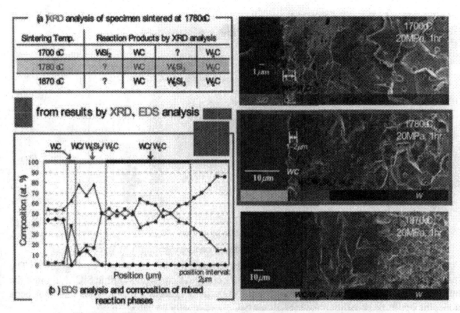

Sintering Temp.	Reaction Products by XRD analysis			
1700 ℃	WSi₂	WC	?	W₂C
1780 ℃	?	WC	W₅Si₃	W₂C
1870 ℃	?	WC	W₅Si₃	W₂C

(a) XRD analysis of specimen sintered at 1780℃

from results by XRD, EDS analysis

(b) EDS analysis and composition of mixed reaction phases

Figures 5: SEM images of the interphase of W coated SiC fabricated at various temperature and the results of XRD and EDS analysis

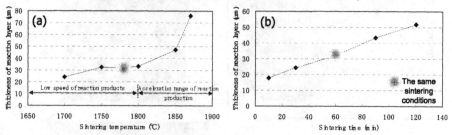

Figures 6: Effect of sintering temperature (a) and sintering time on total thickness of reaction layers

reaction layers and the reaction temperature is show in Figure 6(a). The relation between the total thickness of reaction layers and the reaction time is show in Figure 6(b).

The shear strength of the coating fabricated at various sintering temperature is shown in Figure 7. The specimens fabricated at 1780 °C and 1850 °C showed large strength compared with the other specimens. It was found the weakest link in the reaction layers were the mixed layer of mostly consisted of W_5Si_3 formed at more than 1780 °C. The region included pores. One of the reason of degradation of the shear strength at higher processing temperature was attributed to growth of the pore size at that region. In the case of the coating fabricated at lower temperature (less than 1750 °C), the strength of the SiC itself was smaller than that of SiC fabricated at more than 1780 °C and the shear strengtrh was also smller than that of the coating fabricated at more than 1780 °C.

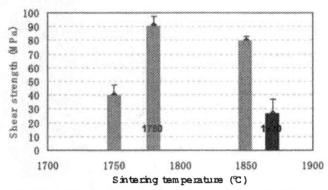

Figure 7: Effect of the sintering temperature on shear strength of W coated SiC

SiC/SiC composites were also successfully coated by W as shown in Figure 8. The microstructure of the reaction layer seemed to very close to that for monolithic SiC. Although the optimization of the processing condition and evaluation of mechanial properties have not been done, the optimized condition for monolithic SiC shold mostly be used.

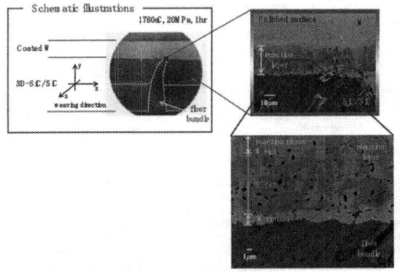

Figures 8: SEM images of W coated SiC/SiC composites

CONCLUSION
1. Silicon carbide was successfully joined applying NITE processing. It showed strong joint even with approximately 40% non-reacted region compared with the other joint for SiC or SiC/SiC composites. Tensile strength was about 40 MPa. Shear strength were not obtained due to strong bonding, while it was more than 90 MPa. By improving the experimental condition and decreasing the porosity, the tensile strength increased to more than 250 MPa.
2. Tungsten coating were formed on both SiC and SiC/SiC composites by hot-pressing. It was found that the reaction layer including W_5Si_3 was the weakest link for the coating formed at more than 1780 °C. Shear strength was improved reducing the reaction layer. The maximum shear strength of 90 MPa was obtained by the specimen fabricated at 1780 °C. The same technique was applied to SiC/SiC composites and similar reaction layers were formed.

ACKNOWLEDGEMENTS
This work was supported by IVNET program of MEXT, Japan.

REFFERENCES
[1] A. Kohyama and Y. Katoh, Ceramic Transactions, **144** (2002) 3-18.
[2] L.L. Snead, R.H. Jones, A. Kohyama and P. Fenici, J. Nucl. Mater., **233-237** (1996) 26-36.
[3] M. Takeda, A. Urano, J. Sakamoto and Y. Imai, J. Nucl. Mater., **258-263** (1998) 1594-1599.
[4] T. Ishikawa, S. Kajii, T. Hisayuki, K. Matsunaga, T. Hogami and Y. Kohtoku, Key Eng. Mater., **164-165** (1999) 15-18.
[5] T. Hinoki, L.L. Snead, Y. Katoh, A. Hasegawa, T. Nozawa and A. Kohyama, J. Nucl. Mater., **307-311** (2002) 1157-1162.
[6] Y. Katoh, S.M. Dong, A. Kohyama, Fusion Engineering and Design 61-62 (2002) 723-731.
[7] P. Colombo, B. Riccardi, A. Donato and G.Scarinci, J. Nucl. Mater., 278 (2000) 127-135.
[8] P. Lemoine, M. Ferraris, M. Salvo, M.A. Montorsi, J. the European Ceram. Soc. 16 (1996) 1231-1236.
[9] Y. Katoh, M. Kotani, A. Kohyama, M. Montorsi, M. Salvo, M. Ferraris, J. Nucl. Mater., 283-287 (2000) 1262-1266.
[10] M. Singh, S.C. Farmer and J.D. Kiser, Ceram. Eng. Sci. Proc. 18[3] (1997) 161-166.
[11] A.R. Raffray, D. Haynes and F. Najmabadi, J. Nucl. Mater., 313-316 (2003) 23-31.
[12] Y. Ueda, K. Tobita and Y. Katoh, J. Nucl. Mater., 313-316 (2003) 32-41.

EFFECT OF ACTIVE FILLERS ON CERAMIC JOINTS DERIVED FROM PRECERAMIC POLYMERS

C.A. Lewinsohn
Ceramatec Inc.
Salt Lake City, UT, 84119

S. Rao
The University of Washington
Seattle, WA, 99352.

R. Bordia
The University of Washington
Seattle, WA, 99352.

Pyrolysis of preceramic polymers has been proposed as a method for obtaining bonds between silicon carbide at relatively low processing temperatures. Shrinkage and evolution of gaseous species during pyrolysis of preceramic polymer-based joints, however, can give rise to strength limiting defects. In this work, the use of various reactive fillers to mitigate shrinkage was investigated. Dilatometry was used to measure shrinkage and x-ray diffraction was used to detect the onset of reactions between the filler phases and the preceramic polymer material. The effects of processing conditions, filler concentrations, and applied pressure on joint strength were also studied. Microscopy was used to complement mechanical testing. This talk will present the findings of this study and directions for additional development.

INTRODUCTION

Many ceramic materials possess properties that make them desirable for specialized applications; fabrication, however, of components with complex shapes is difficult and costly. Therefore, there is a need for satisfactory methods for joining ceramic components to themselves or to components made from dissimilar materials.[1] In some cases, this barrier precludes the selection of ceramic materials, such as silicon carbide (SiC), in applications where their use would result in significant energy savings or environmental benefits. For example, diffusion bonding requires the use of high temperatures and pressures and the shape of components that can be joined is limited. Brazing must be performed under vacuum. Thermal expansion mismatches within brazed joints and the formation of brittle reactants, particularly when silicides and carbide phases can form, can limit durability. Reaction-formed, or reaction-sintered, joints may contain residual silicon that is subject to corrosion and can be a source of lifetime limiting defects. Therefore, there is a need for methods of joining ceramics that are appropriate for applications involving exposure to elevated temperatures. This paper will described recent efforts to join silicon carbide using materials derived from pyrolysis of preceramic precursors.

The pyrolysis of preceramic polymers to join ceramics offers a number of attractive features relative to other joining methods, such as ease of application and low processing temperatures (<1200°C for obtaining a dense ceramic layer).[2-14] During processing, however, evolution of gaseous species and shrinkage during densification can create defects within the polymer derived material that limit the mechanical properties. Work by others has demonstrated that the addition of active fillers that react with preceramic polymers during pyrolysis to form phases with higher molar volumes than the reactants, can be used to improve the density and strength of pyrolysed materials.[15,16] Inert fillers, such as silicon carbide, have also been used to restrict shrinkage and control microstructural development of materials derived from preceramic polymers.

In this paper, methods of incorporating inert and active fillers to preceramic polymer based joining materials, and methods of processing these materials to obtain tailorable, joints for SiC will be described. The results indicate that efforts are required to develop methods for obtaining uniform dispersions of fillers and polymer materials to obtain high strength joints. Furthermore, improved particle packing in the unfired state of the joint may reduce the temperature and pressure required to obtain reliable joints.

METHODS

Mixtures of powder formed by partial pyrolysis of a commercially available preceramic polymer, allyl-hydridopolycarbosilane (aHPCS, Starfire Systems, Watervliet, NY), and reactive fillers were mixed and pressed into pellets. The pellets were heated to various temperatures, in inert atmospheres. The resulting specimens were crushed into powder and analysed using powder X-ray diffraction to determine the phases present. Scanning electron microscopy (SEM) was also performed on the samples, at The University of Washington, to investigate the microstructure.

Experiments were performed to measure the shrinkage, shrinkage rate and weight loss of some of the relevant materials in this study. The effect of inert and/or active fillers on the shrinkage and microstructural evolution of allyl-hydridopolycabosilane (aHPCS) was investigated by heating the specimens to 1200°C at a predetermined heating rate in argon (99.998% pure) in a dilatometer. The concentration of aHPCS matrix was varied in the range of 30-100%. The amount of fillers, active or inert or both, was varied in the range of 5-70%. SiC was used as an inert filler and titanium dislicide (TiSi$_2$) was used as an active filler.

The strength of joints was evaluated using sandwich type specimens of three 25.4 or 12.5 mm^L x 12.5 mm^W x 5 mm^T SiC coupons joined by various mixtures of preceramic polymer derived and filler materials were made (Figure 1). Joining material was made into a paste and applied, by hand, to the 25.4 or 12.5 mm^L x 12.5 mm^W faces of coupons to form a block of three specimens. The specimens were cured at approximately 35°C overnight with a nominal applied load of 1 kg. The specimens were then pyrolysed, using a controlled heating cycle, in an electrically resistance-heated furnace. Load was applied to the specimens by either stacking heavy ceramic plates on top of the specimens or placing the specimens in a furnace attached to an electromechanical test frame[*] and using the crosshead to apply a known load. The shear strength of the joints was measured, using the electromechanical test frame, by applying a compressive load to only the middle coupon of the sandwich specimen and supporting only the outer two pieces of the bottom of the specimen (see Figure 1). The test configuration did not follow a standard, but provided a simple way to measure the effect of joint compositions and process treatments. The shear strength was calculated by dividing the load at fracture by the joining area. Edge effects and residual stresses due to differential shrinkage would be expected to induce shear stresses at the edges of the joints, but finite element, or other, analysis was not performed to estimate these stresses.

RESULTS AND DISCUSSION

Table I summarises the experiments that were performed in order to determine the temperature at which active fillers react with polymer-derived material. Some specimens were

---

[*] Model 5508, Instron Corp., Canton, MA.

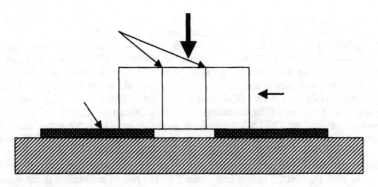

Figure 1   Schematic illustration of joining specimens and testing method.

Table I
Summary of Active Filler Reaction Study

Polymer Precursor Pyrolysis Temperature (°C)	Filler	Reaction Temperature (°C)	Evidence of Reaction
600	Cu	900	N
		1000	Y
		1100	Y
	Fe	900	N
		1000	N
		1100	Y
	Cr	900	N
		1000	(not tested)
		1100	Y
	TiH$_2$	900	N
		1000	N
		1100	Y
	TiSi$_2$	1100	Y
1200	Cr	1000	Y
		1200	Y
		1835	Y
	TiSi$_2$	1000	Y
		1200	Y
		1720	Y

taken to extremely high temperatures to investigate the microstructural stability of the materials. The occurrence of a reaction was determined from the presence of additional peaks in the X-ray diffraction spectra.

A summary of the linear shrinkage, diametral shrinkage, and shrinkage calculated from the sample dimensions is given Table II. The sample containing only aHPCS matrix shrank about 16.5% during pyrolysis. A significant amount of shrinkage occurred in the temperature range 500-1200°C, as shown in Figure 2. The sample was heated to 1200°C in flowing argon. The amount of shrinkage decreased to 3.5% when the sample contained 30%SiC inert filler, as shown in Figure 3. When the filler was changed to $TiSi_2$ active filler while keeping the volume fraction of the filler the same, the sample shrank by about 5.3%, as shown in Figure 4. Although the material containing $TiSi_2$ as an active filler shrank more than material containing inert SiC inert filler, the sample with the inert filler contained one and half times more open porosity (38.2 %) than the one with active filler (24.1 %). In addition, in the material with the active filler, $TiSi_2$ a reaction occured with the aHPCS matrix (mainly carbonaceous material) to form TiC, as shown in Figure 5. For most of the samples, linear shrinkage and diametral shrinkage match closely, which indicates that the shrinkage is isotropic (see Table II). Although this study demonstrated that $TiSi_2$ reacts with the aHPCS matrix and forms TiC, the rate of conversion is not significant enough to bring about zero shrinkage. Nevertheless, the result is very encouraging and the rate of conversion can be increased by decreasing the filler particle size and/or increasing the processing temperature. Unlike SiC, $TiSi_2$ does not appear to hinder the sintering of the matrix. As a result of both the effects mentioned above, $TiSi_2$ is promising from the point of compensating the matrix shrinkage thereby producing zero net-shrinkage SiC-based composites.

Table II

Linear shrinkage and diametral shrinkage of the passive and/or active filler controlled pyrolysis of aHPCS in argon at 1200°C.

Sample	Linear shrinkage (dilatometry), %	Shrinkage from sample dimensions, %	Shrinkage from dilatometry after removing steep portion of length increase, %	Diametral shrinkage from sample dimensions, %	Comments
aHPCS matrix	17.2	16.7	------	17.6	
70%HPCS + 30%SiC	3.5	5.1	-----	5.95	
70%HPCS + 30%TiSi$_2$	5.3	5.2	-----	8.5	
30%HPCS + 70%SiC	-2.7 (expansion)	1.3	0.7	1.1	Formed white sponge like deposit (appears to be amorphous silica)
30%HPCS + 65%SiC+ 5%TiSi$_2$	-1.7 (expansion)	0.3	0.1	1.26	Formed white deposit similar to that of 30%HPCS+70% SiC

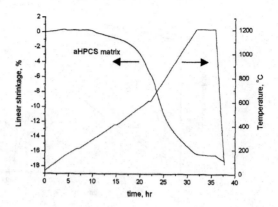

Figure 2   Linear shrinkage of aHPCS matrix.

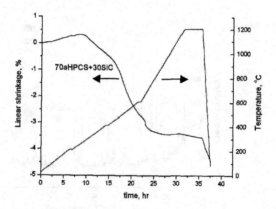

Figure 3   Linear shrinkage of 70% aHPCS+30% SiC.

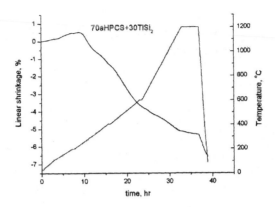

Figure 4   Linear shrinkage of 70% aHPCS+30%TiSi$_2$.

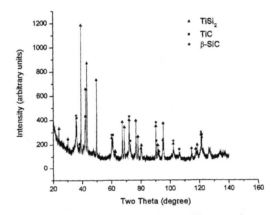

Figure 5   X-ray diffraction pattern of 70% aHPCS+30% TiSi$_2$ sample heated to 1200°C and held for 4 hr in argon.

The shear strength of joints between silicon carbide coupons is given in Table III. In the column labeled "Filler" the volume fraction and composition of filler is given. For aHPCS filler the temperature used to pyrolyse the precursor is shown. After pyrolysis, the precursor was ground into powder. In one case, labeled "milled", additional milling was used to reduce the particle size. It was found that inert filler material derived from pyrolysing the precursor polymer at 1200°C led to very high strengths: Tests # 1-9 in Table III. Consistently high shear strength values were measured for joints processed at 1200°C: Tests # 6-8 in Table III. The prepyrolysed material may bond better to the fresh polymer than the submicron silicon carbide powder.

Table III
Shear strength of joints, derived from aHPCS, between monolithic silicon carbide

	# specimens	Filler	Joining Temp (C)	Joining environment	Joining Pressure (kPa)	Shear Strength (MPa)	95% CI (MPa)
1	1	70% 1200°C aHPCS	1000	argon	736	17.3	na
2	2	70% 1200°C aHPCS	1000	argon	736	23.8	0.9
3	1	60% 1200°C aHPCS	1000	argon	736	15.08	na
4	2	30% Fe	1000	argon	736	22	11.7
5	3	60% 1200°C aHPCS	1050	air	54	32	5.5
6	2	70% 1200°C aHPCS	1200	air	736	49	27
7	4	70% 1200°C aHPCS	1200	air	654	49	13.3
8	4	70% 1200°C aHPCS - milled	1200	air	654	45	4.2
9	2	60% 1200°C aHPCS	1200	air	736	29	4.6

SUMMARY

In summary, temperatures at which active fillers react with preceramic polymer precursors were identified and the addition of certain active fillers was shown to reduce greatly the shrinkage of preceramic polymer materials. Furthermore, the shear strength of joints consisting of preceramic polymer-derived materials were measured and found to be improved by the addition of fillers. The results, however, indicated that pressure was required during joining to obtain acceptable joint strengths.

Analysis of these results suggests that improvement in dispersion and packing of raw materials in the joining materials must be improved for development of a practical and commercially viable joining method. In addition, additional screening of precursor polymers and evaluation of the stability of the microstructure of joint materials after long term exposures at high temperatures are strongly recommended.

REFERENCES

1. J.M. Fragomeni and S.K. El-Rahaiby, Review of Ceramic Joining Technology, Rept. No. 9, Ceramic Information Analysis Center, Purdue University, Indiana (1995).
2. L.M. Ewart, "A study of process variables and bond strength in the use of polycarbosilane to join SiC," pp. 125-132 in *Proc. 19th Army Science Conference*, Orlando FL, June 20-23, 1994, Department of the Army, Washington DC (1994).
3. A. Donato, P. Colombo and M.O. Abdirashid, "Joining of SiC to SiC using a preceramic polymer," pp. 471-476 in *High-Temperature Ceramic-Matrix Composites I: Design,*

413

durability and performance, Ceramic Transactions Vol. 57. Edited by A.G. Evans and R. Naslain. The American Ceramic Society, Westerville OH, 1995.

4. P. Colombo, M.O. Abdirashid, G. Scarinci and A. Donato, "A new method for joining SiC/SiC$_f$ composites," pp. 75-82 in *Fourth Euro-Ceramics, Coatings and Joinings* Vol. 9. Edited by B.S. Tranchina and A. Bellosi. Gruppo Editoriale Faenza Editrice S.p.A., Faenza, Italy, 1995.

5. I. Ahmad, R. Silberglitt, T.A. Shan, Y.L. Tian and R. Cozzens, "Microwave-assisted pyrolysis of SiC and its application to joining," pp. 357-365 in *Microwaves: theory and application in materials processing, III*, Ceramic Transactions 59. Edited by D.E. Clark, D.C. Folz, S.J. Oda and R. Silberglitt. The American Ceramic Society, Westerville OH, 1995.

6. I. Ahmad, R. Silberglitt, Y.L. Tian and J.D. Katz, "Microwave joining of SiC ceramics and composites," pp. 455-463 in *Microwaves: Theory and Application in Materials Processing IV*, Ceramic Transactions 80. Edited by D.E. Clark, W.H. Sutton and D.A. Lewis, The American Ceramic Society, Westerville, OH, 1997.

7. W.J. Sherwood, C.K. Whitmarsh, J.M. Jacobs and L.V. Interrante, "Joining ceramic composites using active metal/HPCS preceramic polymer slurries," *Cer.Eng.Sci.Proc.*, **18**, 177-184 (1997).

8. I.E. Anderson, S. Ijadi-Maghsoodi, Ö. Ünal, M. Nostrati and W.E. Bustamante, "Development of a compound for low temperature joining of SiC ceramics and CFCC composites," pp. 25-40 in *Ceramic Joining*, Ceramic Transactions **77**. Edited by I.E. Reimanis, C.H. Henager and A.P. Tomsia. The American Ceramic Society, Westerville OH, 1997.

9. Ö. Ünal, I.E. Anderson, M. Nostrati, S. Ijadi-Maghsoodi, T.J. Barton, T.J. and F.C. Laabs, "Mechanical properties and microstructure of a novel SiC/SiC joint," pp. 185-194 in *Ceramic Joining*, Ceramic Transactions **77**. Edited by I.E. Reimanis, C.H. Henager and A.P. Tomsia. The American Ceramic Society, Westerville OH, 1997.

10. E. Pippel, J. Woltersdorf, P. Colombo and A. Donato, "Structure and composition of interlayers in joints between SiC bodies," *J.Europ.Ceram.Soc.*, **17**, 1259-1265 (1997).

11. Ö. Ünal, I.E. Anderson and S.I. Maghsoodi, "A test method to measure strength of ceramic joints at high temperatures," *J. Am. Ceram. Soc.*, **80**, 1281-1284 (1997).

12. P. Colombo, V. Sglavo, E. Pippel and J. Woltersdorf, "Joining of reaction-bonded silicon carbide using a preceramic polymer," *J.Mater.Sci.*, **33**, 2409-2416 (1998).

13. A.S. Fareed, C.C. Cropper, "Joining Techniques for Fiber-Reinforced Ceramic Matrix Composites, " Ceram. Eng. & Sci. Proc, 20 [4], 61-70 (1999).

14. C.A. Lewinsohn, P. Colombo, I. Riemanis, and O. Unal, "Stresses Arising During Joining of Ceramics Using Preceramic Polymers," J. Am. Ceram. Soc., [10] (2001).

15. C.A. Lewinsohn, C.H. Henager Jr., and M. Singh, "Brazeless approaches to joining silicon carbide-based ceramics for high temperature applications," pp. 201-208 in Advances in Joining of Ceramics, (C. Lewinsohn, M. Singh, and R. Loehman [Eds.]), Ceramic Transactions Vol. 138, The American Ceramic Society, Westerville (OH, USA), 2003.

16. P. Greil, "Active-filler-controlled pyrolysis of preceramic polymers," J. Am. Ceram. Soc., vol. 78 [4], pp. 835-848 (1995).

17. Margaret Mary Stackpoole, "Processing & mechanical properties of polymer derived silicon nitride matrix composites & their use in coating and joining ceramics and ceramic composites", Ph.D. Thesis, Dept. of Materials Science and Engineering, University of

Washington, Seattle, 2002.

NUMERICAL ANALYSIS OF SINGLE LAP JOINTED CERAMIC COMPOSITE SUBJECTED TO TENSILE LOADING

Daisuke Fujita and Hisashi Serizawa
Joining and Welding Research Institute, Osaka University
11-1 Mihogaoka
Ibaraki, Osaka, 567-0047, Japan

Mrityunjay Singh
MS 106-5, Ceramics Branch, NASA Glenn Research Center
21000 Brookpark Road
Cleveland, OH 44135-3191, USA

Hidekazu Murakawa
Joining and Welding Research Institute, Osaka University
11-1 Mihogaoka
Ibaraki, Osaka 567-0047, Japan

ABSTRACT

As one of the most typical test methods to determine the shear strength of joints, the effect of the joint shape on the tensile strength of lap jointed ceramic composite bonded by ARCJoinT™ was examined by using finite element method with the interface element. Also, the influence of the mechanical properties of the interface element on the tensile strength was analyzed. From the effects of the surface energy and the bonding and shear strength of interface element on the tensile strength, it was found that the tensile strength of beveled lap joint was mainly governed by the surface energy at the interface and almost independent on the bonding and the shear strength. As the results of the joint shape effect, it was revealed that the joint shape slightly affected the tensile strength and this influence was caused by the order of singularity in the stress field at the edge of joint. Furthermore, it was considered that the shear strength measured by the tensile test of lap joint might be different from that obtained by the asymmetrical four point bending test of the butt jointed specimen.

INTRODUCTION

Silicon carbide-based fiber reinforced silicon carbide composites (SiC/SiC composites) are promising candidate materials for high heat flux components because of their high-temperature properties, chemical stability and good oxidation and corrosion resistance [1-3]. For fabricating large or complex shaped parts of SiC/SiC composites, practical methods for joining simple geometrical shapes are essential. As a result of R & D efforts, an affordable, robust ceramic joining technology (ARCJoinT™) has been developed as one of the most suitable methods for joining SiC/SiC composites among various types of joining between ceramic composites [4].

To establish useful design databases, the mechanical properties of joints must be accurately measured and quantitatively characterized. Where the lap joint, in which two sheets are joined together with an overlay, is one of the most common joints encountered in practice and

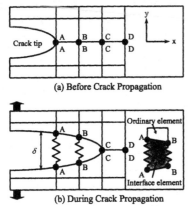

(a) Before Crack Propagation

(b) During Crack Propagation

Fig.1 Representation of crack growth using interface element.

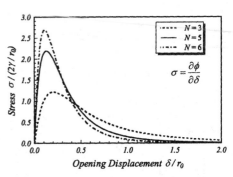

$$\sigma = \frac{\partial \phi}{\partial \delta}$$

Fig.2 Relationship between crack opening displacement and bonding stress.

is the configuration most often used for testing adhesives. Also, in the practical use of lap joint, the edges of joint components are tapered to eliminate stress risers under loading and to reduce the tendency to peel. Although detailed information on the stress field of lap joint under the tensile loading has been reported by using the analytical and numerical techniques [5-7], little information on the criteria of the fracture is available from these types of study. This comes from the fact that the physics of failure itself is not explicitly modeled.

On the other hand, to describe deformation and fracture behavior more precisely, a new and simple computer simulation method has been developed [8-12]. The method treats the fracture phenomena as the formation of new surface during crack opening and propagation. Based on the fact that surface energy must be supplied for the formation of new surface, a potential function representing the density of surface energy is introduced to the finite element method (FEM) using cohesive elements [8] or interface elements [9-12]. This method may have a potential capability not only to give insight into the criteria of the fracture but also to make the quantitative prediction of strength itself. In this research, the effect of the joint shape on the tensile strength of lap and beveled lap jointed ceramic composite bonded by the ARCJoinT™ was examined by using the finite element method with the interface element. Also, the influence of the mechanical properties of the interface element on the tensile strength was analyzed.

INTERFACE POTENTIAL

Essentially, the interface element is the distributed nonlinear spring existing between surfaces forming the interface or the potential crack surfaces as shown by Fig.1. The relation between the opening of the interface δ and the bonding stress σ is shown in Fig.2. When the opening δ is small, the bonding between two surfaces is maintained. As the opening δ increases, the bonding stress σ increases till it becomes the maximum value σ_{cr}. With further increase of δ, the bonding strength is rapidly lost and the surfaces are considered to be separated completely. Such interaction between the surfaces can be described by the interface potential. There are rather wide choices for such potential. The authors employed the Lennard-Jones type potential because it explicitly involves the surface energy γ which is necessary to form new surfaces. Thus, the surface potential per unit surface area ϕ can be defined by the following equation.

$$\phi(\delta_n, \delta_t) \equiv \phi_a(\delta_n, \delta_t) + \phi_b(\delta_n) \tag{1}$$

$$\phi_a(\delta_n, \delta_t) = 2\gamma \cdot \left\{ \left(\frac{r_0}{r_0 + \delta} \right)^{2N} - 2 \cdot \left(\frac{r_0}{r_0 + \delta} \right)^{N} \right\}, \quad \delta = \sqrt{\delta_n^2 + A \cdot \delta_t^2} \tag{2}$$

$$\phi_b(\delta_n) = \begin{cases} \dfrac{1}{2} \cdot K \cdot \delta_n^2 & (\delta_n \le 0) \\ 0 & (\delta_n \ge 0) \end{cases} \tag{3}$$

Where, δ_n and δ_t are the opening and shear deformation at the interface, respectively. The constants γ, r_0, and N are the surface energy per unit area, the scale parameter and the shape parameter of the potential function. In order to prevent overlapping in the opening direction due to a numerical error in the computation, the second term in Eq.(1) was introduced and K was set to have a large value as a constant. Also, to model an interaction between the opening and the shear deformations, a constant value A was employed in Eq.(2). From the above equations, the maximum bonding stress, σ_{cr}, under only the opening deformation δ_n and the maximum shear stress, τ_{cr}, under only the shear deformation δ_t are calculated as follows.

$$\sigma_{cr} = \frac{4\gamma N}{r_0} \cdot \left\{ \left(\frac{N+1}{2N+1} \right)^{\frac{N+1}{N}} - \left(\frac{N+1}{2N+1} \right)^{\frac{2N+1}{N}} \right\} \tag{4}$$

$$\tau_{cr} = \frac{4\gamma N \sqrt{A}}{r_0} \cdot \left\{ \left(\frac{N+1}{2N+1} \right)^{\frac{N+1}{N}} - \left(\frac{N+1}{2N+1} \right)^{\frac{2N+1}{N}} \right\} \tag{5}$$

By arranging such interface elements along the crack propagation path as shown in Fig.1, the growth of the crack under the applied load can be analyzed in a natural manner. In this case, the decision on the crack growth based on the comparison between the driving force and the resistance as in the conventional methods is avoided.

From the results of our previous researches using the interface elements, it was found that the failure mode and the stability limit depend on the combination of the deformability of the ordinary element in FEM and the mechanical properties of the interface element as controlled by the surface energy γ, the scale parameter r_0 and the interaction parameter A in Eq.(2); furthermore, the fracture strength in the failure problems of various structures might be quantitatively predicted by selecting the appropriate values for the surface energy γ, the scale parameter r_0 and the interaction constant A [11,12].

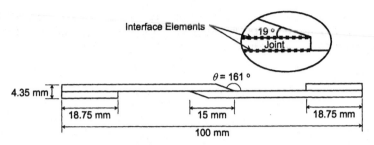

Fig.3   Schematic illustration of beveled lap jointed ceramic composite.

MODEL FOR ANALYSIS

SiC/SiC composite ceramic joint joined by the ARCJoinT™ was selected for this study. Figure 3 shows a model of the beveled lap joint for the tensile test. Joint was made from two SiC/SiC composite plates, whose dimensions were 57.5 mm-long, 12.5 mm-wide and 2.125 mm-thick. The thickness of joint was set to be 100 μm for a typical example of ARCJoinT™ [4] and the angle of the edge, θ, was assumed to be 161 degree according to our ongoing experiments. To prevent a rotation of the joint due to a bending moment, the tabs were also jointed to the ends of the joint via the ARCJoinT™ method. Young's moduli and Poisson's ratios of SiC/SiC composite and the joint were assumed to be 300 GPa, 350 GPa, 0.15 and 0.20, respectively [10,12]. Although the mechanical properties of SiC/SiC composites should be anisotropic, the properties were assumed to be isotropic since the difference between the elastic properties of the composite and the joint material is significantly larger than those due to the composite anisotropy. Because of the brittleness of the ceramic materials, FEM calculations were conducted assuming linear elastic behavior in two-dimensional plain strain. Since the fracture started from the edge of the interface between SiC/SiC composite and the joint where the load had a maximum value in the experiment, the interface elements were arranged along both the interfaces between the composite as shown in Fig.3 and the joint and the mesh division near both the edges was set to fine. The total number of elements and nodes for the joint material were 17162 and 17521, respectively, and the element sizes were decided by continuously refining the mesh until approximate convergence with the numerical solution was achieved.

From the previous studies about the four point and the asymmetrical four point bending tests of a butt jointed SiC/SiC composites via the ARCJoinT™ method by using FEM with the interface element, the surface energy γ and the parameter of the interaction between the opening and the shear deformations A in Eq.(2) were estimated to be 30 N/m and 2.47×10^{-2}, respectively [12]. Also, in this research, a constant K was set to be 5.0×10^{4} N/m. Then, by changing the scale parameter r_0 in the range from 1.0×10^{-4} to 100 μm, the tensile strength of the beveled lap joint was analyzed. Also, the effect of the joint shape on the tensile strength of lap and beveled lap joint was examined by changing the thickness of the joint and the angle of the edge, θ. The shape parameter N was assumed to be 4 according to our previous researches [10-12].

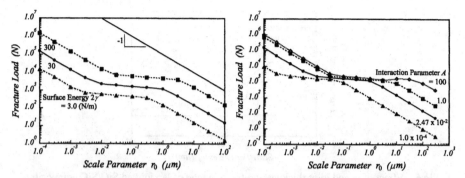

Fig.4   Effect of r_0 and γ on fracture load
          of lap jointed SiC/SiC composite.

Fig.5   Effect of r_0 and A on fracture load
          of lap jointed SiC/SiC composite.

EFFECT OF MECHANICAL PROPERTIES OF INTERFACE ELEMENT

The tensile load was applied to the beveled lap joint through the horizontal displacement given on both the ends of the joint. According to the experimental result, the maximum load obtained was defined as the fracture load. The effect of the scale parameter on the fracture load was summarized into Fig.4 with logarithmic scale where the surface energy was also changed in the range from 3 to 300 N/m to study the effect of the surface energy on the fracture load. As it is clearly seen from this figure, all the curves could be divided into three parts with respect to the size of the scale parameter r_0. When r_0 was in 0.01 and 1.0 μm for $\gamma = 30$ N/m, the fracture load was almost independent on the scale parameter. On the other hand, the slope of the curve became -1 when the scale parameter was smaller or larger than this range. From our previous researches [11], it was found that the results in the middle part, whose slopes were not -1, could be quantitatively compared with the experimental results. Namely, the appropriate value for the scale parameter r_0 was in this range. Also, from this figure, it was found that the tensile fracture strength of beveled lap joint was controlled by the surface energy γ and the strength was almost independent on r_0.

In order to study the effect of the interaction between the bonding and the shear strength of interface element, the influence of the interaction parameter A was analyzed and the result was summarized into Fig.5. This figure suggests that the fracture load was almost independent on the interaction between the bonding and the shear strength of interface element and only the range for the appropriate value of scale parameter was affected. From those results, it was concluded that the tensile strength of beveled lap joint was governed by the surface energy and almost independent on the bonding and the shear strength at the joint interface.

On the other hand, in our previous analyses about the four point bending and the asymmetrical four point bending tests of the butt jointed ceramic composites, which are commonly used to measure the bonding and the shear strength of joint, the bonding and the shear strengths were controlled by not only the surface energy but also the strength at the joint [12]. Then, it can be considered that the shear strength measured by the tensile test of lap joint might be different from that obtained by the asymmetrical bending test of butt jointed specimen.

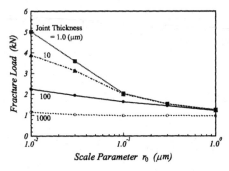

Fig.6 Effect of r_0 and joint thickness on fracture load of lap jointed SiC/SiC composite.

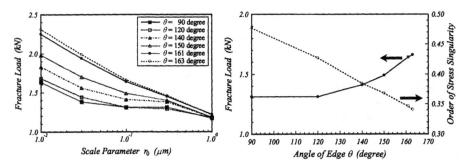

Fig.7 Effect of r_0 and θ on fracture load
of lap jointed SiC/SiC composite.

Fig.8 Effect of θ on fracture load and
order of stress singularity
of lap jointed SiC/SiC composite.

EFFECT OF JOINT SHAPE

To examine the effect of the joint shape on the tensile strength of lap joint, the thickness of joint and the angle of edge were changed in the range from 1 to 1000 μm and from 90 to 163 degree, respectively. The results were summarized into Figs.6 and 7 where the scale parameter was changed in the range from 0.01 to 1.0 μm according to the previous calculation. From those results, it was found that the tensile strength would be affected by the joint shape though its effect was less than that of the surface energy at the joint. As for the joint thickness, the tensile strength monotonically decreases with increasing the joint thickness and this effect would be larger with decreasing the scale parameter, which means that the bonding and the shear strengths at the joint increase as shown in Eqs.(4) and (5). In the case of the angle of edge, its influence on the tensile strength also becomes larger with increasing the bonding and the shear strengths at the joint. Furthermore, the stress distribution at the edge of joint was analyzed and the effect of the angle of edge on the order of stress singularity was summarized into Fig.8. The tensile strengths were also plotted in this figure where the scale parameter was 0.1 μm. From this figure, it was

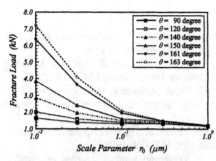

Fig.9   Effect of r_0 and θ on fracture load
of lap jointed SiC/SiC composite.

Fig.10   Effect of θ on fracture load and
order of stress singularity
of lap jointed SiC/SiC composite.

found that there was a strong relationship between the tensile strength and the order of the singularity in the stress field at the interface although the strength was not determined by the order of the singularity.

Practically, the beveled lap joint was prepared by tapering the edge before or after joining SiC/SiC composite plates. In order to examine the effect of this preparation method, the influence of the angle of joint on the tensile strength was studied and the stress field at the edge of joint was also analyzed. Figure 9 shows the effect of the angle of joint on the tensile strength as the function of the scale parameter. The relationship between the tensile strength and the order of stress singularity was summarized into Fig.10 where the scale parameter was 0.1 μm and the results shown in Fig.8 were also plotted. From this figure, it was found that the tensile strength decreased with the increasing order of stress singularity at the edge of interface. However, these changes in the tensile strength were much less than the effect of the surface energy at the interface. Namely, it was concluded that the tensile strength of lap or beveled lap joint was mainly governed by the surface energy at the interface and the effect of the joint of shape were little.

CONCLUSIONS

In order to examine the effect of the joint shape on the tensile strength of single lap jointed ceramic composites, the tensile test of lap and beveled lap jointed SiC/SiC composites plates bonded by the ARCJoinT™ was analyzed by using the finite element method with the interface element. Also, the influence of the mechanical properties of the interface element on the tensile strength was analyzed. The conclusions can be summarized as follows.

(1)   The tensile strength of beveled lap joint was mainly governed by the surface energy at the interface and almost independent on the bonding and the shear strength at the interface.

(2)   The joint shape slightly affected the tensile strength of lap and beveled lap joint and this influence was caused by the order of singularity in the stress field at the edge of joint.

(3)   The shear strength measured by the tensile test of lap joint might be different from that obtained by the asymmetrical four point bending test of the butt jointed specimen.

REFERENCES

[1]H. Serizawa, C.A. Lewinsohn, G.E. Youngblood, R.H. Jones, D.E. Johnston and A. Kohyama, "High-Temperature Properties and Creep Resistance of Near-Stoichiometric SiC Fibers," *Ceramic Engineering and Science Proceedings*, **20** [4], 443-450 (1999).

[2]A. Kohyama and Y. Katoh, "Overview of Crest-Ace Program for SiC/SiC Ceramic Composites and Their Energy System Applications," *Ceramic Transactions*, **144**, 3-18 (2002).

[3]B. Riccardi, L. Giancarli, A. Hasegawa, Y. Katoh, A. Kohyama, R.H. Joines and L.L. Snead, "Issues and advances in SiC$_f$/SiC composites development for fusion reactors," *Journal of Nuclear Materials*, **329-333**, 56-65 (2004).

[4]M. Singh, "Design, Fabrication and Characterization of High Temperature Joints in Ceramic Composites," *Key Engineering Materials*, **164-165**, 415-419 (1999).

[5]M. Goland and E. Reissner, "The stresses in cemented joints," *Journal of Applied Mechanics*, **66**, A17-A27 (1944).

[6]R.D. Adams, J. Comyn and W.C. Wake, *Structural Adhesive Joints in Engineering*, Chapman & Hall, London (1997).

[7]R.W. Messler, Jr., *Joining of Materials and Structures*, ElsevierButterworth-Heinemann (2004).

[8]A. Needleman, "An Analysis of Decohesion Along An Imperfect Interface," *International Journal of Fracture*, **42**, 21-40 (1990).

[9]H. Murakawa, , H. Serizawa and Z.Q. Wu, "Computational Analysis of Crack Growth in Composite Materials Using Lennard-Jones Type Potential Function," *Ceramic Engineering and Science Proceedings*, **20** [3], 309-316 (1999).

[10]H. Serizawa, H. Murakawa and C.A. Lewinsohn, "Modeling of Fracture Strength of SiC/SiC Composite Joints by Using Interface Elements," *Ceramic Transactions*, **144**, 335-342 (2002).

[11]H. Murakawa, H. Serizawa, K. Miyamoto, I. Oda, "Strength of Joint Between Dissimilar Elastic Materials," *Proceedings of 2003 International Conference on Computational & Experimental Engineering & Sciences (ICCES'03)*, **6**, (2003) (CD-ROM).

[12]H. Serizawa, H. Murakawa, M. Singh and C.A. Lewinsohn, "Finite Element Analysis of Ceramic Composite Joints by Using a New Interface Potential," *High Temperature Ceramic Matrix Composites 5*, 451-456 (2004).

Author Index